Suzanne Fyfe

Hyperspectral studies of seagrasses

Ecology, physiology, biochemistry and stress

LAP LAMBERT Academic Publishing

Impressum/Imprint (nur für Deutschland/ only for Germany)
Bibliografische Information der Deutschen Nationalbibliothek: Die Deutsche Nationalbibliothek verzeichnet diese Publikation in der Deutschen Nationalbibliografie; detaillierte bibliografische Daten sind im Internet über http://dnb.d-nb.de abrufbar.

Coverbild: www.ingimage.com

Verlag: LAP LAMBERT Academic Publishing GmbH & Co. KG
Dudweiler Landstr. 99, 66123 Saarbrücken, Deutschland
Telefon +49 681 3720-310, Telefax +49 681 3720-3109
Email: info@lap-publishing.com

Herstellung in Deutschland:
Schaltungsdienst Lange o.H.G., Berlin
Books on Demand GmbH, Norderstedt
Reha GmbH, Saarbrücken
Amazon Distribution GmbH, Leipzig
ISBN: 978-3-8443-3352-7

Imprint (only for USA, GB)
Bibliographic information published by the Deutsche Nationalbibliothek: The Deutsche Nationalbibliothek lists this publication in the Deutsche Nationalbibliografie; detailed bibliographic data are available in the Internet at http://dnb.d-nb.de.

Cover image: www.ingimage.com

Publisher: LAP LAMBERT Academic Publishing GmbH & Co. KG
Dudweiler Landstr. 99, 66123 Saarbrücken, Germany
Phone +49 681 3720-310, Fax +49 681 3720-3109
Email: info@lap-publishing.com

Printed in the U.S.A.
Printed in the U.K. by (see last page)
ISBN: 978-3-8443-3352-7

Contents

Chapter 3 Experimental Methods

S. K. Fyfe & S.A. Robinson

Chapter 4 Spatial and Temporal Variation in the Spectral Reflectance of Seagrass

S. K. Fyfe, S.A. Robinson & A.G. Dekker

Chapter 6 Spectral Indices for the Prediction of Light Stress

S. K. Fyfe

References 338

Appendices 380

Symbols and Abbreviations

ADP	adenosine diphosphate; formed from ATP on hydrolysis releasing usable energy
API	aerial photograph interpretation
ATP	adenosine triphosphate; the major source of usable chemical energy in metabolism
CASI	Compact Airborne Spectrographic Imager (Itres Instruments Inc.)
dGPS	differential global positioning system
ETR	electron transfer rate
F_m	maximum fluorescence
F_o	initial fluorescence
F_v	variable fluorescence
$F_v:F_m$	optimum quantum yield of photochemistry in PSII (photosynthetic efficiency)
$\Delta F:F_m'$	effective quantum yield of photochemistry in PSII (yield; instantaneous photosynthetic performance under ambient PAR)
FOV	field-of-view
FWHM	full-width-half-mean; measure of spectral resolution
g	SI unit for centrifugal force
GE	green edge
GEP	green edge position
I	irradiance; radiant flux per unit area of surface (W m^{-2})
I_c	compensation point; irradiance at P = respiration rate
I_{max}	maximum photosynthetic irradiance (the minimum irradiance required to support P_{max})
IR	infrared

K_d	vertical attenuation coefficient for downward irradiance; light attenuation with depth in water (m^{-1})
LAI	leaf area index
LHCI	light harvesting complex I
LHCII	light harvesting complex II
NADP	nicotinamide adenine dinucleotide phosphate; electron acceptor in photosynthesis
$NADPH_2$	formed from reduction of NADP, supplies usable chemical energy to metabolism
NDVI	Normalised Difference Vegetation Index
NIR	near infrared
NPQ	non-photochemical quenching
P	photosynthetic rate
P_{max}	maximum photosynthetic rate
PAR	photosynthetically active radiation
PMSC	piecewise multiplicative scatter correction
ppt	parts per thousand
PRI	Photochemical Reflectance Index
PSI	photosystem I
PSII	photosystem II
qE	ΔpH-dependent non-photochemical quenching
qI	photoinhibitory quenching
qP	photochemical quenching
R	irradiance reflectance; ratio of the upward to the downward irradiance at a given point in the field ($R = E_u/E_d$)
RE	red edge
REP	red edge position

Rubisco	ribulose-1,5-biphosphate carboxylase/oxygenase; first enzyme catalyst in the dark reactions of photosynthesis, CO_2 acceptor
SNR	signal-to-noise ratio
SVI	simple vegetation index
SWIR	short wave infrared
UV	ultraviolet
W	watt; fundamental unit of optical power or energy flux, defined as a rate of energy of one joule (J) per second
z_{col}	maximum depth to which any water body is colonized by macrophytes (m)
ø	quantum yield (number of CO_2 molecules fixed in biomass per quantum of light absorbed by the plant)
$ø_{max}$	maximum quantum yield; maximum slope of P-I curve
α	slope of P-I curve in the light-limited region (~ $ø_{max}$)
λ	wavelength
$\mu E\ m^{-2}s^{-1}$	equivalent to μmol photon $m^{-2}s^{-1}$
μmol photon $m^{-2}s^{-1}$	SI unit for irradiance
ΔpH	pH gradient

Acknowledgements

I would like to thank Prof. Sharon Robinson of the University of Wollongong for her contributions to several chapters in this book and for her supervision of the research on which this book was based. Thanks also to Prof. Arnold Dekker of the CSIRO for his contributions towards Chapter 4 and for mentoring and advice.

A large number of people provided assistance, friendship and fun in the field and laboratory including Donna Rolston, Simone Dürr, several NSW Fisheries officers, John Marthick, John Reid, Jodie Dunn, Jane Wasley, Nicole Pennel, Jacqui Fyfe, Michelle Paterson and many of the staff and students of the Schools of Geosciences and Biological Sciences at the University of Wollongong. I am grateful to Dr Pam Davies and Assoc. Prof. Andy Davis for statistical support and/or advice and to Julia Slotwinski, Prof. Colin Woodroffe and Prof. Sharon Robinson for proof-reading the manuscript.

Funding and/or logistic support for this project were provided by:

- Australian Postgraduate Award
- Australian Research Council small grant
- Institute for Conservation Biology, University of Wollongong
- GeoQuest, University of Wollongong
- NSW Fisheries Research Institute
- CSIRO Land and Water

I would especially like to thank my family and friends for the support, patience and encouragement offered throughout the development of this work.

Chapter 1
Mapping and Monitoring Seagrass

S.K. Fyfe

1.1 Introduction

Seagrasses are flowering marine plants that form extensive meadows in the shallow waters of sheltered coastal bays, estuaries, lagoons and lakes. These grass-like plants are highly productive primary producers at the base of the estuarine food chain (Phillips and McRoy 1980; Hillman *et al.* 1989). Seagrass meadows are crucial to the maintenance of estuarine biodiversity (Hutchings and Recher 1974; Powis and Robinson 1980) and the sustainability of many commercial fisheries (Bell and Pollard 1989; Coles *et al.* 1993). They not only offer an important food resource, but also provide shelter, support, breeding and nursery grounds to a wide variety of marine vertebrates and invertebrates (Shepherd *et al.* 1989; Klumpp *et al.* 1989; Kirkman *et al.* 1995). In addition, seagrass meadows help trap and stabilise sediments, filter seawater, reduce erosion and maintain coastal water quality and clarity (Ginsburg and Lowenstam 1958; Koch 1996, Komatsu 1996; Terrados and Duarte 2000).

1.2 Threats to seagrass meadows

Over 80% of the population of Australia lives on the coastal fringe in close proximity to waterways supporting seagrasses. In New South Wales alone, there are approximately 133 embayments and estuaries offering habitat for seagrass growth. Unfortunately these waterways have not been well managed under the pressure of agricultural, residential, commercial and industrial development (West 1997). Seagrasses play a pivotal role in coastal ecosystems, yet historical records have consistently demonstrated that most areas of seagrasses investigated in Australia are in a state of decline, and that similar unprecedented declines have been experienced in both temperate and tropical regions throughout the world (e.g. Orth and Moore 1983; Robblee *et al.* 1991; Lee Long *et al.* 1996; Short *et al.* 1996; Pasqualini *et al.* 1999; Kurz *et al.* 2000). Some major NSW estuaries have lost up to 85% of their seagrass beds in the last 50 years (S.P.C.C. 1981; Larkum and West 1983; Larkum and West 1990; Walker and McComb 1992; NSW Fisheries 1999) and seagrass coverage in Cockburn Sound, Western Australia has declined by 77% since 1967 (Kendrick *et al.* 2002).

The factors that have contributed to widespread decline of seagrass meadows have been reviewed by Shepherd *et al.* (1989), Short and Wyllie-Echeverria (1996) and Kirkman (1997). Human disturbance, predominantly through reduced water quality, is considered the primary cause of seagrass loss worldwide (Short and Wyllie-Echeverria 1996). Increased anthropogenic inputs to coastal waters, in particular fine sediments, nutrients and pollutants have been linked to seagrass loss (Shepherd *et al.* 1989; Walker and McComb 1992; Walker *et al.* 1999). Industrial, commercial, rural or residential land uses supply non-point source runoff to estuaries and bays (Cambridge *et al.* 1986; Larkum and West 1990; Short *et al.* 1996) as well as point source outfalls of stormwater and sewage (Pergent-Martini and Pergent 1996). Eutrophication has been blamed for the most devastating disturbances to seagrass meadows in Australia (Kirkman 1996) although seagrasses have, on occasion, been smothered by sediment (Kirkman 1978).

The main effects of eutrophication and sedimentation include a decrease in water quality and stimulation of algal growth (Dennison *et al.* 1993; Short *et al.* 1995; Touchette and Burkholder 2000a), which both effectively reduce the light available for seagrass photosynthesis and reduce their growth rate. Seagrasses are adapted to short periods of light reduction, for example, the high turbidity levels experienced in estuaries after a storm or heavy rain, or when fine sediments are resuspended by wind driven waves. The plants carry sufficient rhizome stores to survive such 'pulsed' events but plant death occurs when the starch reserves are exhausted by extended periods of light reduction (e.g. McComb *et al.* 1981; Longstaff and Dennison 1999; Walker *et al.* 1999). Excessive nutrient loading may also have direct physiological effects that can cause dieback in some species (Touchette and Burkholder 2000a).

Heavy metals and petrochemicals in polluted waters are known to affect seagrass health (Ralph and Burchett 1998a; 1998b; Prange and Dennison 2000); hence, seagrasses growing in ports or bays supporting industry are particularly at risk from toxic pollutants in addition to sedimentation and eutrophication problems (Cambridge and McComb 1984). Direct shading by wharves and jetties block photosynthesis resulting in localised dieback of seagrasses (Kirkman 1989). Seagrass meadows may suffer direct physical damage from boat propellers (Pasqualini *et al.* 1999), mooring and anchor chains (Hastings *et al.* 1995), land reclamation, dredging or mining (Onuf 1994; Long *et al.* 1996; Wyllie *et al.* 1997) and hydrological changes associated with dredging or training wall construction (Larkum and West 1990). Recreational boat usage or construction of boating facilities may result in increased turbidity and sedimentation, erosion, shading, mechanical damage to seagrass beds, nutrient enrichment and other more insidious long term pollution events associated with antifoulants (Kirkman 1996).

Not all seagrass loss can be directly attributed to human activities. Natural pressures include storms and flooding (Clarke and Kirkman 1989), cyclones (Poiner *et al.* 1989), exposure to extreme low tides and weather conditions (Seddon *et al.* 2000), disease (den Hartog 1996), grazing (Cambridge *et al.* 1986; Supanawid 1996) and introduced marine pests (Meinesz 1999). However, human impacts ranging in scale from local changes in hydrology, regional species introductions to global climate change have been recognised as a contributing factor in many instances of 'natural' dieback (e.g. den Hartog 1996).

While fast growing seagrasses such as *Halophila* spp. can recolonise damaged or disturbed areas within months (Supanawid 1996) and *Zostera* spp. within years (Larkum and West 1983; Larkum and West 1990), slow growing species such as *Posidonia* spp. do not respond well to physical damage. The natural recovery in many *Posidonia* spp. is very slow and restoration may not be feasible in many situations (West *et al.* 1989; Kirkman and Kuo 1990; Meehan and West 2000). In addition, the more adaptable, faster growing seagrass species do not always respond well to attempts at restoration or rehabilitation (e.g. West 1995), particularly if the estuarine environment has been irreparably altered. Significant increases in seagrass coverage have, however, been observed in recent years, for example, in Western Australia (Kendrick *et al.* 1999; 2000) and Florida, USA (Kurz *et al.* 2000). These increases were associated with better water quality in response to government legislation and improved waterways and catchment management.

1.3 Seagrass management

Seagrass meadows need to be managed in order to balance their conservation with human usage of estuaries and their catchments. A variety of agencies can be involved in seagrass management at all levels of government (environmental protection authorities, land and water conservation agencies, parks, flora and fauna agencies, land use planning and development control agencies, fisheries managers, port/harbour/airport/defence authorities, etc) (Leadbitter *et al.* 1999). These management authorities need to know which species of seagrasses currently occur where and in what proportions and quantities (McNeill 1996). Other important factors include how seagrasses respond to human induced changes, what the natural seasonal, interannual and longer term dynamics of seagrasses are, and whether damaged beds can potentially be repaired or replanted (Kirkman 1997). In addition, managers need to know where seagrasses are likely to occur for the purposes of recovery and restoration, and to allow for natural spatial dynamics (McNeill 1996). The information is required in formats that can be easily interpreted and shared between management agencies. In short, seagrass managers need maps.

1.4 Mapping and monitoring of seagrass meadows

The first step in management is to provide baseline maps or measurements that document the extent and condition of seagrass meadows. The next step is to establish monitoring programs designed to detect disturbance at an early stage, and to distinguish such disturbance from natural variation in seagrass meadows (Kirkman 1996; Lee Long *et al.* 1996a). Any study of seagrass systems, whether field-based or remotely sensed, needs to recognise the spatial organization of seagrass meadows (Fonseca 1996) and be designed accordingly. Two spatial scales are relevant; the actual extent of the landscape under discussion, and the size of samples being described (Fonseca 1996). Three generalised levels of mapping and monitoring can be considered to provide information that satisfies the spatial scales relevant to seagrass; regional (coarse scale), local (intermediate to fine scale) and individual meadow scales. Mapping at the coarse scale will provide an inventory of the marine habitats in a region and a spatial framework for baseline descriptions and monitoring programs. Mapping at finer scales can provide detail on species composition and biophysical parameters, and can allow detection of small scale change in an area, particularly when combined with comprehensive ground survey (Kirkman 1996).

The three scales are complementary since they provide different kinds of information relevant to managers that can easily be combined in a Geographic Information System (GIS) for an integrated management approach. Such an approach is rarely applied, however, and the scale of mapping that is undertaken is too often determined by the available budget and human resources rather than by the objectives behind producing the map, i.e. the level of information that is required. There is a need to balance accuracy against costs to find the most cost-effective yet statistically robust method of seagrass mapping and monitoring for any particular situation (Kirkman 1997). Digital satellite and airborne remote sensing and aerial photography have been widely, though not always appropriately, applied to the task of seagrass mapping and monitoring (Walker 1988; Table 1.1). A large number of different data collection and processing systems are currently available and many more sensors are planned in the near future to provide panchromatic, multispectral and hyperspectral capacity with spatial resolutions ranging from 1m to 2.5 km (Phinn *et al.* 2000). The success of coastal remote sensing applications would certainly be increased if a set of objective guidelines were used to select the most appropriate data sets and processing techniques to meet a particular objective at a particular site (Phinn *et al.* 2000).

1.4.1 Regional scale: seagrass resource mapping

Coarse scale seagrass maps are typically of low precision and cover 100's – 1000's km^2 at scales usually greater than 1:25 000 and typically 1:100 000. Regional maps provide the spatial information needed to produce inventories of seagrass resources and associated habitats, to describe long term and large scale changes in meadows, to select and define marine parks and reserve boundaries and to identify sensitive areas so that decisions can

Table 1.1. Some examples of sensors and applications for the synoptic mapping of seagrass meadows and other types of submerged aquatic vegetation. Results of accuracy assessment or estimates of error have been included where stated by the authors.

Aim of mapping	Sensor (Resolution)	Accuracy	Reference
MAPPING DISTRIBUTION			
- map seagrass and ocean water colour	Landsat MSS (4 broad bands; 79 m pixels) Nimbus-7 CZCS (6 narrow and broad bands; 825 m pixels)		Claasen *et al.* 1984
- map submerged aquatic vegetation	Landsat MSS Landsat TM (7 broad bands; 30 m pixels)		Ackleson & Klemas 1987
- map intertidal and subtidal macroalgae	SPOT XS (3 broad bands; 20 m pixels)		Ben Moussa *et al.* 1989
- map seagrass meadows	Landsat TM		Lennon & Luck 1990
- map intertidal macroalgae	Itres CASI (8 narrow bands; 2.5m pixels)	65-86%	Zacharias *et al.* 1992
- map marine habitats	Geoscan MK2 (airborne multispectral scanner)		Ong *et al.* 1994
- map seagrass and macroalgal distribution	colour 1:12 000 API (visual interpretation)		Bulthuis 1995
- map seagrass and macroalgae	Itres CASI (13 narrow bands; ~5 m pixels)	85%	Bajjouk *et al.* 1996
- map benthic types, bathymetry and water optical properties	Spectron SE590 spectroradiometer suspended from blimp		Lee *et al.* 1997
- map tropical coastal habitats, e.g. seagrass, coral reefs, mangroves	Itres CASI (8 narrow bands; 1 m pixels)	81-89%	Clark *et al.* 1997 Mumby *et al.* 1998
- map invasive macroalgae *Caulerpa filiformis* in rocky subtidal	Itres CASI (14 narrow bands; 1 m pixels)		Bosma 1998
- map seagrass cover	Landsat TM	65-77%	Macleod & Congalton 1998
- map seagrass distribution	colour 1:20 000 API and side-scan-sonar; (image processing)	92%	Pasqualini *et al.* 1998
- map intertidal saltmarsh and macroalgae	Spectron SE590 spectroradiometer to simulate CASI bandsets	96-100%	Thomson *et al.* 1998

Table 1.1. (cont.)

Aim of mapping	Sensor (Resolution)	Accuracy	Reference
MAPPING DISTRIBUTION (cont.)			
- map macroalgae and seagrass	Daedalus MIVIS (102 narrow bands; 4m pixels)		Alberotanza *et al.* 1999
- map seagrass distribution and cover	HyVista HyMap (128 contiguous narrow bands; 5 m pixels)		Dunk & Lewis 2000
- map 32 ecological themes including seagrass, coral reefs, mangroves and soft substratum	SPOT XS		Chauvaud *et al*. 2001
- map benthic microalgae on coral reefs	Landsat TM	63%	Roelfsema *et al.* 2002
MONITORING / CHANGE DETECTION			
- monitor rapid seagrass dieback	SPOT XS		Robblee *et al.* 1991
- monitor change in seagrass distribution	Landsat TM		Zainal *et al*. 1993
- monitor seagrass, retrospective change detection	colour and B&W 1:5 000 – 1:50 000 API (visual interpretation)	<5% error	Hastings *et al.* 1995
- monitor natural seagrass spatial dynamics	colour and infrared 1:1 500 – 1:3 000 API (visual interpretation)		Turner *et al*. 1996
- map and monitor benthic habitats and cover, integrate data into GIS	Itres CASI (12 narrow bands; 5 m pixels)	overall Baye's error~20-30%	Jupp *et al.* 1995; 1996 Anstee *et al.* 1997
- monitor large area of seagrass, integrate data into GIS	Landsat TM	73%	Ferguson & Korfmacher 1997
- map and monitor sensitive coastal ecosystems	Itres CASI (12 narrow bands; 5m pixels)		Held *et al*. 1997
- feasibility of monitoring seagrass	colour 1:24 000 API (visual interpretation)		Robbins 1997
- monitor seagrass, retrospective change detection	Landsat MSS		Ward *et al.* 1997

Table 1.1 (cont.)

Aim of mapping	Sensor (Resolution)	Accuracy	Reference
MONITORING / CHANGE DETECTION (cont.)			
- monitor seagrass, retrospective change detection	colour and B&W 1:15 000 – 1:40 000 API (visual interpretation)		Wyllie et al. 1997
- retrospective change detection, impact of *Caulerpa taxifolia* on seagrass meadows	Itres CASI (18 narrow bands; 0.3-2.5 m pixels)		Jaubert *et al.* 1998
- monitor seagrass cover, change detection	Landsat TM	55%	Macleod & Congalton 1998
- monitor seagrass, retrospective change detection	colour and B&W 1:10 000 – 1:20 000 API (image processing)	<8% error	Kendrick *et al*. 1999
- monitor environmental impact in coastal zone, especially seagrass meadows	colour 1:20 000 API and side-scan-sonar (image processing)		Pasqualini *et al.* 1999
- monitor seagrass, retrospective change detection	colour and B&W 1:10 000 API (visual interpretation)		Kendrick *et al*. 2000
- monitor seagrass, retrospective change detection in a GIS	colour 1:24 000 API (image processing)		Kurz *et al.* 2000
- monitor seagrass, retrospective change detection	colour 1:20 000 – 1:40 000 API (visual interpretation)		Seddon *et al.* 2000
- retrospective seagrass change detection	colour and B&W (0.5-3 m pixels) API (visual interpretation)		Meehan 2001
- map and monitor seagrass, integrate into GIS	colour, infrared and B&W 1:2 000 – 1:6 500 API (image processing)		Pasqualini *et al.* 2001
- monitor seagrass, baseline for future change detection	colour 1:10 000 API (visual interpretation)		Kendrick *et al*. 2002
- retrospective seagrass change detection	Landsat TM & ETM	54-76%	Dekker *et al.* 2003

Table 1.1. (cont.)

Aim of mapping	Sensor (Resolution)	Accuracy	Reference
ESTIMATING BIOMASS / COVER / DENSITY			
- estimate intertidal seagrass and macroalgal biomass	false colour 1:10 000 – 1: 20 000 API (image processing)		Meulstee et al. 1988
- estimate seagrass biomass (standing crop)	Landsat TM	$r^2 = 0.80$	Armstrong 1993
- estimate harvestable biomass of intertidal macroalgae	SPOT XS	$r^2 = 0.84$	Guillaumont *et al.* 1993
- estimate cover of macroalgal species	Itres CASI (13 narrow bands; 5m pixels)	$r^2 = 0.96$	Bajjouk *et al.* 1998
- estimate seagrass leaf area index and shallow water bathymetry	Ocean PHILLS airborne sensor (128 contiguous narrow bands; 1.25 m pixels)	$r^2 = 0.88-0.98$	Dierrsen *et al.* 2003
SENSOR COMPARISONS			
- compare sensors for monitoring macroalgal biomass, integrate data into GIS	SPOT XS; Positive Systems ADAR (4 broad bands; 2.3 m pixels		Deysher 1993
- compare sensors for mapping impacts of pollution and *Caulerpa taxifolia* on seagrass meadows	Landsat MSS SPOT XS and Pan NOAA AVHRR (5 broad bands; 1.1 km pixels)		Guillaumont *et al.* 1995
- compare sensors for mapping benthic habitats including coral reef, seagrass and macroalgae cover classes	Itres CASI (8 narrow bands; 1 m pixels) SPOT XS Landsat TM and merged TM/SPOT Pan Landsat MSS colour 1:10 000 API (visual interpretation) IKONOS-2 (7 broad bands; 4 m pixels)	81% 37% 30%, 35% 21% 67% 41-64%	Mumby *et al.* 1997a Mumby *et al.* 1999 Mumby and Edwards 2002
- compare sensors for estimating seagrass biomass (standing crop)	Itres CASI (8 narrow bands; 1 m pixels) SPOT XS Landsat TM	$r^2 = 0.81$ $r^2 = 0.79$ $r^2 = 0.74$	Mumby *et al.* 1997b

Table 1.1. (cont.)

Aim of mapping	Sensor (Resolution)	Accuracy	Reference
SENSOR COMPARISONS (cont.)			
- compare sensors for mapping seagrass distribution and abundance	1:24 000 colour API (visual interpretation) Itres CASI (48 narrow bands; 1 m pixels) Itres CASI (20 narrow bands; 2.5 m pixels)		Virnstein *et al*. 1997
- feasibility of detailed benthic habitat mapping	Itres CASI (12 narrow bands; 5 m pixels) colour API (1.8 m pixels; visual interpretation) SPOT XS Daedalus 1268 (many narrow bands; 5 m pixels)	50% 63% 38%	Rollings *et al.* 1998
- compare sensors and field methods for mapping intertidal and subtidal habitats	GPS field mapping at 1:500 1:5 000 colour API (visual interpretation) Itres CASI (spectral resolution ns; 4 m pixels)		Levings *et al*. 1999

Sensors: MSS Multispectral Scanner, CZCS Coastal Zone Colour Scanner, TM Thematic Mapper, XS multispectral, Pan panchromatic, AVHRR Advanced Very High Resolution Radiometer, ADAR Airborne Data Acquisition and Registration (multispectral digital video), CASI Compact Airborne Spectrographic Imager, MIVIS Multi-spectral Infrared and Visible Imaging Spectrometer, HyMap Hyperspectral Mapper, PHILLS API Aerial Photograph Interpretation, B&W black and white.

be made about development in and near estuaries and bays (McNeill 1996; Kirkman 1996; 1997; Thomas *et al.* 1999).

The classes that can be confidently mapped at this scale are typically ‘seagrass’ versus ‘non-seagrass’ (vegetated/non-vegetated), or basic habitat classes such as ‘seagrass’, ‘sand’, ‘coral’, and ‘macroalgae’, which can be mapped to a maximum overall accuracy of around 75-85% using Landsat Thematic Mapper (TM) satellite imagery (Table 1.1).

Satellite imagery is a cost effective approach for mapping and monitoring seagrass in large and remote regions (Ferguson and Korfmacher 1997; Mumby *et al.* 1999) particularly when mapping large and/or continuous meadow forming species such as *Zostera marina* (Ward *et al.* 1997). The spatial resolution of Landsat TM (30 m ground resolution units or pixels) and Systeme Pour l’Observation de la Terra High-Resolution-Visible (SPOT HRV) multispectral sensor (XS; 20 m pixels) are an appropriate compromise between the need for accuracy and large areal coverage with both sensors having been successfully used to map benthic vegetation (Table 1.1). All but the most recent satellite remote sensing systems carry broad band sensors of relatively low spectral resolution, i.e. they gather data in only a few spectral bands that average the signal received across a fairly wide range of wavelengths (see Jensen 1996 for a description of the sensor system characteristics). Seagrass mapping has been attempted with broad band satellite sensors of coarser spatial resolution including Landsat Multispectral Scanner (MSS; pixel size 79 m), National Oceanographic and Atmospheric Administration’s Advanced Very High Resolution Radiometer (NOAA AVHRR; pixel size 1.1 km) and the Nimbus-7 Coastal Zone Colour Scanner (CZCS; pixel size 825 m at nadir) but maps produced were of limited value (Table 1.1). The pixel size of these sensors is too large given the size and patchy nature of most seagrass meadows and may not even be relevant to the scale of many estuaries (Cracknell 1999). The accuracy that can typically be expected when mapping broad benthic habitat classes using Landsat MSS is in the range of only 30-60% (Mumby *et al.* 1997a), except in the unusual case where seagrass meadows are very large and contiguous (Ward *et al.* 1997).

Satellite remote sensing methods have been criticised because a spatial resolution of $\geq$20 m will not discriminate beds that are small, patchy, and/or linear and narrow, all of which commonly occur in temperate estuaries and tropical reef areas (Lehmann and Lachavanne 1997). Hence there is a bias in mapping towards shallow water and large species (Thomas *et al.* 1999). Increasing the spatial resolution of satellite sensors may increase the accuracy of mapping in these circumstances but to date this has not necessarily been the case. Mumby and Edwards (2002) noted that the IKONOS-2 multispectral satellite with a spatial resolution of 4 m mapped coarse benthic habitat classes (seagrass, sand, coral, macroalgae) to an overall accuracy of 75% - only 2% better than the accuracy gained using Landsat TM for the same task and at an enormously greater cost. Merging the SPOT HRV 10 m pixel size panchromatic (Pan) band with Landsat TM and SPOT XS imagery actually decreased the accuracy of classification of these benthic classes (Mumby *et al.* 1997a). Mumby *et al.* (1997a) found 1:10 000 aerial photographs significantly less accurate than Landsat TM for coarse scale seagrass mapping. To map a large region with aerial photographs requires a huge investment in time and money, especially if the photographs need to be georeferenced and mosaiced into a single map (O'Neill *et al.* 1997; Mumby *et al.* 1999). For example, Landsat TM imagery was used to delineate underwater reefs and shoals and define park boundaries for the 2 000 km long Great Barrier Reef Marine Park because aerial photograph mapping would have taken several decades to complete (Claasen *et al.* 1986).

The spectral resolution (i.e. the wavelength location and width of spectral bands) and signal-to-noise ratio (SNR) of Landsat TM is somewhat better for the discrimination of benthic vegetation than that of the SPOT XS. The radiometric resolution (i.e. number of brightness levels that can be resolved) of the TM and XS sensors are also well suited for this task (Lubin *et al.* 2001) and both have significantly better resolution than that of Landsat MSS (Ackleson and Klemas 1983; Mumby *et al.* 1997a). Three of the seven broad wavebands of Landsat TM are useful for water penetration and discrimination of submerged aquatic vegetation; these bands are centred in the visible blue (450-520 nm), green 520-600nm and red (630-690 nm) wavelength regions. Such spectral resolution appears adequate for benthic mapping where the level of discrimination is simply vegetated from non-vegetated, however Landsat TM has been less successful at distinguishing reef or soft bottom habitat classes (e.g. Mumby *et al.* 1997a; 1998).

The selection of a good satellite scene displaying favourable atmospheric, weather and sea surface conditions, and the availability of accurate bathymetry data, will greatly enhance seagrass discrimination (Ferguson and Korfmacher 1997). Turbid water, a hazy or cloudy atmosphere, a rough ocean surface, and error in the knowledge of the mean ocean depth for a given scene, make satellite mapping of benthic habitats much more difficult (Lubin *et al.* 2001). The effects of atmospheric interference and attenuation must be corrected for in all remote sensing data but particularly when images are collected from a satellite platform (Duggin 1987). Attenuation of the benthic signal by an overlying water column leads to depth-dependent effects on spectral reflectance that can be accounted for using modelling or masking procedures (e.g. Lyzenga 1981; Zainal *et al.*

1993; Jupp *et al.* 1996; Tassan 1996; Malthus *et al.* 1997b; Plummer *et al.* 1997; Bosma 1999).

1.4.2 Local scale: habitat mapping and change detection

Local scale seagrass maps (of typical scale 1:5 000 to 1:25 000) usually attempt to map seagrass species, associations or habitat classes within a single estuary or bay, giving some indication of their cover or density as well as their distribution. At the finer scale (generally around 1:10 000), mapping will assist managers in deciding whether seagrass meadows are stable or declining, whether particular developments have, or will impact on meadows, or whether mitigation of damage in polluted areas has occurred and seagrasses are recovering (Kirkman 1997). Mapping can also be used to monitor natural spatio-temporal patterns and dynamics in seagrass meadows for landscape ecology studies or to assess their value as fisheries habitat (e.g. Irlandi *et al.* 1995).

Fine scale mapping

The synoptic overview provided by remote sensing offers many advantages over purely field-based mapping of seagrass even at this scale. Field methods may involve huge investments of time and labour (Mumby *et al.* 1999), and even the more sophisticated underwater videography (e.g. Norris *et al.* 1997; Norris and Wyllie-Echeverria 1997) and acoustic (e.g. Lee Long *et al.* 1998; McCarthy and Sabol 2000) transect methods, carry an inherently large capacity for mapping error because of the necessity for spatial interpolation between data points. Field data are, however, a crucial component of any mapping exercise (Mumby *et al.* 1999). Transect, point or quadrat data collected in conjunction with a differential Global Positioning System (dGPS) for accurate location data will provide detailed information that can be used in the development of classes prior to classification, or as independent samples for post-classification accuracy assessment.

Side-scan sonar has the potential to produce spatially descriptive benthic maps (Sabol *et al.* 1997; McCarthy and Sabol 2000). Pasqualini *et al.* (1998; 1999) found this method an excellent adjunct to aerial photographs in water too deep or turbid for optical penetration. For example, *Posidonia oceanica* grows to a reported 50 m depth in the Mediterranean Sea (Duarte 1991), well below the effective range of most satellite and airborne sensors that can penetrate to a depth maximum of around 20 m in very clear water (e.g. Pasqualini *et al.* 1998; Chauvaud *et al.* 2001). Most seagrass species do not grow this deep (generally to around 2-3 m in estuaries; Kirkman 1997, and to 10 m in open water bays; Kirkman *et al.* 1995) so the distribution of most species is well within the scope of optical remote sensing methods.

Aerial photograph interpretation (API) is still the most widely used method for mapping seagrasses (Thomas *et al.* 1999) and submerged aquatic vegetation in general (Lehmann and Lachavanne 1997). This may be partly due to the fact that aerial photographs are readily available in a variety of scales and a range of formats (colour, black and white, IR) (Watford and Williams 1998) and can be flown as required temporally. Unlike digital remote sensing, visual interpretation of aerial photographs does not necessarily require highly trained staff and specialist software and many researchers still use subjective visual interpretation methods for API because of the advantages that contextual and textural knowledge provide to the mapping (e.g. Meehan 2001; Kendrick *et al.* 2000). These methods are very time consuming, however, and they do require specialist knowledge of species and their habitat preferences. Image processing techniques have recently gained appeal because they are objective and make the production of maps much more time efficient and economical (e.g. Pasquilini *et al.* 2001; Kendrick *et al.* 2002). The disadvantage of these techniques is that the semi-automated segregation of phototones usually requires accurate knowledge of bathymetry (Pasqualini and Pergent-Martini 1996; Hart 1997; Pasqualini *et al.* 2001) and is successful only where seagrass canopies are dense (Kendrick *et al.* 2000). In addition, there is much scope for confusion between seagrass meadows and other 'dark' benthic substrates, e.g. kelp beds and accumulations of organic detritus.

API supported by extensive ground survey has been used to map detailed benthic habitat and/or cover classes of seagrass and macroalgal species (Table 1.1). Although accuracy assessment was not performed in the studies mentioned here, the accuracy of such maps is probably very high when classes are distinct in colour, texture, density or habitat preference and grow in mainly monospecific stands in clear shallow water. Mumby *et al.* (1997a) found that 1:10 000 API mapped eight habitat classes (assemblages of corals, seagrass, macroalgae, sponges and associated substrata) to significantly greater accuracy (67% overall accuracy) than satellite sensors (Landsat TM, SPOT XS and IKONOS 2 enhanced by a textural filter; overall accuracy of each was around 50%). However, this level of accuracy is low compared to the 81% overall accuracy achieved by a digital airborne imaging spectrometer CASI (Compact Airborne Spectrographic Imager, Itres Instruments Inc.) when applied to the same task (Mumby *et al.* 1997a; 1998). Some authors have suggested that different seagrass species cannot be distinguished in aerial photographs (Pasqualini and Pergent-Martini 1996; Robbins 1997). This is likely to be the case where species of similar colour and canopy structure occur in the same estuary or where they grow together in mixed assemblages (Kendrick *et al.* 2002). Problems with species discrimination will be aggravated if seagrasses grow in deeper or turbid water.

Aerial photographs are difficult to georectify and that is one reason they are rarely accuracy assessed. Despite this, aerial photographs and maps produced from API (that were never field checked themselves) are often used to 'ground truth' remote sensing classifications or subsequent API maps (e.g. Meehan 2001). There seems to be a general assumption that, because aerial photographs have high spatial resolution and can map the boundaries of benthic classes in detail, these boundaries are geographically correct and the classes have been identified correctly. Aircraft provide sensor platforms that are far less stable than satellites, particularly over turbulent coastal areas. Hence, the precision of aerial photography is seldom better than 5m, even after meadow boundaries have been positioned geographically, surveyed and rectified for such aberrations as roll, twist and yawl of the plane and curve of the sensor's lens (Kirkman 1996; Mumby *et al.* 1999). Some quantification of the level of classification accuracy and geometric error should be included with all maps, no matter how they are produced (Thomas *et al.* 1999). Accuracy should be verified after classification has been completed by statistically valid field checking of the mapped classes (Congalton 1991).

Traditional satellite remote sensing instruments do not have the spatial or spectral resolution to readily distinguish between different benthic plant types (Guillaumont *et al.* 1993; Jernakoff and Hick 1994). For example, Landsat TM imagery could not adequately discriminate between 13 coral reef and seagrass habitats in the clear, shallow water of the Caribbean (Mumby *et al.* 1997a). Mixed pixels were a problem in this classification because the pixel size of 30 m was of similar magnitude to the size of habitat patches in this region, with many seagrass and reef habitats dominated by photosynthetic organisms (coral zooxanthellae, macroalgae, seagrasses) (Mumby *et al.* 1999). Spectral reflectance differences between these organisms are subtle, and require high spectral resolution data and possibly spectral derivative analysis for class separation (e.g. Holden and LeDrew 1999; Myers *et al.* 1999; Hochberg and Atkinson 2000; Kutser *et al.* 2001). As the current satellite sensors have only three to four water penetrating bands with bandwidths around 70 nm, their capacity for spectral discrimination of benthic vegetation and corals is limited. However, fine scale habitat mapping with SPOT XS has been reported; Chauvaud *et al.* (2001) mapped 32 complex ecological communities including coral, mangroves and 12 cover classes of the seagrass *Thalassia testudinum* on a variety of substrates in the West Indies. Presumably these habitat classes were large and quite distinct in their spectral response.

Hyperspectral remote sensing can provide the spectral resolution needed to detect subtle reflectance differences in the pigment content of plants. Airborne hyperspectral scanners (imaging spectrometers) and high spectral resolution multispectral ('superspectral') scanners provide images with high spatial resolution in many, specific, narrow spectral bands (Jensen 1996). True hyperspectral systems typically offer around 200 contiguous bands of ca. 2-10 nm width through the visible, NIR and mid-IR portions of the spectrum. NASA's Airborne Visible-Infrared Imaging Spectrometer (AVIRIS) was one of the first of these sensors but more recent instruments such as the HyMap (HyVista Corp.) have improved on sensor design features including the SNR. The CASI has hyperspectral potential when in 'spectral mode' but is most often used in 'spatial mode', in which it is programmed to record 10-20 spectral channels of desired location and bandwidth.

High spectral resolution and band placement may be as important as high spatial resolution for detailed vegetation classification (Mumby *et al.* 1997a; 1997b). It is desirable to measure the spectral signatures of the target species with a ground spectroradiometer before image acquisition and select or program an appropriate bandset based on this information (Bajjouk *et al.* 1996). Thomson *et al.* (1998) found that more information could be extracted from many, narrow bands of ca. 10-15 nm widths than from fewer, broad bands. These authors did note that band placement was less important in the classification of intertidal seaweeds if many narrow bands were available (Thomson *et al.* 1998). It would be preferable, however, to program bands with the most appropriate bandwidth and wavelength position for any particular application so as to avoid redundancy in the data set.

The CASI scanner has been applied in the detailed mapping of coral reef and seagrass habitats, subtidal and intertidal macroalgal communities (Table 1.1) and to identify and quantify phytoplankton blooms (e.g. Jupp *et al.* 1994). Alberotanza *et al.* (1999) mapped seagrass and macroalgal beds using the airborne Daedalus Multi-spectral Infrared and Visible Imaging Spectrometer (MIVIS). Promising results have also been obtained using HyMap imagery to map beds of *Zostera* spp. and *Posidonia* spp. in South Australia (Dunk and Lewis 2000). In a pilot study to determine the sensor and methods most suitable for mapping benthic assemblages in a relatively large and turbid estuary (Port Phillip Bay in southern Australia), Rollings *et al.* (1998) found that colour aerial photographs gave better overall accuracy (63%) than CASI (50%), Daedalus 1268, and SPOT XS (38%). The superior accuracy of API in this study must be related to a large degree to the higher spatial resolution of the aerial photographs (1.8 m pixels compared to 5 m for CASI and Daedalus, and 20 m for SPOT) and the fact that the area mapped was small and well known to the interpreter. When the spatial resolution of image data is roughly equivalent, CASI will generally perform better than API (Mumby *et al.* 1997a). Rollings *et al.* (1998) noted that the skill and time required for API would be prohibitive over a large area and recommended CASI for the production of benthic maps across the whole of Port Phillip Bay. A very important feature of the CASI data set was that it could penetrate the estuarine water column and map to a depth of 12 m while aerial photographs could only penetrate to around 6 m, SPOT XS to 5 m and Daedalus to < 2m (i.e. insufficient for mapping benthic vegetation) (Rollings *et al.* 1998).

While airborne remote sensing studies appear to have been very successful at mapping benthic vegetation into detailed classes, like API, few of these studies have reported post-classification accuracy assessment. Geometric correction of any airborne image is difficult and georectification problems are exacerbated by the fact that ground control points are usually non-existent in aquatic scenes (Dunk and Lewis 2000). CASI has recently overcome this problem to some extent with new onboard instruments that collect pixel-by-pixel dGPS and irradiance data for significantly better geometric and radiometric correction (Levings *et al.* 1999).

With the recent launch of the earth orbiting EO-1 Hyperion (220 narrow spectral bands at 30 m spatial resolution), and with several more hyperspectral satellites planned for launch in the near future, the question is whether hyperspectral resolution will improve the accuracy of seagrass mapping from satellite platforms. To investigate the potential for detailed species or habitat class discrimination by satellite sensors, Lubin *et al* (2001) collected the spectral signatures of four coral species, sand and three benthic algal habitats *in situ* using an underwater spectroradiometer. The spectra were modelled through an 'ideal' atmosphere and water column using a radiative transfer algorithm to examine how the reflectance might appear after attenuation by the water column and atmosphere. It was concluded that many of the spectral features used to discriminate coral species and associated benthic algal habitats using surface reflectance (Holden and LeDrew 1999) or aircraft data (Hochberg and Atkinson 2000) would be obscured in satellite data. This suggests that the new generation hyperspectral satellites may not achieve a higher level of seagrass species or habitat discrimination than currently afforded by broad band satellite sensors. At this stage, airborne imaging spectrometry appears to offer the most accurate solution for seagrass species and habitat mapping.

Change detection

The historical archive of aerial photographs dating back at least to the 1960's has allowed for retrospective study of historic changes in the extent and distribution of seagrass meadows (e.g. Larkum and West 1990; Meehan 2001; Kendrick *et al.* 2002). Opportunistic use of archival material provides valuable historic information that could not otherwise be obtained. For example, Short *et al.* (1996) were able to establish a relationship between increased housing development and a 32-year decline in *Z. marina* in Ninigret Pond, Rhode Island, USA using colour and thermal IR aerial photographs. However, the variable radiometric quality and scale of archival photographs introduces error (Hastings *et al.* 1995), and when combined with image rectification errors, allows only relatively large magnitude changes in the extent or distribution of seagrass meadows to be detected with any degree of certainty. The lack of identifiable ground control points for georectification of multidate aerial photographs is not the least of these problems (Meehan 2001). Thomas *et al.* (1999) warn against quantitative assessment of seagrass change when using aerial photographs that are not of identical format and not specifically commissioned for the purpose of identifying and monitoring seagrass beds.

The geometric repeatability of satellite data is high but the precision of the coarse scale maps that can be produced is low and largely unquantified. Satellite remote sensing is a cost effective method for detecting large changes in seagrass distribution or extent (Robblee *et al.* 1991; Zainal *et al.* 1993; Ward *et al.* 1997; Macleod and Congalton 1998) but local change may be indistinguishable from map error. If satellite data are used for local scale monitoring then significant real change may be missed (Thomas *et al.* 1999).

All airborne instruments allow the user to specify the spatial resolution to suit the aims of the project and scale of the targets being mapped. Very high resolution of less than 0.5 m is possible and limited only by the flying speed of the aircraft platform (e.g. see Jaubert *et al.* 1998) although there is always a trade off between accuracy and areal coverage. Airborne instruments also allow for flexibility in the timing of data collection, for example, in response to a significant event such as an oil spill. In addition to high spatial and temporal resolution, modern digital airborne remote sensing provides high spectral and radiometric resolution, repeatability in image geometry and processing methodology, digital format for ease of integration into GIS and the rapid turnaround time required to monitor vegetation for small scale change. For example, Smith *et al.* (1998) mapped relatively subtle seasonal and longer term changes in the vegetation cover of a coastal wetland using multitemporal CASI images. In the Mediterranean, the recent distribution of the introduced alga *Caulerpa taxifolia* has been mapped with CASI and the effect of its invasion on *P. oceanica* beds has been deduced using aerial photographs from as far back as 1988 (Jaubert *et al.* 1998). While multidate airborne scanner data has not yet been used to monitor seagrass meadows, several baseline maps have been produced for the specific purpose of future monitoring (e.g. Jupp *et al.* 1996; Virnstein *et al.* 1997; Pasqualini *et al.* 2001).

1.4.3 Meadow scale: Monitoring seagrass health

Seagrass research to date has concentrated on meadow scale studies of ecological and physiological factors using field-based methods predominantly carried out by SCUBA divers (Kirkman 1996; Duarte 1999). Permanent transects or random quadrats have been used to record shoot density, biomass and other factors such as epiphyte cover, distance from shallow to deep edge of beds, position of blowouts, etc (e.g. King 1986; King and Holland 1986; Gordon *et al.* 1994; 1996; Mumby *et al.* 1997c; Kirkman and Kirkman 2000; Morris *et al.* 2000). At the meadow level, managers are primarily interested in the 'health' of the seagrass, as well as seagrass productivity, nutrient cycling and ecological interactions between seagrasses and associated communities of flora and fauna. It has become particularly relevant to monitor meadow condition under the current levels of stress imposed by human development and usage of estuaries and their catchments. While field measurements provide detailed information and are precise methods for detecting small scale change at the meadow level, they are very localised (Kirkman 1996; Wilzbach *et al.* 2000). Remote sensing methods are more likely to detect longer term and larger scale trends in seagrass decline, particularly if regular monitoring programmes are established.

Monitoring seagrass canopy parameters

Stress-related dieback can be observed in seagrass meadows through a loss of standing crop, reduced shoot density or reductions in meadow extent (Kirkman 1996). For example, the maximum depth distribution in which a certain species can survive decreases rapidly with declining water quality (Duarte 1991; Dennison *et al.* 1993). Growth and productivity can be severely impaired by shading (e.g. Dawes and Tomasko 1988; Kirkman 1989; Czerny and Dunton 1995), and the result will be reflected in a loss of seagrass biomass (Duarte and Chiscano 1999) and reduced shoot density (e.g. Kirkman 1989). Short, stunted growth forms can be observed for all seagrass species when exposed to persistent environmental stress (Walker *et al.* 1999).

Shoot or stem density is possibly the best biophysical measure of seagrass stress because it responds rapidly and predictably to disturbances such as shading (Bulthuis 1984; Neverauskus 1988; Alcoverro *et al.* 2001) while productivity may give confusing results (Kirkman 1996). Shoot dieback typically occurs over a time frame of months or years in stressed seagrass beds (Pergent-Martini and Pergent 1996; Kirkman 1989). Meehan (2001) found that monitoring shoot density was a particularly effective method of distinguishing anthropogenic impact from natural variation in *Posidonia australis* meadows. However, it may be logistically difficult to count the shoots of small species such as *Halophila* and *Zostera* spp. Since shoot density is often closely correlated with biomass (Kirkman and Manning 1993) which also responds quickly and measurably to perturbation (Kirkman 1996), then biomass may be a better general measure of meadow condition.

Armstrong (1993) and Mumby *et al.* (1997b) evaluated the potential for remote sensing measurement of seagrass biomass by establishing empirical relationships between vegetation indices derived from image data and field measurements of the standing crop. A crucial processing step was to remove the confounding effects of variable water depths from the scene. Mumby *et al.* (1997b) used Lyzenga's (1981) method of creating single depth-invariant bottom indices from pairs of spectral bands (i.e. ratios of two bands whose product cancels out the affects of depth on the brightness of the image bands) using only those wavelengths suitable for water penetration (400-650 nm). The best estimation of seagrass biomass (r^2 = 0.81; 95% confidence interval) was obtained from a ratio of narrow CASI wavebands at the green vegetation reflectance peak and red chlorophyll absorption features in which reflectance differences between sand and vegetation were maximised (Mumby *et al.* 1997b). The level of predictive confidence obtained from remote sensing was similar to that achieved by *in situ* quadrat sampling of seagrass biomass at medium densities, and better than quadrat sampling for dense seagrass stands (Mumby *et al.* 1997b).

Plummer *et al.* (1997) and Malthus *et al.* (1997a) used an alternative approach by developing invertible radiative transfer models to estimate the leaf area index (LAI) or biomass of seagrass canopies from spectral reflectance at a remote sensor. Plummer *et al.* (1997) combined the SAIL (Scattered by Arbitrarily Inclined Leaves; Verhoef 1984) model of radiative transfer within a vegetation canopy and PROSPECT model of leaf reflectance and transmittance (Jacquemoud and Baret 1990) with a water propagation model (based on Sathyendranath and Platt 1989) to simulate the reflectance from seagrass canopies with various LAI. These model inversions do not appear to have been tested for submerged aquatic vegetation on imagery acquired from satellite or airborne platforms as yet, although they have been successfully applied in remote sensing the biophysical properties of terrestrial crops (e.g. Jacquemoud *et al.* 1995). There is much potential for future work in this area.

Monitoring the physiological status of seagrass plants

While canopy condition is an excellent indicator of a seagrass meadow's ultimate response to stress, the immediate health of seagrass leaves can be directly assessed by measuring aspects of their biochemistry and photosynthetic performance. Seagrass species have demonstrated the ability to photoacclimate their photosynthetic apparatus to changing environmental conditions by adjusting their total pigment concentration and/or the relative proportions of their light harvesting and photoprotective pigments (e.g. Dennison and Alberte 1985; Abal *et al.* 1994; Major and Dunton 2002; see Chapter 5). Chlorophyll *a* and other photosynthetic pigments determine the photosynthetic potential of a plant and therefore their concentrations should be strongly related to physiological status (Chapelle *et al.* 1992). Declines in photosynthetic capacity due to environmental stress are often accompanied by changes to the photosynthetic pigments such as reductions in chlorophyll, changes in the ratio of chlorophyll *a* to chlorophyll *b*, and increases in various carotenoids (particularly ß-carotene and the xanthophyll cycle pigments). In some plant species there may also be increases in nonphotosynthetic pigments such as anthocyanins (Barker *et al.* 1997; Stone *et al.* 2001; Coops *et al.* 2004).

Many field- and laboratory-based measures of seagrass physiology or growth have been used to examine the response of seagrasses to various forms of stress. The pigment content of seagrass leaves can be determined by laboratory analysis of leaf samples collected in the field using spectrophotometric methods (e.g. Pérez-Llorens *et al.* 1994; Major and Dunton 2000) or high performance liquid chromatography (e.g. Ralph *et al.* 2002). The rate of oxygen evolution into leaf chambers (e.g. Goodman *et al.* 1995; Alcoverro *et al.* 2001) or graduated pipettes (lacunal-gas technique; Pollard and Greenway 1993) provides a measure of the actual photosynthetic rate of seagrasses. Submersible pulse-amplitude modulated fluorometers or 'diving-PAMs' have also been used to measure the instantaneous photosynthetic efficiency of seagrass leaves from their chlorophyll fluorescence (e.g. Dawes and Dennison 1996; Ralph *et al.*

1998; Major and Dunton 2002). Other techniques include leaf marking to examine growth rates (e.g. Zieman and Wetzel 1980), estimates of photosystem density and size (Major and Dunton 2002), and measures of leaf nitrogen content, amino acid concentrations and stable isotope ratios (e.g. Abal *et al.* 1994; Udy and Dennison 1997a).

There is some potential for remote sensing methods to detect stress in seagrass meadows before actual reductions in biomass or areal extent occur, but no such studies have been attempted to date. Stress-induced changes in the physiology and morphology of terrestrial plants cause changes in spectral reflectance that have been detected using satellite or airborne remote sensing (e.g. Rock *et al.* 1988; Lelong *et al.* 1998; Curran *et al.* 1998; Qiu *et al.* 1998 and see review of Field *et al.* 1995). Numerous laboratory (leaf level) or field spectrometry (canopy level) studies have shown that spectral indices, derivatives or algorithms derived from hyperspectral data can predict the physiological status of vegetation from indicators such as chlorophyll content, the relative concentration of carotenoids to chlorophylls, and photosynthetic performance (e.g. see review of Peñuelas and Filella 1998 and see Chapter 6). Relatively fewer studies have actually applied narrow band indices or algorithms in the detection of plant stress from airborne or satellite remote sensing data (e.g. Zarco-Tejada *et al.* 1999; 2000; Coops *et al.* 2003; 2004). The advantage of remote sensing over field based methods for monitoring the health of seagrasses is that it is possible to acquire a synoptic overview of a large meadow or whole estuary in a fraction of the time and effort it would take to measure the same area by boat. Broad scale patterns and trends in photosynthesis or stress can only be identified using spatially relevant data.

1.5 Image processing and the application of spectral libraries

Effective classification of benthic vegetation requires that the different species, habitats or communities to be mapped can be discriminated by their spectral reflectance (Hochberg and Atkinson 2000). The spectral separation of benthic classes is a function of the inherent reflectance differences between these features, how strongly their reflectance is attenuated by the overlying water column and atmosphere, and how sensitive the remote sensing instrument is to the reflectance differences that are preserved after attenuation (Ackleson and Klemas 1983). The spectral signatures of benthic classes must therefore differ, after atmospheric and water column attenuation of the signal has occurred, by a magnitude that can be detected by the sensor. In addition, these differences must occur in wavelength regions that can be resolved and are not averaged out by the sensor.

If remote sensing mapping techniques are to be generally applied, then the spectral reflectance differences that separate vegetation species or habitats must also be consistent from place to place, and over time. There is intrinsic natural variability in the spectral signature of a healthy seagrass due to the influence of water quality, nutrient availability, light levels, temperature, water movement, and associated epiphyte communities, etc, that all vary considerably both spatially and temporally under natural conditions (Kirkman *et al.* 1995). Hence, it is necessary to characterise the degree of variability and range of signatures expected for any species or habitat class under natural conditions if the signatures are to be used in spectral libraries for quantitative spectral analysis (Price 1994) or as a baseline for the detection of spectral change in response to stress. Seagrass managers, in particular, need to know the extent of natural changes in seagrass meadows so that human impacts can be separated from normal background variation (Lee Long *et al.* 1996a).

Jupp *et al.* (1995) and Malthus and George (1997) have identified the need to build marine spectral libraries as the basis for developing routine methods of applying remote sensing to aquatic vegetation surveys. Spectral libraries are a fundamental part of many hyperspectral image classification and radiative transfer modelling procedures including atmospheric and water column correction, spectral unmixing and endmember mapping (e.g. Jupp *et al.* 1996; Anstee *et al.* 1997; Held *et al.* 1997; Kruse *et al.* 1997; Malthus *et al.* 1997a; Plummer *et al.* 1997). The success of these image processing techniques depends on the comprehensiveness and quality of the spectral library that is used (Taylor *et al.* 1998). Spectral libraries also provide information needed to make an informed choice about the spectral resolution required of a sensor and the wavebands that will be most useful for a particular application.

1.6 Why spectral studies of seagrasses?

The urgent need to manage estuarine ecosystems and to minimise future impacts has been recognised in government environmental policy whose major themes include the conservation of biodiversity and Ecologically Sustainable Development. The State of the Marine Environment Report (Zann 1995) identified the loss and deterioration of temperate seagrasses as one of the most serious environmental issues facing the nation's coastal waters. Resource managers and policy makers often assume scientists already have a good knowledge of the biological composition and function of coastal ecosystems and can therefore predict the impact of management decisions on an estuary. In fact, little information is available on what constitutes 'healthy' estuarine habitats or the extent to which existing and proposed activities might impact on estuarine communities (West 1997). Managers need a rapid means of assessing estuarine condition so that decisions can be made about the type and extent of development, human activity or resource use that should be allowed to take place in and around an estuary. Ecologically sustainable development can only be realised in coastal regions if rapid feedback on the impacts associated with development becomes available. Airborne remote sensing is an obvious tool for this kind of application.

Seagrass meadows have been demonstrated to be an important bioindicator of the condition of the whole estuarine ecosystem (Dennison *et al.* 1993; Bearlin *et al.* 1999; Walker *et al.* 2001) because they show measurable and timely responses to environmental impacts (Lee Long *et al.* 1996a). Eelgrasses (*Zostera* species), for example, respond dynamically to the environment by changes in growth form, spatial pattern, biomass (Kirkman *et al.* 1995). Once seagrass decline has been expressed through a change in biomass, plant distribution or morphology, however, the effects are nearly always irreversible. Increases in our understanding of seagrass physiology may provide an opportunity to detect stress in seagrasses before anthropogenic impacts lead to the dieback of meadows (Major and Dunton 2000).

The ground-based spectral reflectance studies presented in this book were undertaken for the purpose of generating baseline data that would advance the development of remote sensing procedures for the mapping and monitoring of seagrass meadows. Chapter 2 provides background on the ecology, growth, morphology and physiology of the Australian seagrass species used in the studies; eelgrass *Zostera capricorni*, strapweed *Posidonia australis* and paddleweed *Halophila ovalis*. Leaf structure and photosynthesis have a particular influence on the spectral reflectance measured from seagrass leaves. Chapter 3 describes the field and laboratory experiments and data collection and analysis methods that were applied in this research.

In Chapter 4, patterns of seagrass reflectance are characterised using detailed field spectroradiometer measurements made on seagrass leaves taken from different estuaries and habitats, across seasons and years. Chemical analysis of the photosynthetic and accessory pigment composition of concurrent leaf samples explains the differences in spectral reflectance observed between and within seagrass species. The result is a comprehensive spectral library of the three most common seagrass species of south eastern Australia, which is used to characterise 'healthy' reflectance signatures for the species by defining the levels of spectral variability associated with these seagrasses over a range of natural conditions in which they grow in the field. The results were then used to determine whether there were consistent and detectable spectral reflectance differences between the species.

The spectral reflectance and physiological changes that occur as eelgrass *Z. capricorni* decline in response to light stress were measured in laboratory growth experiments that are described in Chapter 5. These studies were carried out to identify the pigment changes characteristic of light stressed *Z. capricorni* and to relate these to changes in the spectral reflectance of the seagrass leaves, and hence assess the potential for remote sensing of pigment-based 'light stress indicators'.

The research reported in Chapter 6 develops and tests the effectiveness of a range of existing and new hyperspectral reflectance indices suitable for the remote detection and monitoring of stress in seagrasses. Data from the light stress experiments are applied to examine the correlation between reflectance indices and the pigment content or physiological status of *Z. capricorni* leaves. A subset of field-measured spectral reflectance and pigment data has been used to validate the effectiveness of the best performing of these indices.

Finally, Chapter 7 examines the practical implications of this research towards maximising the potential for remote sensing mapping and monitoring of seagrass meadows. These studies lay important groundwork for the hyperspectral monitoring of physiological stress in seagrass meadows, which may provide the means for early remote sensing detection of environmental problems in estuaries.

Chapter 2
Seagrasses – Species, Structure and Photosynthesis

S.K. Fyfe

2.1 Introduction

There are less than 70 species of seagrass worldwide but over half of these species, and all but one genus, occur in Australia (den Hartog 1970). 'Seagrass' is a functional group referring to marine angiosperms that live entirely submerged and produce flowers that are pollinated under water. It appears that the earliest seagrasses developed the osmoregulatory capacity to exist in marine waters sometime during the Cretaceous approximately 100 million years ago (Larkum and den Hartog 1989). They evolved from three lineages of land plants in separate evolutionary events to produce the Cymodoceaceae complex, Zosteraceae and Hydrocharitaceae (Waycott and Les 1996). These seagrasses display a number of convergent morphological and physiological characteristics (Larkum and den Hartog 1989) that allow them to survive in marine environments despite having basically the same physiology and biochemistry as terrestrial angiosperms. Such adaptations include lacunae or internal gas spaces, epidermal chloroplasts, a lack of stomata, rapid leaf turnover and a specialised carbon metabolism (Walker *et al.* 1999; Touchette and Burkholder 2000b).

Seagrasses generally occur in shallow (< 30 m), sheltered shore habitats and may form meadows of considerable extent. The largest beds in Australia are found in Spencer's Gulf and the Gulf of St Vincent, South Australia, and along the Western Australian coast north to Exmouth including vast areas of Shark Bay (McComb *et al.* 1981). The high wave energy environment of the New South Wales (NSW) coast restricts seagrass growth to sheltered situations in approximately 130 estuaries and a small number of semi-enclosed embayments that occur along the coastline (West *et al.* 1989). Fast growing species such as *Halophila* and *Zostera* spp. can rapidly colonise even small, intermittent estuaries where they tolerate extremes in water temperature and salinities as low as 2 parts per thousand (ppt) or as high as 46 ppt for short periods of time (McComb *et al.* 1981). Since many of the smaller estuaries in NSW are relatively turbid, seagrasses are commonly restricted to depths of only 1-2 m, the exception being in open bays and estuaries with wide entrances allowing for unlimited tidal exchange. Seagrasses draw nutrients from both the water column and sediments, although sediments are thought to be the primary source (Udy and Dennison 1996). The colonization and extension of most seagrass meadows is by vegetative propagation, though many species do produce large quantities of viable seed (Tomlinson 1974; McComb *et al.* 1981; Turner *et al.* 1996). Propagation by seed is rarely seen for some species (e.g. *Cymodocea serrulata, Halodule uninervis*) but it is the dominant method for reproduction in others (e.g. *H. ovalis*) (McComb *et al.* 1981). Seagrass seedlings are often short lived, however, and their presence does not indicate successful settlement of bare sediments (den Hartog 1970).

Although seagrass plants can tolerate considerable environmental stress, the stability of meadows can be markedly affected by environmental change. Zonation of species can occur, as do patterns of community development speculated to involve cyclic processes of succession mainly driven by wave, currents and substrate deposition (McComb *et al.* 1981). Walker *et al.* (1999) discussed the concept of a seagrass form/function model to explain the different response of seagrass species to changing environmental conditions. For example, small seagrass species (e.g.

Halophila spp.) have thin leaves and rhizome turnover is rapid, on the order of weeks to months. Large species (e.g. *Posidonia* spp.) have thick leaves, large rhizomes and rhizome turnover that can take months to years. The leaf turnover rate for large species is also much slower than for smaller species and epiphyte load on the leaves is correspondingly greater, as is the build up of secondary compounds in the leaves to reduce their palatability to grazers (e.g. Preen 1995). The leaf turnover rate for *P. australis* is 30-40 days and epiphytes begin to build up significantly after about two weeks. In contrast, *Zostera* spp. are in the smaller 3rd of seagrass genera so leaf turnover may be on the order of days (Bearlin *et al.* 1999). Smaller seagrasses tend to respond more quickly and significantly to environmental changes. The photosynthetic responses of *Halophila* spp. are particularly rapid (e.g. Drew 1979; Ralph *et al.* 1998; Dawson and Dennison 1996) and plants survive only weeks when light deprived (Longstaff and Dennison 1999; Preen 1995) but recover quickly and can readily colonise new areas by seeding (Duarte 1991). Large seagrasses such as *P. australis* display lower photosynthetic rates (Ralph *et al.* 1998) and slower photosynthetic responses (Dawson and Dennison 1996) but they survive much longer when deprived of light (Duarte 1991; Gordon *et al.* 1994; Czerny and Dunton 1995) because they develop greater below ground biomass (Duarte and Chiscano 1999). However, large seagrass species do not recover well or recolonise readily if die back occurs (Duarte 1991). In addition, biomass turnover tends to be faster for temperate seagrasses and they devote more to below-ground storage than do tropical seagrasses (Duarte and Chiscano 1999).

2.2 Seagrass species of south eastern New South Wales

Australian seagrasses include 30 species in 11 genera, extending across all areas from tropical to temperate waters (Kuo and McComb 1989). Den Hartog provided the first comprehensive treatment of the group in 1970 but the taxonomy of several species and groups has since been reviewed. For example, Jacobs and Williams (1980) stressed that the anatomical characters used by den Hartog (1970) to separate *Zostera* and *Heterozostera* at generic level were not convincing and subsequently, after genetic analysis the genera were grouped (Waycott and Les 1996). However, the species are here referred to by their original genus names, *Zostera* and *Heterozostera*.

2.2.1 Eelgrass *Zostera capricorni* Aschers.

(Order Potamogetonales, Family Zosteraceae, Subgenus Zosterella)

Zostera capricorni (Plate 2.1) is the most common seagrass in NSW waters (den Hartog 1970). It is a dynamic and adaptable seagrass that can tolerate a wide range of environmental conditions (e.g. Kirkman *et al.* 1995) and is therefore found in almost all estuaries along the NSW coastline including

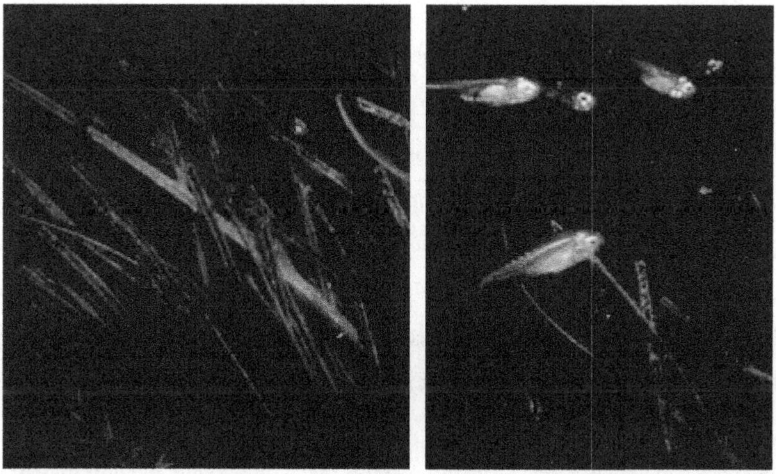

Plate 2.1. Eelgrass *Zostera capricorni* in tidal channel, Sussex Inlet, NSW. These specimens show a strong reddish-bronze colouration attributed to anthocyanins in the leaves. The leaves of *Z. capricorni* may also be entirely green or various shades of green, bronze and red.

very small and intermittently open tidal inlets, lakes and lagoons. The seagrass has an extensive distribution across Queensland, NSW, Victoria, South Australia, Tasmania, Lord Howe Island and New Zealand. The leaves of *Z. capricorni* are straplike, 7-50 mm long x 2-5 mm wide, with a truncate, rounded or denticulate apex, 3-5 lateral veins and 6-8 air canals either side of the midvein (Ashton 1973; Harden 1993). Leaves are carried on horizontal branching rhizomes with two groups of roots at each node. Male and female flowers (7-10 of each) are carried on a spadix usually between September and March, with fruits being released as they mature.

Zostera capricorni was originally considered to be a tropical-subtropical species and *Z. muelleri* a temperate species that overlapped in range in south eastern NSW (e.g. Mills 1995), although *Z. capricorni* was by far the most common species found north of Sussex Inlet (Jacobs and Williams 1980). Their taxonomic distinction was difficult to determine without chemotaxonomic analysis (Jacobs and Williams 1980) so the species were generally identified on the basis of the geographic location from which they were collected. Les *et al.* (2002) recently re-examined the four species members of the subgenus Zosterella (*Z. capricorni*, *Z. muelleri*, *Z. mucronata* and *Z. novazelandica*) and suggested that all should be merged within the single species, *Z. capricorni*. The *Zostera* spp. encountered in the current studies were subsequently named according to Les *et al.* (2002) although all specimens sampled would almost certainly have belonged to the previous species definition of *Z. capricorni*.

Zostera capricorni tends to grow in small, dense meadows in shallow water but can also appear as sparse, scattered beds in deeper water. Eelgrasses typically line the sublittoral zone and grow on shoals and banks, and in tidal creeks where they tolerate disturbance, exposure and changing salinity (Larkum and West 1983; West *et al.* 1989). These plants have the ability to recolonise denuded areas very quickly by seedling or rhizome growth (West 1995) and can grow prolifically, particularly in the warmer months. Often *Z. capricorni* displays seasonal leaf growth during spring and summer with some dieback over winter (Kirkman *et al.* 1982).

2.2.2 Strapweed *Posidonia australis* Hook. f.

(Order Potamogetonales, Family Posidoniaceae)

Posidonia australis (Plate 2.2.) is endemic to the temperate waters of southern Australia, extending in range from southern NSW, across Victoria, Tasmania and South Australia to Western Australia. This species has large, bright green, linear and sometimes falcate straplike leaves to 60 cm long x 6-15 mm wide, with rounded or truncate apices and a distinct ligule at the junction of the leaf blade and sheath. Short erect stems rise from short 6-8 mm thick horizontal branching rhizomes and both are densely covered with the persistent fibres of decaying leaf sheaths. Flowers and fruit are produced in spikes on leafless stems in Spring (Ashton 1973; Harden 1993).

Plate 2.2. Strapweed *Posidonia australis* in tidal channel, Sussex Inlet, NSW.

Posidonia australis is a long-lived and very productive seagrass and in some regions fruiting is prolific, but it is a poor coloniser that spreads very slowly by rhizome or seedling growth (Kirkman and Kuo 1990). Its distribution is restricted to permanently open lagoons and bays where there is good exchange of marine waters since it does not withstand brackish conditions (West *et al.* 1989) though it can tolerate hypersaline conditions and periodic decreases in salinity (den Hartog 1970). Strapweed is generally only found on sandy bottoms in clear non-turbid waters from low tide mark to around 10 m depth (Ashton 1973).

2.2.3 Paddleweed *Halophila ovalis* (R.Br.) Hook.

(Order Hydrocharitales, Family Hydrocharitaceae)

Halophila ovalis (Plate 2.3) is a small, fast growing, highly fecund and adaptable seagrass, capable of living in a wide range of salinities, sediment types and water temperatures and well adapted to unstabilised substrates and low light conditions (den Hartog 1970; Lee Long *et al.* 1996b). It occurs in all regions of Australia and extends throughout Indo-Pacific tropical waters to Japan, Hawaii, Tonga, Tahiti, Eastern Africa and Malaysia. This species can be found on most substrates (den Hartog 1970), surviving across the full range of water depths at which seagrasses occur (West 1995) from the shallow intertidal to very deep (up to 50 m) in tropical reef areas (Lee Long *et al.* 1996b). *Halophila ovalis* has broad, ellipsoid leaves with entire margins, 1-4 cm long, held in pairs at the nodes of narrow creeping rhizomes. Male flowers are produced from August to April on membranous tepals while female flowers are carried on prostrate horizontal stolons. *Halophila ovalis* is often a pioneer in denuded or unstable areas and is known to be the fastest coloniser of seagrass meadows in NSW (West 1995). Due to its fecund nature and fast growth, *H. ovalis* may behave as an annual and die back over winter, or it can persist in dense, monospecific clumps throughout the year (Lee Long *et al.* 1996b).

Plate 2.3. Dense patches of paddleweed *Halophila ovalis* (A) in tidal channel, Sussex Inlet, NSW, and (B) exposed during an extreme low water Spring tide growing in gaps in a *P. australis* meadow at Plantation Point, Jervis Bay, NSW.

2.2.4 Minor species

There are a number of other seagrass species that occur in south eastern Australia but they are less common or have a more restricted distribution than *Z. capricorni, P. australis* and *H. ovalis.* For example, *Heterozostera tasmanica* is a temperate seagrass previously thought to occur in NSW only within Jervis Bay (Jacobs and Williams 1980), although it may have a more widespread distribution and has since been reported in sheltered oceanic bay sites near Eden and as far north as Port Stephens (West *et al.* 1989). Mills (1995) considered *H. tasmanica* to be "rare in southern NSW or rare over the species whole range".

Halophila decipiens and *H. australis* have a very similar growth form to *H. ovalis,* and they can also be found across a wide range of habitats through SE NSW, but they are less commonly encountered. While

H. decipiens occupies a similar range to *H. ovalis*, *H. australis* is considered to be more of a southern Australian species (West *et al.* 1989).

Ruppia spp. (sea tassels) can not strictly be considered as seagrasses since these angiosperms can also grow and reproduce in freshwater ecosystems but they do occupy very similar estuarine habitats as the seagrasses and are often found growing in association with them, particularly in the brackish backwaters of coastal lagoons. These aquatic plants are often transient and usually proliferate during periods of low salinity. Several species occur in the region but identification is difficult. *Ruppia megacarpa* is thought to be the dominant species in SE NSW (West *et al.* 1989).

2.3 Seagrass structure and morphology

Seagrasses show surprisingly similar structure given the different evolutionarily origin for the three major taxonomic groups. Tomlinson (1980), McComb *et al.* (1981) and Larkum *et al.* (1989) have described the structure and morphology of a range of seagrass taxa, the main features of which have been summarised below for the genera *Zostera*, *Halophila* and *Posidonia*.

The roots of seagrasses are adventitious and generally arise from the nodes of the rhizomes as in all monocotyledons. The rhizomes of seagrasses are usually herbaceous, branched, and are typically buried under the sediment. Erect stems, if present, are similar in structure to the rhizomes which are modified stems. The leaf sheaths that enclose, support and protect developing leaves are persistent on the rhizomes of many seagrass genera including *Posidonia* and *Zostera* spp. because they contain lignified fibres, but leaf sheaths are not present in *Halophila* spp.

The leaves of seagrasses are surprisingly uniform in their main structural features, but these features are unlike those of terrestrial plants and most other aquatic plants;

- photosynthesis occurs almost exclusively in the epidermis while the hypodermis often lacks pigmentation,
- the epidermis is covered by a very thin cuticle,
- there are no stomates,
- instead of intercellular spaces, the mesophyll usually encloses a system of well-developed air canals (lacunae) segmented by transverse diaphragms of small compact cells at regular intervals,
- leaves are supported longitudinally by a mechanical system, mostly of fibres,
- longitudinal veins (vascular bundles) with reduced xylem extend down the length of the leaf and are interconnected by transverse veins, and
- idioblastic secretory cells (tannin cells) are present in many species.

Since a functional stomatal system is absent, the reduction of the cuticle to a thin, electron-transparent layer in some genera including *Zostera* and *Halophila*, probably allows for direct inorganic carbon and nutrient diffusion between the open water and photosynthetic epidermal cells. The cuticle of *Posidonia* spp. has a distinctly porous texture that may act as an ion-water exchange column that regulates nutrient flow between the seagrass leaf, epiphytes and surrounding medium (Kuo 1978). The high desiccation tolerance observed for seagrass leaves when exposed at low tide is probably due to their overlapping leaves rather than any protection offered by the thin cuticles (e.g. Clough and Attiwell 1980). However, the epidermal cell walls of seagrasses are thickened with pectin, protein, cellulose and other polysaccharide compounds (but are never lignified) and are highly hydrated, and therefore they take water up rapidly after being dried (Tomlinson 1980).

The leaf anatomy of seagrasses does not conform closely to either C_3 or C_4 terrestrial plants (reviewed in McComb *et al.* 1981; Kuo 1983; Figure 2.1). Epidermal chloroplasts are rare in terrestrial plants and occur in only a few shade-adapted species. In seagrasses, the leaf blade epidermis is the major site of photosynthesis and consequently the single layer of epidermal cells are densely packed with chloroplasts. There are no visible differences between epidermal cell layers (i.e. no obvious adaxial or abaxial surfaces; Abal *et al.* 1994).

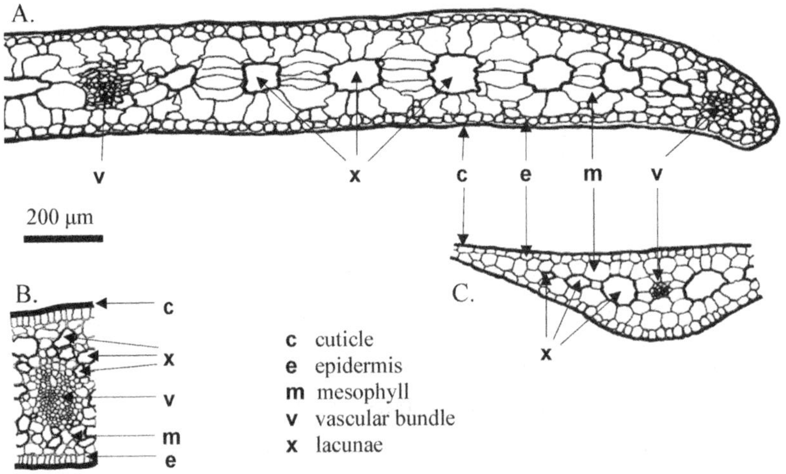

Figure 2.1. Leaf cross-sections displaying internal structure of A. *Zostera capricorni* (midvein to margin), B. *Posidonia australis* (midvein area) and C. *Halophila ovalis* leaves (mid-section of leaf). Sections were redrawn from the figures and photographs of A. McComb *et al.* 1981, B. Kuo and McComb *et al.* 1989 and C. Tomlinson 1980.

The size of the epidermal cells is relatively consistent for most seagrass species at ~10 μm wide although *Enhalus* sp. reach 20-30 μm. Seagrass epidermal cells are generally cubic in shape although shape may differ considerably between genera and species, for example within the genus *Posidonia* (see examples in Larkum *et al.* 1989).

In *Posidonia* spp., the epidermal cells contain tannin (in the vacuoles) in addition to chlorophyll but in *Zostera* spp. they do not. In contrast, large secretory cells are interspersed with the chlorenchymatous cells in the epidermis of *Halophila* spp. Flavonoids such as anthocyanins and related secondary compounds including other kinds of UV- blocking pigments will also be contained within or just below the epidermal cells (Chalker-Scott 1999). Anthocyanins are water-soluble pigments that occur in the vacuoles of higher plants and are responsible for the red, blue, purple, violet and scarlet colours of leaves, stems, flowers and fruits (Mancinelli 1985). These pigments have been attributed a wide range of functions, one of which is photoprotection from UV or visible light (Chalker-Scott 1999).

The mesophyll tissue in seagrass leaves is composed of homogeneous, thin-walled parenchyma cells with large vacuoles. There are fewer chloroplasts in the mesophyll cells; those that are present contain small starch grains and are dispersed in a thin, peripheral cytoplasm. The mesophyll surrounds air lacunae, which in *Zostera*, *Heterozostera* and some other species are large and regularly arranged across the leaf cross-section in a single layer separating longitudinal vascular bundles. *Z. capricorni* has 15-20 large lacunae across the leaf blade, positioned one cell depth below the photosynthetic epidermis, and with 4-5 lacunae between each vascular bundle (Grice *et al.* 1996). Lacunae are smaller in the leaves of *Posidonia* spp. but occur in several layers through the thick leaves of this genus. A single layer of small lacunae is present only in the mid-vein region of the thin leaves of *Halophila* spp. In all species however, the lacunae form continuous air tubes from the roots to the leaves, interrupted at regular intervals by perforated septa made up of small parenchyma cells. Lacunae play an important role in seagrass photosynthesis (e.g. Roberts and Moriarty 1987) because they store and transfer gases such as carbon dioxide, oxygen and nitrogen throughout the plant, allowing for recycling of the gaseous products of photosynthesis and respiration (Larkum *et al.* 1989). Interestingly, earlier authors had previously suggested their primary function was for support of the leaf (Williams and Barker 1961).

Seagrass leaf blades are pliable and elastic because fibre cells rather than lignified tissues provide mechanical support. Fibre cells with thickened but not lignified walls are abundant in the leaves of most genera where they occur near the leaf margins and veins as small strands associated with the longitudinal vascular bundles. Seagrasses do not display the ability to form leaf surface outgrowths although marginal features may be present, e.g. hairs are a diagnostic feature of *H. decipiens* while marginal teeth restricted to the leaf apex are a feature of many *Zostera* spp.

Leaf turnover is important to seagrasses in order to shed the older leaves that have been covered with epiphytes and to provide new, unfouled leaves for continued photosynthesis. In some taxa, leaf abscission appears to be a programmed process associated with the formation of regular abscission layers at the nodes. In other species, the leaf blade breaks off without the formation of abscission layers since the junction between the leaf blade and sheath is mechanically weak.

2.4 Photosynthesis in seagrasses

Photosynthesis is the oxidation-reduction reaction by which plants capture light energy and convert it into chemical energy, storing the energy as carbohydrate molecules built from inorganic carbon and water. The 'light reactions' of photosynthesis are the initial photochemical reactions in which light energy captured by photosynthetic pigments is used to oxidise water and release electrons. The light energy absorbed by pigment molecules results in the modification of their electronic structure to the extent that electrons can be physically transferred to acceptor molecules within the photosynthetic reaction centres. Hydrogen ions released during the oxidation of water are passed along a series of hydrogen carriers (the electron transport chain), reducing NADP to $NADPH_2$. Associated with these processes is the conversion of ADP and inorganic phosphate to the energy rich ATP. The $NADPH_2$ is subsequently utilised in the Calvin-Benson cycle, reducing CO_2 to produce the carbohydrates that fuel plant growth and metabolism. Relevant aspects of the process of photosynthesis in aquatic plants have been summarised below but more detail can be found in the reviews of Kirk (1994), Falkowski and Raven (1997) and Touchette and Burkholder (2000b).

2.4.1 Photosynthetic pigments

The photosynthetic pigments contained within the chloroplasts of seagrasses are the same as those found in terrestrial higher plants; i.e. the chlorophylls *a* and *b*, and a range of xanthophylls and carotenes that constitute the carotenoids (eg. Beer *et al.* 1998). All chlorophyll pigments have two major absorption bands; B or Soret bands in the blue to blue-green wavelengths, and bands in the red wavelengths called Q bands (Falkowski and Raven 1997). Chlorophyll *a* plays the major role in light-harvesting in all photosynthetic green plants. Chlorophyll *b* supplements light-harvesting in all higher plants and some algal groups by increasing the absorption window provided by chlorophyll *a* to longer wavelengths in the blue and shorter wavelengths in the red. The molar ratio of chlorophyll *a*:*b* is about 3:1 in higher plants (Kirk 1994), although this ratio will vary with environmental conditions.

Carotenoids occur in green plants in the molar ratio of around 1:3 total carotenoid:total chlorophyll ($a + b$), although again, this ratio is variable. The carotenoids are a diverse group of light absorbing molecules that can demonstrate a wide range of spectral characteristics by means of subtle structural changes (Falkowski and Raven 1997). Carotenoids display blue and blue-green absorption bands that overlap the Soret bands and, depending on their structure, either facilitate the transfer of excitation energy to, or remove excess excitation energy from chlorophyll *a*. They therefore function in both light-harvesting and photoprotection. The xanthophylls (oxygenated carotenoids) lutein, violaxanthin and neoxanthin assist in light-harvesting by extending the range of light absorption provided by the chlorophylls even further in the short (high energy) wavelengths. However, another essential function of carotenoids is in photoprotection (reviewed in Owens 1994). Carotenoids limit damage to the photosynthetic apparatus resulting from the photochemical generation of oxygen radicals by taking up the excess excitation energy from chlorophyll and dissipating it as heat. Photoprotection occurs in green plants through the regulation of absorbed light energy by the xanthophylls antheraxanthin and zeaxanthin (see Chapter 5). In addition, carotenes in very high concentrations may act as a photoprotective colour filter, reducing blue light absorption e.g. in the algae *Dunaliella salina* where a large amount of β-carotene is not complexed in the photosynthetic system (Loeblich 1982).

2.4.2 The light reactions of photosynthesis

In the light reactions of photosynthesis, electron transfer occurs within the reaction centres or photosystems after the pigment molecules absorb light energy and are raised to an excited electronic state. There are two photosystems; Photosystem I (PSI) and Photosystem II (PSII), both of which consist of photochemical reaction centres coupled to Light Harvesting Complexes (LHCI and LHCII) or 'antennae'. The reaction centres for the photosystems (termed P_{700} for PSI and P_{680} for PSII because of spectral changes at these wavelengths accompanying the loss of electrons from chlorophyll) are composed of a special form of chlorophyll *a* complexed with a specific protein. PSI is involved in the reduction of NADP while PSII brings about the oxidation of H_2O. A functional photosynthetic unit consists of PSI and PSII together with their associated antennae of light-harvesting pigment-protein complexes, interconnected by electron transport chain intermediates.

The pigment composition of the antennae varies for each photosystem and between species but includes most of the chlorophyll *a* in the plant, plus chlorophyll *b* and various carotenoids. Only xanthophylls such as lutein, violaxanthin and neoxanthin are present in LHCI and LHCII, carotenes are absent (Young 1993). LHCI contains a higher proportion of chlorophyll *a*:*b* than LHCII (Young 1993; Falkowski and Raven 1997) and in some higher plants LHCII may contain all the chlorophyll *b* (Schiefthaler 1999). LHCII initially transfers most (but not all) of the energy it harvests to PSII but then 'spillover' of energy from PSII to PSI takes place, ensuring that both photoreactions continue at the same rate (Myers and Graham 1963). Photosynthesis is therefore a very efficient process because the photosynthetic pigments work together in assemblages to maximise the probability of photon interception and ultimately electron transport.

After the chlorophyll *a* molecules in reaction centres have absorbed light there are four possible de-excitation pathways from the excited singlet state back to the ground state (see Kirk 1994; Björkman and Demmig-Adams 1995). The most important of these is photochemistry where the absorbed energy results in electron transfer. If all reaction centres are full and electron transfer cannot occur, then excess energy may be dissipated to the environment as heat in the processes of non-photochemical quenching (see Section 5.1.3). Excess excitation energy may also be dissipated through decay of the chlorophyll triplet state. In addition, a small amount of excess energy (3-4% of the total) may be emitted in slightly longer wavelengths as chlorophyll fluorescence (Kraus and Weis 1991). The fluorescence emission spectrum of chlorophyll *a* is a mirror image of the absorption spectrum but shifted about 10 nm to the red (Stokes shift) as a consequence of energy losses as vibrational energy during de-excitation of the excited singlet state. While photosynthetic pigments other than chlorophyll *a* can be observed to fluoresce when extracted in solvent, only fluorescence from chlorophyll *a* can be detected *in vivo*. This is because the energy absorbed by other light harvesting pigments is efficiently transferred to chlorophyll *a* as a result of the organization of the pigment-protein complexes in the thylakoid membranes. Since the de-excitation pathways are competitive processes, chlorophyll fluorescence can be used to monitor photochemistry and non-photochemical heat dissipation.

2.4.3 Limitations on photosynthesis

When light is limiting but CO_2 is not, the rate of photosynthesis is limited by the availability of NADPH and ATP, the products of the light reactions. At saturating light intensities carbon fixation may be operating at a rate below maximum capacity if the CO_2 concentration is too low to saturate the Rubisco enzyme (ribulose-1, 5-biphosphate carboxylase/oxygenase; the first enzyme in the Calvin-Benson Cycle). In this case, carbon fixation will be the rate limiting step in the photosynthesis as these reactions will not be able to consume the products of the light reactions as fast as they are being formed (Falkowski and Raven 1997; Kirk 1994). This is of particular importance in the case of aquatic photosynthesis because dissolved CO_2 occurs in marine waters in very low concentrations. In general, seagrasses appear well-adapted to utilise the inorganic carbon supply available in their environment though the mechanisms by which they do so whilst maintaining such high productivity are not well understood (see reviews of Beer 1996; Touchette and Burkholder 2000b). Many researchers believe that submersed seagrasses are potentially carbon limited for growth despite their adaptive mechanisms (Beer 1996; Larkum and James 1996). Indeed, photosynthetic activity is relatively low, on a chlorophyll basis, in seagrasses (Larkum *et al.* 1989).

Chapter 3
Experimental Methods

S.K. Fyfe & S.A. Robinson

3.1 Field experiments

Seagrass leaves, both with and without their characteristic leaf epibionts, were sampled at marine and brackish habitats within three temperate Australian estuaries to study spatial and temporal variation in the spectral reflectance and foliar chemistry of seagrasses. This destructive method for sampling seagrass was carried out once during each season over a two year period. *In situ* field spectra of seagrass meadows were also measured from above the water surface over a range of canopy, substrate and water column conditions.

3.1.1 Study sites

Three estuaries on the southeast coast of NSW were selected to represent the range of conditions under which seagrasses grow in this region (Figure 3.1). Port Hacking on the southern edge of Sydney (E 151°07′, S 34°04′) is a drowned river valley with a deep, permanently open entrance. Much of the catchment for this estuary lies in the Royal National Park that adjoins its southern shores although the northern side of the estuary is heavily urbanised. The entrance to the estuary opens into a wide bay and on the seaward side of the tidal delta, seagrass meadows are restricted by wave energy to small transient patches (West *et al.* 1989). Costen's Point (E 151°06′30″, S 34°04′50″) is located on the tidal channel some 5 km inland from the coast but the area receives free tidal exchange with ocean water and has an oceanic tidal range. At Costen's Point, a large sandbar of clean yellow marine sands supports an extensive and stable *P. australis* meadow with fringing beds of *Z. capricorni* and occasional patches of

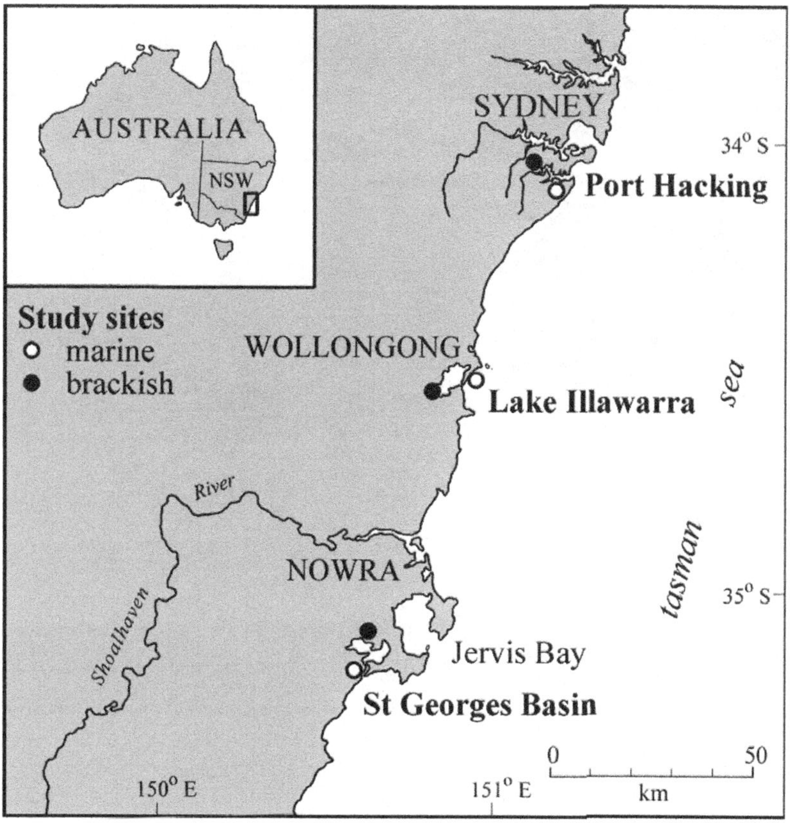

Figure 3.1. Location of the field study sites in marine and brackish habitats within three estuaries in south eastern NSW (Fyfe 2003).

H. ovalis (Plate 3.1). Gray's Point (E 151°04′35″, S 34°03′45″) lies several kilometres further upstream in the north western arm of Port Hacking. The Hacking River is still tidal at this point but freshwater inputs lower the salinity considerably during periods of rain. The sediments in the fluvial delta at Gray's Point are black organic silts, sand and mud that support extensive mangrove growth and linear *Z. capricorni* beds at the littoral fringe.

Plate 3.1. The sandbar at Costen's Point, Port Hacking, NSW, supports an extensive and stable *P.australis* meadow with fringing *Z. capricorni* beds (darker coloured seagrass on the shallow edge of the lighter green *P. australis*) and occasional patches of *H. ovalis* in meadow gaps.

Lake Illawarra is situated approximately 80 km south of Sydney near Wollongong (E 150°50′, S 34°32′). It is a shallow, partially infilled barrier estuary with a mud basin that shows little tidal influence because the entrance channel is relatively small and closes intermittently. The catchment of Lake Illawarra is heavily urbanised and has a history of intensive agricultural and industrial landuse prior to residential development. By the 1980's, significant environmental problems had been recognised in the estuary; in particular, eutrophication resulting in excessive algal growth and sedimentation causing accelerated infilling of the lake (Harris 1977; King 1988). While measures have been taken to implement catchment management and improve water quality since this time, Lake Illawarra remains one of the more polluted estuaries in NSW.

The seagrass meadows in Lake Illawarra are limited to shallow water depths of generally less than 1-1.5 m because of the poor water quality. *Zostera capricorni* occurs throughout Lake Illawarra on barrier sands, mudbanks and in littoral situations, covering possibly 30% of the area of lagoon (King 1988). *Ruppia* and *Halophila* spp. are also present but they occupy a smaller, more variable area and their distribution is often patchy and/or intermittent (King 1988; West *et al.* 1989). At times they may be completely absent from the lagoon (Harris *et al.* 1980). Sedimentation rates and entrance condition have a great deal of influence on the presence and area of each seagrass species occurring in the estuary at any one point in time (West *et al.* 1989). Seagrasses were sampled in Lake Illawarra from a fully tidal site of marine dominated sediments on the main entrance channel adjacent to Windang bridge (E 150°52′05″, S 34°32′00″) and from Koona Bay (E 150°48′10″, S 34°33′40″), a back lagoon site on the western edge of the mud basin where no tidal influence is apparent.

St Georges Basin is a barrier estuary of much younger geological age than Lake Illawarra (West *et al.* 1989) with a permanently open and strongly tidal entrance. This estuary lies approximately 200 km south of Sydney (E 150°37′, S 35°08′) within a relatively pristine forested catchment area. Residential development of the foreshores has increased over recent years and includes some canal development along the south side of the entrance channel. However, the northern shoreline of the channel lies in a Commonwealth Territory protected zone and the greater proportion of the mud basin foreshores have not yet been cleared or developed. Consequently, the quality of the estuary water is relatively good and the seagrass meadows do not show any obvious effects of human usage of the estuary and its catchment. All three seagrass species (*P. australis*, *Z. capricorni* and *H. ovalis*) occur both in the entrance channel at Sussex Inlet and in the mud basin of St Georges Basin. At Lion's Park, Sussex

Inlet (E 150°35′25″, S 35°10′20″) linear meadows of intertidal and subtidal *Z. capricorni* line the edges of the channel and contain small, patchy beds of *H. ovalis* and *P. australis*. The largest meadows of *P. australis* occur on tidal delta sands at the junction of the basin and the entrance channel. Tidal range is minimal (0.1 m) on the landward side of St Georges Basin and at Island Point (E 150°35′40″, S 35°06′05″), sparse and/or patchy meadows of *H. ovalis* and *Z. capricorni* grow on shallow sandy mudflats, while *P. australis* is absent.

3.2 Manipulative laboratory experiments

Three laboratory experiments were carried out in order to observe changes in the spectral reflectance of *Z. capricorni* associated with changes in photosynthetic vigour, pigment content and growth of the seagrass when subjected to light stress over a period of time;

1. a preliminary experiment – for the purpose of establishing the experimental setup
2. low light stress experiment – using paired control (unshaded) and treatment (shadecloth covered) replicates exposed to light levels below the compensation point for seagrass growth.
3. high light stress experiment – using paired control (shadecloth covered) and treated (unshaded) replicates exposed to illumination at sufficient intensity to induce photoinhibition.

Paired samples were used within each aquarium tank to ensure that environmental factors other than light availability, such as carbon availability and water temperature, did not differ between controls and treatments. Experiments were run for at least three months each to examine both the short and longer term responses of the plants to the light treatments, i.e. long enough to observe either acclimatization to the stressful light regime or significant dieback. The frequency of sampling, however, emphasises the responses of the seagrass plants in the first two weeks of treatment since the main objective of this research was to determine whether seagrass decline could be detected in the spectral reflectance of *Z. capricorni* prior to the onset of dieback.

3.2.1 Preliminary growth experiments

Zostera capricorni were transplanted from Botany Bay and Lake Illawarra in September 1998 and grown in 24 plastic tanks in the laboratory for approximately seven months to ensure that this seagrass could be successfully cultivated in relatively basic laboratory facilities. The experiment was used to establish the experimental conditions required for maintaining healthy, growing *Z. capricorni* plants in the laboratory; e.g. the appropriate light quality and intensity, aeration and stirring of tank water, maintenance of salinity, maintenance of water temperature, algal control, etc. The seagrass samples were also used to trial experimental procedures such as laboratory measured spectral reflectance, fluorimetry and pigment extractions.

3.2.2 Collecting and transplanting seagrass

The *Z. capricorni* samples for all further laboratory experiments were collected from the western side of Wegit Point, Lake Illawarra (E 150°52′10″, S 34°29′35″) in water depths ranging from 0.3-0.6 m. In each case, samples were obtained from a dense meadow of healthy and actively growing plant material and planted into 28 L transparent, rectangular plastic tanks (420 mm x 315 mm x 230 mm).

To minimise disturbance to the seagrass, intact turfs were dug with approximately 5 cm sediment attached. Turfs were planted compactly into the tanks without lifting the plants out of the lake water. Consequently, many associated seagrass epibionts and benthic flora, fauna and infauna were also retained in the tanks for the duration of the experiment. The tanks were sealed and transported back to the laboratory immediately after planting had been completed. Sample tanks were randomly allocated to a position on a laboratory bench and topped up with clean ocean water.

The seagrass samples used in the low light stress experiment were held in the laboratory for one month at PAR 168 $\pm$ 49 μmol photon m^{-2} s^{-1} prior to experimental treatment to allow them to acclimate to laboratory conditions. It was not possible to stabilise the samples used in the high light stress experiment for more than one day prior to treatment but these *Z. capricorni* samples suffered minimal disturbance during transplanting and adjusted rapidly to laboratory conditions. The lack of acclimation time did not appear to influence the results of the experiment as evidenced in the chlorophyll fluorescence measurements, which indicated that the seagrass experienced no decline in vigour over the initial weeks of the experiment (see Section 5.4.2; Figure 5.8).

3.2.3 Laboratory setup and maintenance

Benches

Two fibreglass benches, each comfortably holding 12 plastic tanks, were set up to provide controlled environmental conditions for the growth of the seagrasses (Plate 3.2). All 24 tanks were used for the high light stress experiment, with replicate tanks randomly distributed over both of the benches and the results pooled. The low light experiment was run on a single bench in the following year utilising only 12 tanks.

Lighting

The photosynthetically active radiation (PAR) required for plant growth was provided by 400W metal halide lamps of the type used in commercial

*Plate 3.2.*_Bench setup with aerated plastic tanks and artificial lighting for growing seagrass in the laboratory. Chilled water is continually pumped through the tray holding the tanks to maintain tank water temperatures.

hydroponics. Each bench was illuminated by six equally spaced metal halide lamps plus two centrally located fluorescent bars. The spectral output of these lights over the laboratory benches was compared with direct solar irradiance measured at two of the field study sites using the remote cosine receptor (RCR) attachment on the Fieldspec-FR spectroradiometer (Figure 3.2). The artificial lighting provided particularly high levels of far green and near red illumination. Irradiance in the region of the red chlorophyll absorption trough (660-685 nm) was similar to levels found in natural sunlight but in the blue chlorophyll/carotenoid absorption region (420-480 nm), light levels were well below that which a plant would experience in the field. Nevertheless, the lighting proved more than adequate for healthy seagrass growth in the laboratory. The lamps were set to a 12 hour light/dark cycle (6 am to 6 pm) which supplied unshaded tanks with an average daily integrated dose of photosynthetically active radiation (PAR) approximately equivalent to the insolation received by seagrasses

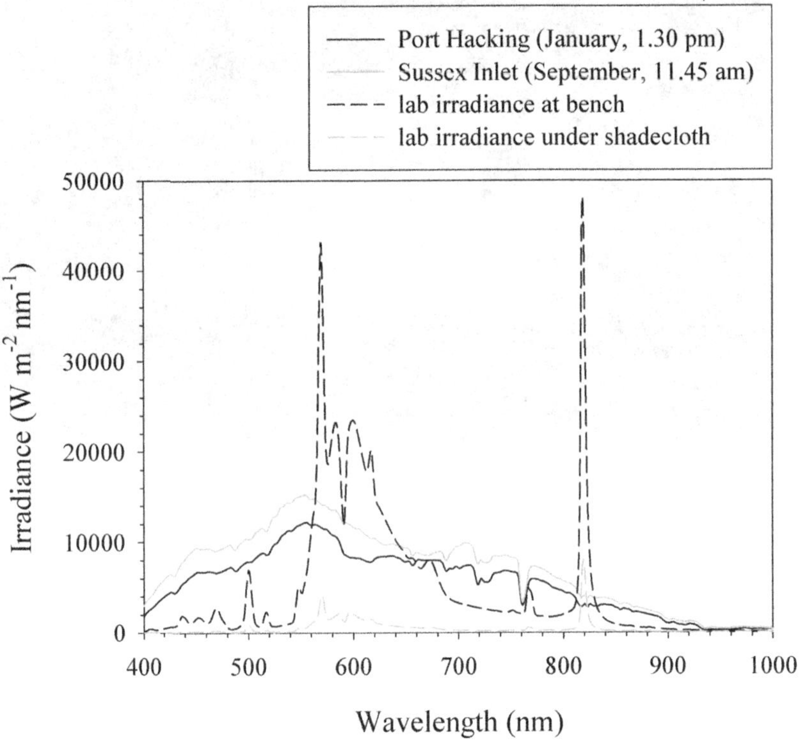

Figure 3.2. The quantity and quality of irradiance received by seagrass grown under artificial lights in laboratory experiments compared with natural illumination at two field study sites.

growing in the field during autumn in south eastern NSW (Table 3.1). The field values presented here are PAR measurements taken on cloud free days but since coasts tend to be cloudy areas, the average daily insolation at field sites may be much lower than this. Despite the fact that the maximum PAR received during the course of a day in the field and laboratory differed substantially, average insolation is the factor that must be taken into account in studies of marine production or of ambient light levels required for photosynthesis (Kirk 1994).

Table 3.1. Comparison of daily insolation received by seagrasses growing under natural illumination in the field at sites in south eastern NSW and in the laboratory under artificial illumination. Daily insolation (total daily integrated PAR) was calculated as (2 x daylength x mean monthly irradiance at solar noon)/π for field sites (Kirk 1994) and as daylength x mean irradiance in the laboratory. Maximum PAR is the maximum insolation received at midday.

	Maximum PAR	Total daily PAR
FIELD SITES		
Port Hacking, autumn	1200 μmol photon $m^{-2}s^{-1}$	28.9 mol photon $m^{-2}d^{-1}$
Lake Illawarra, summer	2000 μmol photon $m^{-2}s^{-1}$	64.2 mol photon $m^{-2}d^{-1}$
LABORATORY SAMPLES		
High light, treatment	661 μmol photon $m^{-2}s^{-1}$	28.6 mol photon $m^{-2}d^{-1}$
High light, control	238 μmol photon $m^{-2}s^{-1}$	10.3 mol photon $m^{-2}d^{-1}$
Low light, control	168 μmol photon $m^{-2}s^{-1}$	7.3 mol photon $m^{-2}d^{-1}$
Low light, treatment	39 μmol photon $m^{-2}s^{-1}$	1.7 mol photon $m^{-2}d^{-1}$

Temperature and salinity

Preliminary work showed that the photosynthetic vigour of *Z. capricorni* was quite sensitive to temperature changes. There was potential for overheating of tank water because a large number of metal halide lamps were required to illuminate experiments within an enclosed laboratory. This problem was overcome by recirculating cooling water around each tank, pumped constantly through the benches from a holding tank chilled to a constant temperature of 15 °C. Tank temperatures were generally maintained at between 17-22 °C for the low light experiment and 19-24 °C for the high light experiment. Estuary water temperatures measured in the field in September at Port Hacking were 17-19 °C.

Tanks were initially filled with clean ocean water (35 ppt) and kept as close as possible to this salinity by topping up with reverse osmosis water to the volume marked on each tank whenever water levels fell due to evaporation. Tank salinities were checked weekly using simple hygrometers.

Nutrients

It was not considered necessary to add fertiliser to the experimental tanks because the seagrasses were growing in lake sediments considered to be highly eutrophic (e.g. Harris 1977) and therefore the seagrass had sustained access to a considerable sediment-stored nutrient pool. Seagrass growth in the preliminary growth trials did not appear to be limited in any way after seven months without addition of nutrients.

Aeration and stirring

Tanks were individually aerated to provide dissolved carbon dioxide and oxygen, and more importantly, to induce water movement around the seagrass leaves. Poor stirring significantly limits leaf uptake of carbon and other dissolved nutrients as well as the uptake of oxygen for respiration and the release of photosynthetic oxygen (Brix and Lyngby 1985; Larkum *et al.* 1989).

If the CO_2 concentration passing into the seagrass leaves from the tank water is too low to saturate the Rubisco enzyme then carbon supply may limit the photosynthetic rate (Falkowski and Raven 1997; Kirk 1994). While all seagrasses are probably carbon limited at low water flow rates, some are more efficient at active carbon transport across the boundary layer than others and therefore their photosynthesis is less restricted by the boundary layer (Beer 1996). *Zostera capricorni* appears to utilise a variety of carbon concentrating mechanisms, including the active uptake and use of bicarbonate ions for photosynthesis (reviewed in Touchette and Burkholder 2000b). In this way the species ensures adequate inorganic carbon supply for high growth rates even under slow moving water conditions (e.g. see Raven 1995). It is unlikely that the supply of inorganic carbon would be a major limiting factor to photosynthesis in the current experiments.

Cleaning

Seagrass leaves were only lightly fouled at the time of collection and epibiont foulers did not persist under laboratory conditions. However it was necessary to clean tank walls, the water surface and leaves at approximately weekly intervals to maintain the tanks free of bloom algae and malcroalgal species such as *Enteromorpha intestinalis* which proliferated in a few of the tanks. Dead seagrass leaves were also removed from the tanks at weekly intervals by 'combing' with the hands to prevent excessive amounts of dead material from shading growing leaves. Under field conditions, dead seagrass leaves do not usually accumulate in the standing crop because they are rapidly broken down into suspended and deposited detritus by microbial activity and/or removed from the seagrass meadows by wind, waves and currents (e.g. Klumpp *et al.* 1989)

3.2.4 Experimental treatments

The illumination provided to the laboratory benches by the combination of metal halide and fluorescent lamps was of sufficient intensity to induce photoinhibition in *Z. capricorni* samples exposed directly to this light. The 'treatment' in the high light stress experiment therefore involved leaving one half of each replicate tank uncovered while the 'control' half of the tank was supplied with a shadecloth cover to reduce illumination to a level suitable for healthy seagrass growth and avoiding the effects of photoinhibition. The covers were made from 70% knitted green shadecloth and shaped so as to extend down to sediment level on all four sides of the tank half (Figure 3.3). Shadecloth covers were applied to tank halves at random which resulted in a relatively wide range of PAR values for the control (238 $\pm$ 127 μmol photon m^{-2} s^{-1}) and treated replicates (661 $\pm$ 190 μmol photon m^{-2} s^{-1}) depending on whether the cover was located on the end of the tank facing the outer edge or centre of the bench.

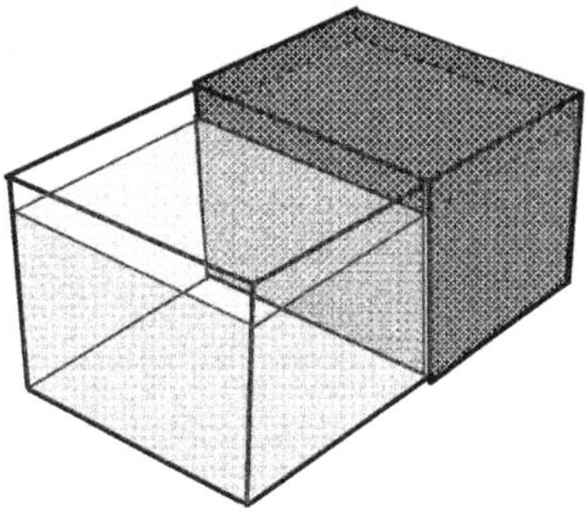

Figure 3.3. Experimental replicates for seagrass shading experiments. Shade treatment was provided to half of each tank by covers sewn from 70% green, knitted shadecloth. Covers extended down each side of the tank to the base, including a flap which divided the interior of the tank to sediment level but did not impede water movement. The shadecloth cover represented the low light treatment in the low light stress experiment and the control in the high light stress experiment.

For the low light stress experiment, an additional layer of shadecloth was suspended between the lights and the tanks to reduce the overall light intensity incident on the tanks to an average PAR of 168 $\pm$ 49 μmol photon m^{-2} s^{-1}. Hence controls were supplied with sufficient light for continued photosynthesis and growth with little risk of photoinhibition. The half-tank shadecloth covers used on controls in the high light stress experiment were employed to provide the low light treatment in this experiment. Shadecloth covers were placed over the outer ends of the tanks, i.e. furthest from the light source, to ensure that the treatment halves received considerably less light than the control halves. Low light treatments received an average PAR of 39 $\pm$ 12 μmol photon m^{-2} s^{-1}. In all cases, the PAR of treated replicates was below the compensation point, i.e. seagrass biomass rapidly declined in the treated tank halves because growth could not be sustained.

Shadecloth is commonly used in shading experiments with seagrasses (e.g. Dawson and Dennison 1996; Longstaff *et al.* 1999). It has been noted that shadescreens do not change the spectral distribution of the irradiance (Longstaff *et al.* 1999) and this can be observed in the general reduction of all visible wavelengths of light to about 65-70% of their unshaded values when a horizontal layer of shadecloth is suspended over the laboratory bench (Figure 3.4).

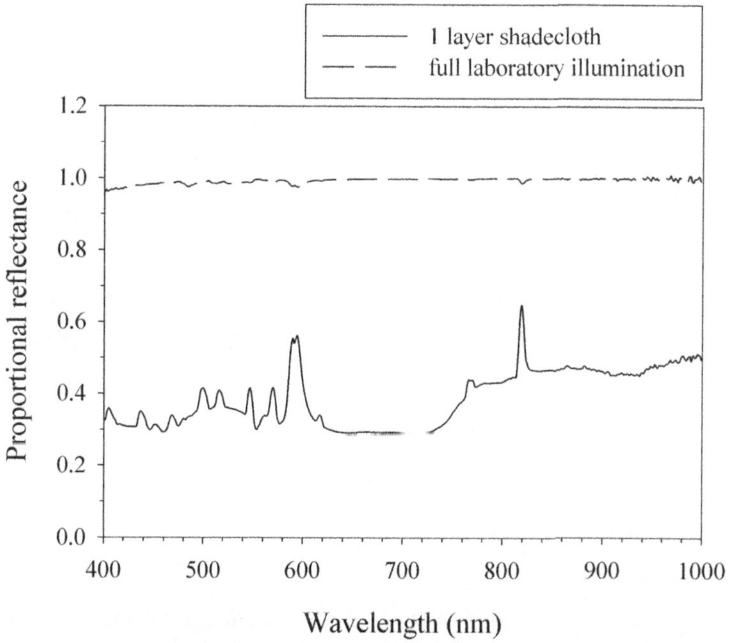

Figure 3.4. Spectral reflectance collected from the surface of a white reference panel with and without a layer of green, 70% knitted shadecloth suspended below the laboratory lights. Note that the spikes in the reflectance are due to noise in the white reference prior to division of the sample spectrum by the reference spectrum.

3.3 Sampling procedures

Leaf samples taken from the field or from tanks during laboratory experiments were primarily composed of the top ½ - ⅔ of the blades and contained leaves of all ages, excluding any obviously dead or diseased leaves. Some authors have noted the importance of comparing leaves of the same age in physiological studies, for example, the second youngest leaf on a seagrass shoot has a fully developed pigment content but not excessive epiphyte cover (Beer *et al.* 2001). Others have shown that photosynthetic parameters and pigment content changes with distance along a leaf blade (Enríquez *et al.* 2002). Alcoverro *et al.* (2001) used only the upper-middle 10 cm of the second youngest leaf to ascertain photosynthetic capacity, chlorophyll and nitrogen content, although Ralph *et al.* (2002) found no difference in a range of fluorescence parameters, chlorophyll and carotenoid content along healthy, epiphyte-free blades of *Z. marina*. In the current study the emphasis is on the remote sensing detection of seagrass physiological parameters so leaf samples represent the mix of leaves present in the part of the seagrass canopy that would be visible to a sensor. While this potentially introduces more variability into the physiological and biochemical results, the method of sampling is more appropriate to the nature of the study.

3.3.1 Spectroradiometry

Background

Spectroradiometers quantitatively measure radiance, irradiance, reflectance or transmittance of light from a remote target (not in direct contact with the instrument) by translating light energy into electrical current. The Fieldspec-FR (ASD, Inc.) is a field portable instrument that applies three integrated spectrometers to collect light in the visible–NIR–SWIR spectral range between 350-2500 nm. In the visible-NIR region (350-1050 nm), the bandwidth of each channel is 1.4 nm but the spectral resolution (FWHM;

full-width-half-mean of a single emission line) is approximately 3 nm at around 700 nm (ASD Inc. User Manual 1997). The sampling interval for the SWIR regions (900-1850 and 1700-2500 nm) is 2 nm but spectral resolution (FWHM) varies between 10-12 nm. The spectral information from the three spectrometers is subsequently corrected within the software for baseline electrical signal (dark current), and then interpolated to a 1 nm sampling interval over the wavelength range. The Fieldspec-FR collects light passively by means of a fiberoptic cable with a conical view subtending a full angle of around 25°. Foreoptics may be attached to the cable; e.g. to limit the lens angle (1° foreoptic), or to measure downwelling irradiance by integrating the light flux from all directions intercepted by a hemispherical surface (RCR).

Spectral reflectance (R) is the actual fraction of light reflected from the surface of a target after the material has interacted with the downwelling irradiance. Light energy may be absorbed, transmitted or reflected by a target. In addition, absorbed and transmitted light may be re-emitted at slightly different wavelengths after interacting with the material. Software used in the spectroradiometer enables relative reflectance to be estimated, i.e. the light reflected from a surface divided by a reference standard assumed to reflect 100% of the irradiance. This reference will be a Lambertian reflector of known radiance, in this case a 99% Spectralon ® diffuse reflectance target (Labsphere Inc.), which should be measured under exactly the same irradiance and geometry conditions as the target. It is ideal to measure downwelling irradiance and target radiance simultaneously to avoid error in the reflectance value caused by irradiance changes that arise from atmospheric fluctuations (Duggin 1980). Since this is not possible with the single cabled Fieldspec-FR, the white reference and target spectra are instead acquired with as short a time delay as possible between the two measurements.

Each reflectance spectrum is produced by averaging 20-60 target spectra collected consecutively from the sample to minimise random noise and therefore increase the signal-to-noise ratio (SNR). An equivalent number of dark current and white reference measurements are also averaged for each spectrum.

Timing of spectral reflectance sampling

Sampling dates for each study were determined by the availability of a consortium owned Fieldspec-FR instrument, which limited usage. In particular, this determined the length of the laboratory experiments and the timing of resampling of seagrasses growing in the laboratory, but it also influenced the dates on which field sampling were carried out.

Field spectroradiometry methods

For each seagrass species, leaf samples were haphazardly collected from an area of approximately 200 m^2 within each field study site and sorted into leaves with epibionts ($n = 5$) and those occurring naturally without epibionts ($n = 5$). The spectral reflectance of each sample was immediately measured in the field with the Fieldspec-FR spectroradiometer under natural sunlight. Leaves were kept wet with seawater throughout the sorting and measurement processes. To achieve a pure signal, leaves were piled on a matt black background in multiple layers (at least seven leaf layers deep (Lillesaeter 1982; O'Neill *et al.* 1990) but usually several times that) to overfill the field of view of approximately 5 cm diameter. Leaf samples were scanned from a height of approximately 5 cm, and from as close as possible to nadir above the sample without shadowing it, using a handheld 25° FOV (field-of-view) foreoptic pointed toward the direction of the sun whenever off-nadir (zenith angle 0-30°, azimuth angle 180°) (Plate 3.3). Sample reflectance was calculated immediately after measurement of radiance from a 99% white Spectralon ® panel (Labsphere, Inc.) under the same viewing and illumination geometry. Spectral averaging of 10-30 spectra per sample was performed to ensure adequate SNR. Sampling was carried out on cloud free days wherever possible within three hours of solar noon. Given that spectra were collected on a moving boat under changing field conditions, every practical effort was made to standardise the viewing and illumination geometry while maximizing the signal from the sample in each case. The correction procedures described below compensate for less than perfect measurement conditions when applied to a homogenous target such as the leaves of a single seagrass species.

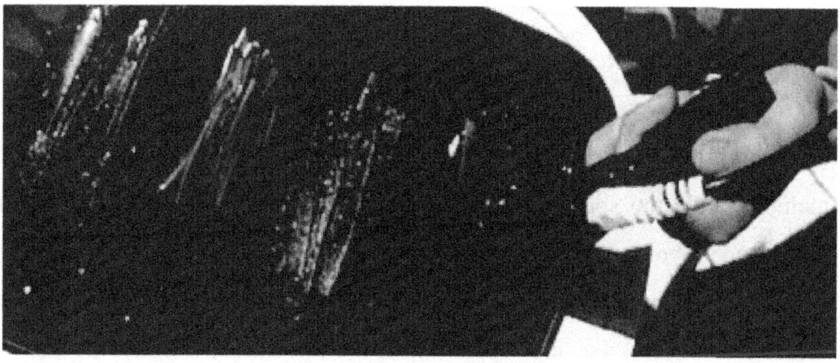

Plate 3.3. Field spectroradiometry of fresh *Posidonia australis* leaf samples sorted into piles of leaves with and without the epibiont foulers that are characteristic of mature seagrass leaves in the field.

Laboratory spectroradiometry methods

During each experiment, leaf samples were cut with scissors from the treatment and control halves of each replicate tank and the leaves were immediately placed in sample jars filled with tank water. The sample jars were placed in a sealed esky for dark adaptation and transport to the spectroradiometer laboratory. The spectral reflectance of each leaf sample was measured under controlled laboratory conditions in a sealed dark room immediately after chlorophyll fluorescence measurements had been made in the dark (see Section 3.3.2). Illumination was provided by three equidistant unfiltered (i.e. no glass cover shield), aluminium backed 50 W halogen lamps fixed at 600 mm distance and 30° zenith angle from the sample. Spectra were collected using the 1° foreoptic suspended at nadir at a constant height 350 mm above the sample. Samples were placed on a matt black background to fill the FOV of the foreoptic in layers at least seven leaves deep in order to saturate the signal. Sample reflectance was once again calculated after referencing to the 99% white Spectralon ® panel with spectral averaging of 60 spectra per sample to ensure optimal SNR.

Correction of spectra

To focus on the wavelengths most useful for the remote sensing of benthic plants, only the visible–NIR wavelengths in the range 430-900 nm were included from the spectral output available from the Fieldspec FR. Both field and laboratory data were converted from relative to absolute reflectance by multiplying the reflectance spectrum of each sample by the actual calibrated reflectance spectrum of the Spectralon ® reference panel. Each sample spectrum was then corrected for nonlinear additive and multiplicative scatter effects using a piece-wise multiplicative scatter correction technique (PMSC) (Isaksson and Kowalski 1993). Multiplicative scatter correction (MSC) applies a simple mathematical correction to the reflectance value at each wavelength of the sample spectrum using offset and slope values estimated by linear least-squares regression of that sample spectrum against a standard spectrum (Isaksson and Kowalski 1993). In PMSC, linear regressions are fitted to local wavelength regions using a moving window of specified length (Isaksson and Kowalski 1993). The 'standard' spectrum applied in PMSC to sample spectra of each seagrass species, was the mean spectrum of all samples of that species collected during the study. A window size of 100 nm was selected after investigating the influence of a range of window sizes (20, 30, 50, 75, 100, 150, 200 nm and 235 nm ~ linear MSC) on the correction of a test data set of 30 field spectra of *Z. capricorni* (Lake Illawarra, January 1999). The purpose of PMSC correction was to ensure that the field measured reflectance spectra were comparable regardless of sampling date, illumination conditions or sample geometry (Datt 1998). The PMSC correction also allowed laboratory measured spectra to be compared directly with field measured spectra.

3.3.2 Chlorophyll fluorescence

Background

Pulse amplitude modulated (PAM) fluorometer studies of the photosynthetic reaction centre, photosystem II (PSII), have been shown to be a useful tool for the detection of various stress responses in terrestrial and marine plants, including seagrasses (e.g. Ralph and Burchett 1995; Dawson and Dennison 1996; Ralph and Burchett 1998b). The first response of a plant to environmental stress is to increase non-radiative energy dissipation in an attempt to avoid photoinhibitory damage, and this is readily observed as chlorophyll fluorescence (Schreiber *et al.* 1995). Fluorescence emissions compete with photochemistry and heat dissipation in the quenching of absorbed light energy (discussed in Chapter 5). The amount of excitation energy that is dissipated as fluorescence is small (see Section 2.4.2) but measurable and provides a means of investigating the efficiency of energy conversion or degree of reaction centre closure in PSII in a non-intrusive manner (Kraus and Weis 1991; Björkman and Demmig-Adams 1995).

The amount of light emitted as fluorescence rapidly increases as the photosynthetic efficiency of the plant decreases (Kraus and Weis 1991). The maximum quantum yield for photochemistry (and hence minimal fluorescence) occurs when the large majority of reaction centres are oxidised or open (F_o). The maximum quantum yield for fluorescence occurs when all reaction centres are reduced or closed (F_m), so absorbed excitation energy can not be directed to photochemistry and no electron transfer can occur. Variable fluorescence (F_v), is the difference between F_o, the initial fluorescence observed in plants that have been dark adapted and F_m, the maximal fluorescence obtained after illumination with a saturating light ($F_v = F_m - F_o$). The optimum quantum yield of photochemistry in PSII, F_v:F_m, is positively correlated with the quantum efficiency with which plants produce oxygen (Björkman and Demmig 1987). F_v:F_m averages around 0.83 for healthy green plants of many different species and ecotypes (Björkman and Demmig 1987) since the basic processes of photosynthesis are the same for most plants (Schreiber *et al.* 1995). Decreases in this ratio have been associated with a wide range of environmental stressors that directly effect PSII efficiency (e.g. Kraus and Weis 1991; Schreiber and Bilger 1987). F_v:F_m, is therefore a quick and easy measure of PSII photosynthetic efficiency.

The effective quantum yield of photochemistry in PSII, ΔF:F_m' is a measure of instantaneous photosynthetic performance of a plant under ambient PAR. This measure can be used to calculate the current electron transport rate (ETR) of PSII, which under light limiting conditions is directly related to O_2-evolution, i.e. the rate of photosynthesis (Maxwell and Johnson 2000).

PAM fluorometers work by measuring F_o prior to illumination, using a measuring light that has no actinic effect, and immediately afterwards applying a strong saturation pulse that completely closes all reaction centres enabling measurement of F_m. F_v:F_m is calculated from this change in fluorescence by $(F_m - F_o)/F_m$ (Krause and Weis 1988; Krause and Weis 1991). More recently, submersible ('diving') PAMs have been applied to measure photosynthetic activity of seagrasses *in situ* (e.g. Beer *et al.* 1998; Ralph *et al.* 1998; Beer and Björk 2000; Macinnis-Ng and Ralph 2001). In the current study all seagrasses were destructively sampled for out of water spectral reflectance measurement, and since it was important for chlorophyll fluorescence measurements to be collected under exactly the same conditions as the spectra, a PAM-2000 (Walz, Germany) was used for this purpose rather than a diving PAM.

Methods

All fluorescence measurements in these studies were collected with a PAM-2000 fluorometer from detached seagrass leaves immediately after their removal from the water and immediately before measurement of their spectral reflectance. Measurements were made on the samples of layered leaves (as described in Section 3.3.1 Field and Laboratory Spectroradiometry Methods) to ensure that the seagrass sample filled the FOV of the fiberoptic cable, the end of which was held equidistant from the leaf samples by means of a 'distance clip'. Photosynthetic efficiency (F_v:F_m), F_o and F_m were measured in the dark after at least 10 minutes of dark adaptation, during which samples held in sample jars of tank water were covered by black plastic or placed in a darkened area (e.g. esky) to ensure all PSII reaction centres were fully oxidised. Initial tests indicated that for *Z. capricorni*, 10 minutes was a sufficient minimum period for dark adaptation. The slightly longer periods used in some experiments were for logistic reasons and did not alter the measurements.

Determinations of apparent photosynthetic electron transport rate (ETR) were madc by measuring the effective quantum yield (ΔF:F_m') of samples held in a leaf clip holder which simultaneously collects PAR incident on the leaf sample plus leaf temperature. ETR was calculated as ETR = Yield x PAR x 0.5 x 0.84 which assumes that 84% of the light incident on the leaf surface is absorbed by the leaf (Heinz Walz 1993). Plant species vary in the quantity of light absorbed into the leaves because of differences in leaf surface reflectance, chlorophyll content, etc. The actual leaf absorbance of *Z. capricorni* was not determined so the ETR values presented in this study are relative values only. This was not considered to be a problem because ETR comparisons were only made on the one species.

3.3.3 High Performance Liquid Chromatography

Background

Chromatography is widely used for the separation, identification and determination of the chemical components in complex mixtures. The components of a mixture are carried through a stationary phase by the flow of a gaseous or liquid mobile phase and separation occurs because of differences in the migration rates among the components (Skoog *et al.* 1992). High performance (or high pressure) liquid chromatography (HPLC) offers high resolution in the separation of complex biological macromolecules such as the chlorophyll and carotenoid pigments (Hancock and Sparrow 1984). HPLC is the preferred method for separating molecules of high polarity, high molecular weight, those containing a number of ionic groups or those displaying thermal instability (Hancock and Sparrow 1984). The most commonly used technique (Skoog *et al.* 1992) and the preferred technique for photosynthetic pigment separation (Wright and Shearer 1984) is reversed-phase HPLC, which utilises a nonpolar (hydrocarbon) stationary phase and a relatively polar solvent for the mobile phase. Since liquids are viscous and have relatively low diffusion rates, the microparticulate silica-based separation columns used in HPLC are highly selective but must operate under high pressure.

The retention rates of the components within the stationary phase are predominantly a function of their polarity and the functional groups carried by each compound (Skoog *et al.* 1992). Since the compounds are eluted from the column one at a time, they can be individually identified by the position of their absorbance peak along a time axis in a chromatogram produced using a spectrophometric detector located at the end of the column. The observed retention times are compared with those from a standard of known composition, previously run on the column under identical conditions. In addition, the height of, and area under each absorbance peak provides a quantitative measure of the amount of each component eluted.

HPLC analysis has been previously been used to determine the chlorophyll *a* and *b* content (van Lent *et al.* 1995) and xanthophyll content (Dawson and Dennison 1996; Ralph *et al.* 2002) of seagrass leaves.

Methods

Seagrass leaf samples for pigment analysis were frozen at -80 °C as soon as practical after harvesting and stored at this temperature until analysis. The chlorophyll and carotenoid pigment concentrations of samples were analysed by reverse-phase HPLC using a method adapted from Gilmore and Yamamoto (1991) as described in Dunn *et al.* (2004). Small, 1-2 cm long sections taken from 2-5 leaves to achieve 50-100 mg fresh weight (FW) per sample were ground to a paste in a mortar and pestle with liquid nitrogen and sand. Ground samples were then extracted in 1.5 mL of 100% acetone and transferred to a microcentrifuge tube containing approximately

1 mg sodium bicarbonate. After 20 mins storage on ice in the dark these samples were centrifuged (13 000 g) for 5 min and the supernatant was poured off into a covered test tube stored on ice. The pellet was re-extracted in 0.5 mL of 80% acetone by grinding with a polypropylene tissue grinder (Crown Scientific) for a minimum of 30 seconds. After a further 5-10 min on ice in the dark, the samples were again centrifuged for 5 min and the supernatant combined into the collection test tube. The remaining pellet was re-extracted, as above, in 0.5 mL of 80% acetone and combined with the previous fractions. Preliminary HPLC testing had shown a third extraction step to be necessary for adequate quantification of the photosynthetic pigments in the leaves of *Z. capricorni*. The combined extract was then made up to 3 mL in the test tube using 80% acetone and approximately 1 mL of this solution was filtered through a 45 μm polypropylene filter into an amberglass vial. Samples (40-60 μL) were automatically injected into a Shimadzu HPLC system at a flow rate of 2 mL min^{-1}. Solvent A (acetonitrile:methanol:Tris HCl buffer 0.1 M, pH 8.0; 79:8:3) ran isocratically from 0 to 4 min followed by a 2.5 min linear gradient to 100% Solvent B (methanol:hexane; 4:1) which then ran isocratically from 6.5 to 15 min. Flow rate was decreased from 2 to 1.5 mL min^{-1} between 8 and 12 min and then ran at 1.5 mL min^{-1} until 15 min to maintain stable pressure. The column was re-equilibrated with Solvent A between samples. Solvents were degassed using an inline degasser. Pigments were separated on a Allsphere ODS1 analytical column (Alltech) preceded by a guard column and were quantified by integration of peak areas detected at 440 nm by a photodiode array detector (Shimadzu Model SPD-M10AVP) relative to pure chlorophyll (Sigma) and carotenoid (Extrasynthase and VKI) standards.

Water content of seagrass leaves

Leaf pigment concentrations are more typically quantified according to leaf surface area in remote sensing studies where the pigment concentrations are related to the spectral reflectance recorded from the surface of a single leaf (e.g. see Blackburn 1998). However, it was not practical to punch replicate sized leaf discs from the narrow, fibrous leaves of *Z. capricorni*. Since all spectral reflectance measurements in this study were collected from a saturating pile of stacked leaves, and because seagrass leaves are bifacial and vary considerably in their thickness, the calculation of pigment content according to fresh weight (FW) may actually be a better measure. Determining pigment composition on the basis of fresh weight may, however, lead to inaccuracies in the absolute values that are calculated because the water content of leaves can vary across species and environmental situations. It may be reasonable to assume that aquatic plants will always be fully turgid and therefore, percent water content of leaves will not differ significantly. However, variations in leaf morphology, surface features and internal characteristics may vary the amount of water that a leaf can hold.

To establish whether pigment analysis results would need to be corrected for differences in leaf water content, the percentage weight of water in the leaves of *Z. capricorni*, *P. australis* and *H. ovalis*, fouled with leaf epibionts and unfouled, were compared from a number of field sites and dates. Samples were sourced from leftover field samples that had not been used in pigment analysis and could therefore be considered equivalent to the samples actually used in pigment analysis. Water content was determined by weighing samples, drying them at 60 °C for several days in a drying oven until stable weight was achieved, then reweighing. The percent water content of samples was calculated as the weight lost by drying x 100/FW of sample.

3.3.4 Spectrophotometry

Background

Optical spectroscopy is the qualitative and quantitative analysis of the chemical composition of a sample on the basis of the characteristic absorption properties of each chemical species in the sample (Skoog *et al.* 1992). Most spectrophotometric instruments pass a stable source of radiant energy through a solution of the sample and through a standard containing only the solvent, which are held in paired transparent sample cuvettes. A radiation detector records the amount of light energy transmitted through the sample within a narrow wavelength region and converts it to a measurable signal. The measurement waveband is selected to represent the region of maximum absorption by the chemical of interest. Spectrophotometric methods have been widely applied in the determination of pigment concentrations in plants, including the anthocyanin content of seagrass leaves (Dawson and Dennison 1996).

Methods

The anthocyanin concentration of seagrass leaves was determined using the differential pH method of Fuleki and Francis (1968) and Francis (1982). After measuring the weight of the sample, approximately 50-100 mg of frozen seagrass leaves (samples from the field and high light laboratory experiment) were ground to a paste in a mortar and pestle with liquid nitrogen and sand. For the dried samples (those from the low light laboratory experiment), around 10-20 mg of dried leaf were ground using sand only. Each sample was then extracted in 1.5 mL of 1% concentrated hydrochloric acid (pH 1) in methanol using a polypropylene tissue grinder and stored on ice in the dark for 10 minutes. Samples were centrifuged (13 000 g) for 6 min. The absorbance of the supernatant was measured against a standard at 529 nm (A_{529}) using a Shimadzu UV-1601 spectrophotometer. One mL of this solution was then mixed thoroughly with 80 μL of sodium acetate buffer (pH 5.25) in an microcentrifuge tube, stored for 5-10 minutes in the dark, and centrifuged for 10 min at 13 000 g. The A_{529} was measured a second time. The relative anthocyanin content of samples was calculated by subtracting the second absorbance value from the first, and dividing the difference by the weight of the sample multiplied by the volume extracted.

3.3.5 Elemental analysis of Carbon and Nitrogen content

Background

Organic compounds can be decomposed to determine the elemental composition of the sample by oxidative methods that involve either the conversion of carbon to CO_2 and hydrogen to water, or heating of the sample with a potent reducing agent (Skoog et al. 1992). The dry-ashing oxidation procedure involves ignition of the organic sample in air or in a stream of oxygen until all carbonaceous material is converted to CO_2. Many important elements including carbon and nitrogen are converted to gases during this procedure but the gases can be trapped and quantified. Automated combustion-tube analysers such as the one applied in the current study and described below simplify the procedure.

Methods

Carbon and nitrogen elemental content was determined with a Carlo Erba NCS1500 elemental analyser. Leaf samples were dried to constant weight at 60 °C and ground in an agate mortar and pestle. Aliquots of sample were then weighed into tin cups, typically 0.3 mg for carbon determination and 4-5 mg for nitrogen determination. Using an autoloader the tin cups and sample were dropped into the combustion column of the elemental analyser, which was maintained at 1050 °C, while oxygen was injected into the helium carrier gas stream. The tin and sample 'flash' combusted and the gaseous combustion products were swept down the combustion column over chromium oxide and silvered cobaltous cobaltic oxide catalysts, to convert all CO to CO_2, and then through the reduction column packed with high purity copper, to remove excess oxygen and reduce all nitrous oxides to N_2, and finally through a water trap. The purified gas was then passed through a chromatograph column to separate the gases, whose quantity was determined with a TCD (thermal conductivity detector). Calibration was achieved using pure urea and an internal laboratory standard (ground plant material) before the initial sample and after every 10 samples.

Chapter 4

Spatial and Temporal Variation in the Spectral Reflectance of Seagrass

S.K. Fyfe, S.A. Robinson & A.G. Dekker

4.1 Introduction

Recent developments in sensors and processing methodologies have led to an increase in the use of hyperspectral and high resolution multispectral remote sensing techniques for environmental applications such as the mapping and monitoring of the coastal zone. High spectral and spatial resolution imagery provides researchers with the potential to map vegetation to species level, provided that the plant species under study are spectrally distinct.

A basic assumption of mapping by remote sensing is that the features of interest in an image reflect or emit light energy in different and often unique ways (e.g. Harrison and Jupp 1990; Lillesand and Kiefer 1994). Remote sensing classification procedures are usually based on numerical techniques that group the image pixels by their spectral response across the available wavebands, i.e. by their multivariate statistical parameters (Jensen 1996). There would be no point in trying to map objects that are not spectrally different at the most basic level, no matter how much time is spent on processing or preprocessing, or how advanced the classification or image extraction techniques are.

The assumption that individual plant species have unique spectral signatures has however, been questioned (Price 1994). First, there is an overall qualitative similarity in the spectral response curves of green plant species (Kleshnin and Shul'gin 1959), although quantitative spectral reflectance differences between species have been observed for plant leaves (e.g. Gates *et al.* 1965; Gausman 1982; Datt 2000) and canopies (Gong *et al.* 1997; Kumar and Skidmore 1998; Yu *et al.* 1999). Second, reflectance from vegetation is governed by a small number of physical and physiological parameters and these may vary for a species, or within an individual plant, over space and time. It is therefore useful to determine whether the observed reflectance differences between plant species are not only statistically significant, but consistent (over different seasons or in different habitats for example) before they can be generally applied with success in remote sensing species mapping.

4.1.1 Spectral reflectance of plant leaves and canopies

The reflectance curve for healthy green vegetation is characterised by the absorption of blue (400-500 nm) and red (600-700 nm) wavelengths and by the reflectance of green radiation (500-600nm), and very strong reflectance of near-infrared (NIR; 700-1300 nm) and mid-infrared (MIR; 1300-2600 nm) radiation (e.g. Billings and Morris 1951; Kleshnin and Shul'gin 1959). The steep rise usually observed in a vegetation curve at about 700 nm, the 'red edge' (Horler *et al.* 1983) is unique to the spectral signatures of green vegetation and indicates the limit of chlorophyll absorption by plant tissue. While the visible reflectance of green light gives plants their characteristic green colouration, plants absorb blue and red light as the energy source for photosynthesis. Chlorophyll *a* absorbs around 70-90% of the radiation entering a plant in wavelengths centred at 430 nm and 660 nm (Gausman 1982; Curran 1989). Chlorophyll *b* absorbs slightly longer blue wavelengths (460 nm) and slightly shorter red wavelengths (640 nm) than chlorophyll *a*, while a variety of carotenoids absorb maximally at various points across the blue wavelengths (centred around 450 nm). The effects of individual carotenoids on blue light absorption may be small but the combined effect of the chlorophylls and carotenoids together results in the characteristic broad blue absorption trough observed in the spectral signatures of plants (Thomas and Gausman 1977).

While the absorption of light by plants is governed by foliar chemistry (Curran 1989), the magnitude of reflectance of light from plant leaves will depend primarily on plant physiognomy. The amount of visible light reflected from the leaves of different plant species will be affected by leaf thickness and morphological complexity (Gausman and Allen 1973), leaf surface quality (Shul'gin *et al.* 1960; Woolley 1971; Vogelmann 1989), surface adaptations such as leaf hairs (Billings and Morris 1951) and leaf internal structure (Gausman *et al.* 1969; Woolley 1971; Buschmann and Nagel 1991, Vogelmann 1989; Vogelmann *et al.* 1996). The internal structure of leaves controls the magnitude of reflectance and transmittance across the whole spectrum, but the effects are most apparent in wavelengths where absorption is low (Jacquemoud and Baret 1990). Consequently, reflectance differences between plant species in the NIR wavelengths are of much greater magnitude than those observed in the visible. The high NIR plateau observed in the spectral signature of plants is predominantly a function of internal light scattering, determined by the size, shape and arrangement of intercellular air spaces in the leaf tissue (Gausman *et al.* 1975; Gausman *et al.* 1984; Vogelmann 1989; Guyot 1990). Other anatomical and physiological features will also influence the degree to which light is scattered within a leaf; e.g. variations in cell size, shape and packing, cell wall structure and hydration, stomata, plastids, nuclei, crystals, cytoplasm and other cell organelles (Gates *et al.* 1965; Gausman 1977; Guyot 1990).

When remote sensing at the canopy scale, the spectral signatures recorded from seagrass meadows will be influenced by the density and geometry of the seagrass canopy. Shoot density, leaf length and width, leaf inclination angle (which may vary considerably with tides and currents), non-random clumping of leaves and associated patterns of shadowing can influence reflectance, particularly in the NIR region (Li and Strahler 1986; Heilman and Kress 1987; Curran and Wardley 1988). An increase in plant canopy density, usually expressed as the leaf area index (LAI), leads to increased red light absorption and NIR reflectance (Curran and Wardley 1988). The spectral reflectance signal will be solely controlled by the seagrass canopy in dense meadows where the projected foliage cover is greater than 70% (Lillesaeter 1982; Horler *et al.* 1983; Heilman and Kress 1987). However, spectral reflectance from a sparse vegetative canopy is significantly influenced by background reflectance (Jupp *et al.* 1986; Clevers 1988; Spanner *et al.* 1990). In the case of seagrass meadows, the background is likely to include sediment, benthic microalgae (BMA) and associated macroalgae and marine fauna (e.g. corals, fanworms, sponges, etc). Reflectance differences between plant species or community types are therefore a consequence of the combined spectral response of individual leaves, the vegetative canopy as a whole and non-vegetative elements beneath or adjacent to the canopy.

4.1.2 Spectral discrimination of plant species

In studies investigating the discrimination of terrestrial plant species, the largest differences in reflectance have been recorded in the near infrared (NIR) and short wave infrared wavelengths (e.g. O'Neill *et al.* 1990; Borregaard *et al.* 2000). Remote sensing of benthic aquatic plants, however, is limited to the visible wavelengths where light penetrates the water column and can be reflected back to a sensor. Although pure water absorbs light to some extent in shorter wavelengths, significant attenuation occurs at wavelengths beyond 680 nm (Kirk 1994). In coastal waters, spectral scattering and absorption by phytoplankton, suspended organic and inorganic matter and dissolved organic substances further restrict the light passing to, and reflected from, the benthos (Dekker *et al.* 1992). Hence, efforts to discriminate between aquatic plant species must concentrate on pigment related spectral features within the visible wavelengths.

Because all angiosperms share basically the same physiology and biochemistry, the wavelength position of pigment absorption features does not differ significantly between species (Curran 1989). However, the relative concentrations of photosynthetic pigments and the presence of accessory pigments do vary among taxa (Young *et al.* 1997). The depths and widths of pigment absorption troughs and the position and magnitude of reflectance peaks can be quite different among species (Gates *et al.* 1965; Gausman *et al.* 1975; Thomas and Gausman 1977; Gausman 1982; Horler *et al.* 1983; Ustin *et al.* 1993). Different arrangements of photosynthetic pigments into pigment-protein complexes in the thylakoid membranes, plus the 'package effect' of pigment molecules within chloroplasts, within cells and within plant leaves, will also modify the magnitude and wavelength position of the absorption features in different ways for different plants (Kirk 1994). Hence, differences in the spectral reflectance of terrestrial plant species have been recorded in the visible wavelengths (e.g. Gong *et al.* 1997; Kumar and Skidmore 1998; Yu *et al.* 1999; Datt 2000). Remote sensing has been even more successful in identifying and quantifying phytoplankton taxa in algal blooms (Richardson *et al.* 1994; Richardson 1996; Aguirre-Gómez *et al.* 2001b).

The relative concentrations of photosynthetic and accessory pigments will also vary within a seagrass species because of genetic variation or chromatic adaptation to seasonal cycles, stage of growth, health or environmental conditions (Pérez-Llorens *et al.* 1994; Dawson and Dennison 1996; Longstaff and Dennison 1999; Ralph 1999; Alcoverro *et al.* 2001). Chromatic acclimation of pigments can occur in an individual aquatic plant grown under changing conditions of water depth or clarity (Jeffrey 1981; Dennison, 1987; Duarte, 1991; Kirk 1994). Spatial and temporal variations in light and nutrient availability, as well as water temperature, salinity and the degree of water movement around the plant will influence growth, photosynthesis and therefore the spectral response of the seagrass. Spectral reflectance changes in aging or diseased plants have been related to changes in the intercellular air spaces (Tageeva *et al.* 1961; Gausman *et al.* 1971), chlorophyll breakdown (Tageeva *et al.* 1961; Gausman 1982; Brakke *et al.* 1989) and the subsequent unmasking of carotenoids and other leaf pigments (Thomas and Gausman 1977; Gausman 1982; Peñuelas *et al.* 1995b; Adams *et al.* 1999) including the synthesis of anthocyanins and tannins (Boyer *et al.* 1988). In addition, the epibionts that grow on the surface of aquatic plants may mask their reflectance to some extent while contributing their own absorption and reflectance features to the spectral response. Seagrass epibionts include a diverse array of microalgae, bacteria, juvenile macroalgae and sessile invertebrates such as tubeworms, bryozoans, hydroids and sponges. However, the algae are the most prominent epibionts and have the greatest impact on remote sensing. The major light harvesting pigments and therefore, the wavelengths where maximal absorption occurs, vary widely between angiosperms and the different algal classes (Table 4.1). May *et al.* (1978) recorded fifty six algal species in a survey of the leaf epiphytes of *Z. capricorni* and *P. australis* in two estuaries in New South Wales. The biomass and species composition of epiphytes on a single seagrass species will change with location and may vary over time. Spectral reflectance response in the visible wavelengths is therefore variable for any seagrass species over space and time. Intraspecific variability may increase the chance of spectral reflectance overlap with other species and make spectral discrimination using remotely-sensed image data difficult or impossible.

Table 4.1. The chlorophylls and major carotenoids that characterise angiosperm plants and some selected classes of algae often found growing as epiphytes on seagrass leaves (adapted from Kirk 1994).

	Angiosperms e.g. seagrass	Chlorophyta green algae	Phaeophyta brown algae	Rhodophyta red algae	Baccilliariophyta diatoms	Dinophyta dinoflagellates	Cyanobacteria blue-green algae
CHLOROPHYLL							
a	+	+	+	+	+	+	+
b	+	+	-	-	-	-	-
c_1	-	-	+	-	+	+/-	-
c_2	-	-	+	-	+	+	-
CAROTENOID							
α-carotene	+	-	-	+	-	-	-
β-carotene	+	+	+	+	+	+	+
echinenone	-	-	-	-	-	-	+
lutein	+	+	-	+	-	-	-
zeaxanthin	+	+	-	+	-	-	+
neoxanthin	+	+	-	-	-	-	-
violaxanthin	+	+	+	-	-	-	-
fucoxanthin	-	-	+	-	+	+/-	-
diatoxanthin	-	-	-	-	+	-	-
diadinoxanthin	-	-	-	-	+	+	-
peridinin	-	-	-	-	-	+	-
myxoxanthophyll	-	-	-	-	-	-	+

NB. predominant chlorophyll or carotenoid is + present; - absent; +/- either present or absent. In the majority of dinoflagellates, peridinin is the major carotenoid and only chlorophylls *a* and c_2 are present. In a few dinoflagellates, fucoxanthin replaces peridinin as the major carotenoid and chlorophyll c_1 is present in addition to c_2.

4.1.3 Are seagrasses spectrally distinct?

Hyperspectral image data offers the high spectral resolution required to detect visible reflectance differences in the pigment content of plants. High spatial resolution airborne spectrometer imagery has been applied in the delineation of seagrass beds, the estimation of seagrass standing crop and in mapping subtidal and intertidal seagrass and macroalgal communities (see Table 1.1). While these studies have been relatively successful in mapping the extent of taxonomically quite different groups of plants, few published studies have been able to accurately delineate co-occurring species.

If species are to be mapped accurately by remote sensing, differences in the spectral signatures of seagrass species need to be distinct and consistent over seasons, geographic locations and changing environmental conditions. Hence, it is necessary to characterise the range of signatures expected for any species under natural conditions if the signatures are to be used in spectral libraries for quantitative spectral analysis (Price 1994).

The main aim of this field-based spectroradiometer study was therefore to determine whether consistent differences were apparent in the spectral reflectance of the three common seagrass species of south eastern Australia, and to explain any observed differences in terms of the photosynthetic and accessory pigment composition of their leaves. Patterns of reflectance associated with epibiont cover and the year, season, estuary and habitat of sampling were characterised to compare the magnitude of interspecific versus intraspecific variability in the spectral signatures of the seagrass species. Within species variability in reflectance was explained by examining pigment differences between samples obtained from different seasons, habitats and estuaries. Accurate remote sensing of seagrasses to species level will only be possible if the species prove to be spectrally distinct despite variation within each species.

A further objective of this work was to determine the location and width of wavebands that may be practically applied in the remote sensing of benthic aquatic plants. These wavelengths can subsequently be targeted to improve the efficiency and accuracy of image data collection and processing. The information may influence decisions about the type of remote sensing scanner best suited to seagrass mapping.

4.2 Study methods

4.2.1 Field sampling of spectral reflectance

The spectral reflectance response of the three common species of seagrass that occur in the estuaries of south eastern Australia, eelgrass *Zostera capricorni*, strapweed *Posidonia australis* and paddleweed *Halophila ovalis*, was investigated over a two year period between May 1999 and January 2001. To assess the range of spectral variability that may be found in each species, reflectance from fresh seagrass leaves, both with and without their characteristic leaf epibionts, was sampled during each season from Port Hacking, Lake Illawarra and St Georges Basin (see Section 3.1.1; Figure 3.1). In addition, samples were examined from marine dominated habitats near the mouths of each estuary and from brackish habitats in the backwaters of each estuary.

Field sampling of spectral reflectance was carried out with a spectroradiometer according to the procedures described in Section 3.3.1. The total number of samples collected for each species was not equal because all species did not occur at all sites and also because leaves with epibionts were occasionally difficult to locate in brackish habitats. Hence, the total sample sizes were; 230 unfouled *Z. capricorni*, 209 fouled

Z. capricorni; 79 unfouled *P. australis*; 78 fouled *P. australis*; 104 unfouled *H. ovalis* and 89 fouled *H. ovalis*.

4.2.2 Analysis of spectral data

Figure 4.1 summarises the data preparation, correction (see Section 3.3.1) and analysis procedures undertaken for the spectral reflectance data sampled in this study. A Model I (fixed effects) single factor analysis of variance (ANOVA) was applied at each wavelength over the range 430-900 nm to indicate the wavelengths where significant reflectance differences between seagrass species occurred. ANOVA was only performed on fouled leaves since these are more abundant in the upper canopy of seagrass meadows and are therefore more likely to dominate the field of view when imaged by an airborne or satellite sensor. For the purpose of assessing the merit of PMSC correction, and to demonstrate that significant results were not simply an artefact of the correction procedure, ANOVAs were performed on both the raw uncorrected and PMSC-corrected data sets. Prior to each analysis, the data at each wavelength were visually assessed for normality, although in general it is reasonable to assume a normal spectral reflectance distribution for any homogenous target material (Lillesand and Kiefer 1994). The data at each wavelength were also examined for homoscedasticity using Cochran's C test (Winer 1971). At wavelengths where variances were heterogeneous, a square root transformation was applied to the data. If transformation failed to homogenise the variances, analysis was performed on the untransformed data. Where significant differences were found in the ANOVA, multiple comparisons were made using the Tukey test for unequal sample sizes (Zar 1984). All univariate analyses involving spectral reflectance data were performed in S-Plus (MathSoft).

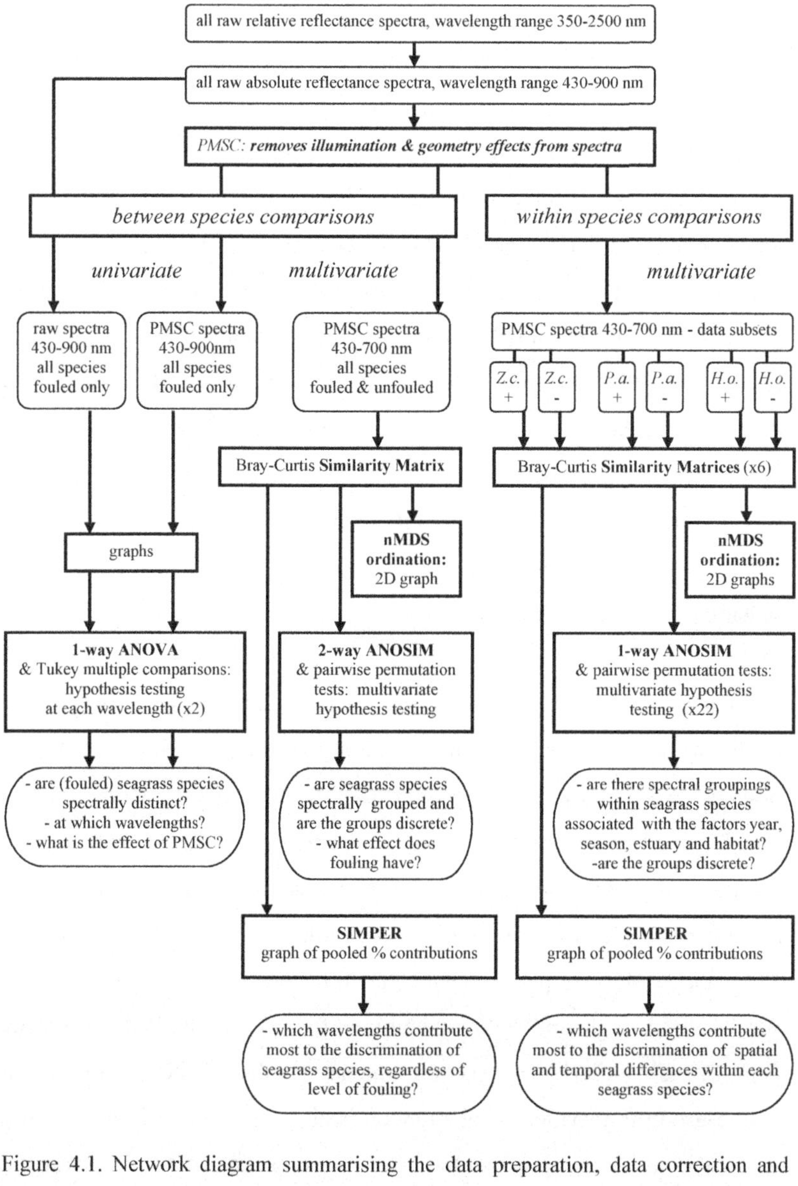

Figure 4.1. Network diagram summarising the data preparation, data correction and statistical analysis procedures undertaken in this study. PMSC = piece-wise multiplicative scatter correction, nMDS = non-metric multidimensional scaling, ANOVA = analysis of variance, ANOSIM = analysis of similarity, SIMPER = similarity percentages; *Z.c. Zostera capricorni*, *P.a. Posidonia australis*, *H.o. Halophila ovalis*; + leaves fouled by epibionts, - unfouled (Fyfe 2003).

Multivariate techniques were used to compare the magnitude of intraspecific variation in spectral reflectance with interspecific variation. Reflectance differences within each seagrass species associated with the estuary, habitat, year and season of sample collection were also examined by multivariate statistical analysis. Analysis was performed on the PMSC-corrected data using all samples collected during the study, i.e. all species both with and without epibionts, and on individual species subsets of the fouled and unfouled leaf samples. The data sets were reduced to include only the 430-700 nm range after preliminary analyses suggested that NIR wavelengths overpowered the contribution of visible wavelengths in the similarity matrix. Since NIR wavelengths are absorbed by water, it was important to ensure that any reflectance differences detected between factors were based only on those wavelengths that can be applied to remote sensing through a water column.

An association matrix was generated for each data set using Bray-Curtis similarity and applied in non-metric multidimensional scaling (nMDS) to produce two dimensional ordination plots (Clarke 1993). Data were not transformed or standardised prior to generation of similarity matrices. To test the hypotheses that the spectral reflectance of seagrass species differs, a non-parametric two-way analysis of similarity (ANOSIM) (Clarke and Warwick 1994) was performed on the full data set of samples from all species, both with and without epibiont foulers. ANOSIM utilises the Bray-Curtis similarity matrix to compare ranks of between-group to within-group similarities using a randomization test of significance in order to test the null hypothesis that the *a priori* groups do not differ (Clarke and Warwick 1994; Anderson 1999). One-way ANOSIM tests were used to investigate differences within each species of seagrass associated with the factors; year, season, estuary and habitat. Pairwise permutation tests followed each ANOSIM to examine where the significant differences between the levels of the factors lay.

Similarity percentages (SIMPER) procedures were also applied to each data set to identify the wavelengths that contributed most to the significant differences between species or factor groups (Clarke and Warwick 1994). The SIMPER algorithm determines the relative contribution that each wavelength (in this case) has made to the average similarity within a group and the average dissimilarity between groups. If a wavelength consistently contributes to within group similarity between pairs of samples, as well as to between group dissimilarity between pairs of samples, then its percentage contribution to similarity is high and it can be considered a good discriminating wavelength (Clarke and Warwick 1994). All multivariate analyses were performed using PRIMER 5 software (Plymouth Marine Laboratory, UK).

4.2.3 Field sampling of pigment content

Seagrass leaf samples were collected during the field sampling of spectral reflectance and frozen at –80 °C until later used in pigment analysis. A number of fresh leaf pieces from the upper central area of the FOV sampled by the spectroradiometer were wrapped in foil and snap frozen in liquid nitrogen immediately after their spectral signature had been measured.

The time and costs associated with pigment extraction, identification and quantification restricted subsequent analysis to a subset of the available samples. Four fouled and four unfouled samples were randomly selected from the five replicates available for each seagrass species collected from each habitat in each estuary. Leaf pigments were analysed for samples collected during October 2000 (to represent the colder months as water temperatures are still low in spring) and January 2001 (representing seasonally warm water). The chlorophyll and carotenoid content of the seagrass leaves were quantified by HPLC using the methods described in Section 3.3.3 for fouled *Z. capricorni* (n = 44), unfouled *Z. capricorni* (n = 52), fouled *P. australis* (n = 17), unfouled *P. australis* (n = 18), fouled *H. ovalis* (n = 21) and unfouled *H. ovalis* (n = 25).

The anthocyanin content of field samples was determined for samples collected during October 1999 and July 2000 since these were the only remaining dates that could provide relatively complete sets of samples for each seagrass species from each estuary (since some sets of samples degraded or were lost during transport and storage). The anthocyanin concentration per gram fresh weight of leaf was determined by spectrophotometer using the differential pH method (see Section 3.3.4) for four randomly selected fouled and four random unfouled samples of each seagrass species from each habitat within each estuary. The total sample sizes were; fouled *Z. capricorni* (n = 39), unfouled *Z. capricorni* (n = 52), fouled *P. australis* (n = 18), unfouled *P. australis* (n = 16), fouled *H. ovalis* (n = 17) and unfouled *H. ovalis* (n = 21).

4.2.4 Percent water content of seagrass leaves

The percentage water content of seagrass leaves was determined for the subset of leaf samples collected during field sampling but not used in pigment analysis (see Section 4.2.3). The percent water content of samples for all three seagrass species obtained from a range of sites and on a number of dates, was determined according to the procedure described in Section 3.3.3.

Results were analysed with a Model II (random effects) one-way ANOVA in the JMP 5 software package (SAS Institute Inc.). Quantile plots were visually examined to confirm the normality of the data but Cochran's tests suggested that there was significant heterogeneity in the variance of the sample groups. As square root transformation did not lead to homoscedasticity, and since the experimental design was also not balanced, it was necessary to apply caution when interpreting a significant result due to the likelihood of a Type I Error occurring. The data were divided into subsets for further investigation of the effects of species and fouling on marine samples, and of habitat or estuary and fouling on *Z. capricorni* samples, using more balanced designs (with proportional replication; sensu Zar 1984) and Model I two-way ANOVA tests using JMP 5 software.

4.2.5 Analysis of pigment data

Fouled and unfouled seagrass leaf samples of *Z. capricorni*, *P. australis* and *H. ovalis* were analysed for differences between the species in their total pigment (chlorophyll and carotenoid) content, in their concentrations of individual pigments normalised to total pigment content, and in their anthocyanin content, using a series of Model I ANOVA tests in JMP 5. The data were visually assessed for normality and Cochran's tests were applied to check that variances were homogeneous. Square root, fourth root, log or arcsine transformations were applied where necessary to improve the homoscedasticity of the data. Since transformation did not always result in homoscedasticity, a significant result achieved from heteroscedastic data was considered justified only if the level of significance was very strong. That is, if variances were homogeneous, then ANOVA results were considered significant at $\alpha \leq 0.05$, but if variances were heterogeneous at $0.01 < p < 0.05$ then ANOVA results were considered significant at $\alpha \leq 0.01$, and if variances were heterogeneous at $p < 0.01$ then ANOVA results were considered significant at $\alpha \leq 0.001$).

The concentrations of individual seagrass chlorophylls and carotenoids normalised to total pigment content were also applied in a multivariate analysis using Primer 5 software to compare the grouping of species by similarities in their pigment content with those observed when species were grouped on the basis of their spectral reflectance. The brown algal pigment fucoxanthin was included in the data set because it was present in large enough quantities to have an important effect on the spectral reflectance of leaves. In particular, it would presumably contribute a great deal toward spectral differences between fouled and unfouled samples. An ordination plot was generated by nMDS based on a Bray-Curtis similarity matrix of untransformed data. Weighting the individual pigments by transforming the data would have overemphasised the less abundant pigments and since the more abundant pigments probably contribute most to spectral reflectance, their dominance of the analysis would be realistic. Standardization was unnecessary since the data had already been normalised to relative pigment content. A two-way ANOSIM was applied to examine the data for significant differences between species and between fouled and unfouled leaf samples. The data set was subsequently analysed by SIMPER to examine which pigments contributed the most to the similarities and differences within and between species groups.

Within species differences in chlorophyll and carotenoid composition associated with the estuary, habitat, or season of sample collection were examined using one-way ANOSIM after a Bray-Curtis similarity matrix and nMDS plot had been generated for each individual species. Significant ANOSIM results were followed by pairwise permutation tests where required to ascertain the levels of those factors that differed significantly. SIMPER was subsequently applied to the data set for each seagrass species to determine the influence of individual pigments on the similarities and differences between season, estuary and habitat groups observed within each species.

Intraspecific differences in the anthocyanin content of fouled and unfouled seagrass leaves related to the estuary, habitat or season of sampling were examined using one-way, two-way, or three-way ANOVAs in JMP 5. The ANOVA design used for each analysis was dependent on samples being available within each factor to allow for a balanced design. The assumptions of normality and homoscedasticity were checked prior to each analysis as described previously in this section.

4.2.6 Influence of epibionts on leaf pigment content

To examine the influence that epibionts have on the pigment composition of seagrass leaves, leaf samples of *Z. capricorni* were collected haphazardly from marine situations in Sussex Inlet and Lake Illawarra in May 2000 and immediately frozen in liquid nitrogen in the field. The samples were stored at –80 °C until later use in pigment analysis. Five Sussex Inlet samples were analysed by HPLC for chlorophyll and carotenoid content according to the methods described in Section 3.3.3. The anthocyanin content of five Lake Illawarra samples was determined by spectrophotometer according to the methods described in Section 3.3.4. For each replicate sample, one fouled leaf was selected and divided longitudinally in half along the midvein. The epibionts were carefully scraped from the surface of one half of this leaf using a razor blade immediately before pigments were extracted from the fouled section, the scraped section and from an unfouled leaf selected from the same sample.

Differences in the chlorophyll, carotenoid or anthocyanin concentrations per gram fresh weight of fouled versus scraped leaf sections were examined for normality and analysed using paired t-tests. Two-sample t-tests were applied to investigate differences in the pigment content of fouled leaf sections versus unfouled leaves after examining the data visually for normality and checking for homogeneity of variances with Cochran's test.

Data that were found to be heteroscedastic were tested using a t-test for unequal variances. All statistical analyses were performed in JMP 5.

4.3 Results: Spectral reflectance of seagrass leaves

4.3.1 Reflectance differences between species

The potential for spectral discrimination of unfouled seagrass species was apparent even before correction of the reflectance data to remove the effects of illumination and sample geometry (Figure 4.2A). Differences were apparent in the relative magnitude of reflectance of green and NIR wavelengths by each species. It was somewhat more difficult to establish reflectance differences between species when the seagrass leaves were fouled by epibionts (Figure 4.2B). Despite large overlaps in the standard deviations of the raw spectral reflectance curves, there were significant differences in the reflectance of fouled leaf samples (Figure 4.2C; $F_{(1), 2, 378} > 6$, $p < 0.002$ for all wavelengths, Figure 4.2D; Cochran's C is NS for wavelengths 530-900 nm, $0.01 < p < 0.05$ for wavelengths 472-529 nm, $p < 0.01$ for wavelengths 430-471 nm). However, there were only a few short wavelength regions in which all three species were significantly different from each other (Figure 4.2C). *Posidonia australis* consistently reflected more light than the other two species in the shorter (430-520 nm) and NIR (700-900 nm) wavelengths while *Z. capricorni* was significantly less reflective than the other two species over much of the green to red wavelength range (545-580 nm, 605-700 nm). PMSC greatly reduced the variability in the spectral signatures of each species by removing albedo effects caused by the varying illumination and geometry of samples measured under constantly changing field conditions. The PMSC-corrected spectral reflectance curves (mean $\pm$ SD) for the three seagrass species were clearly separated across the green (500-600 nm), near red (600-650 nm) and NIR (700-900 nm) wavelengths for leaf samples without (Figure 4.2E) and with epibionts (Figure 4.2F). There was a highly

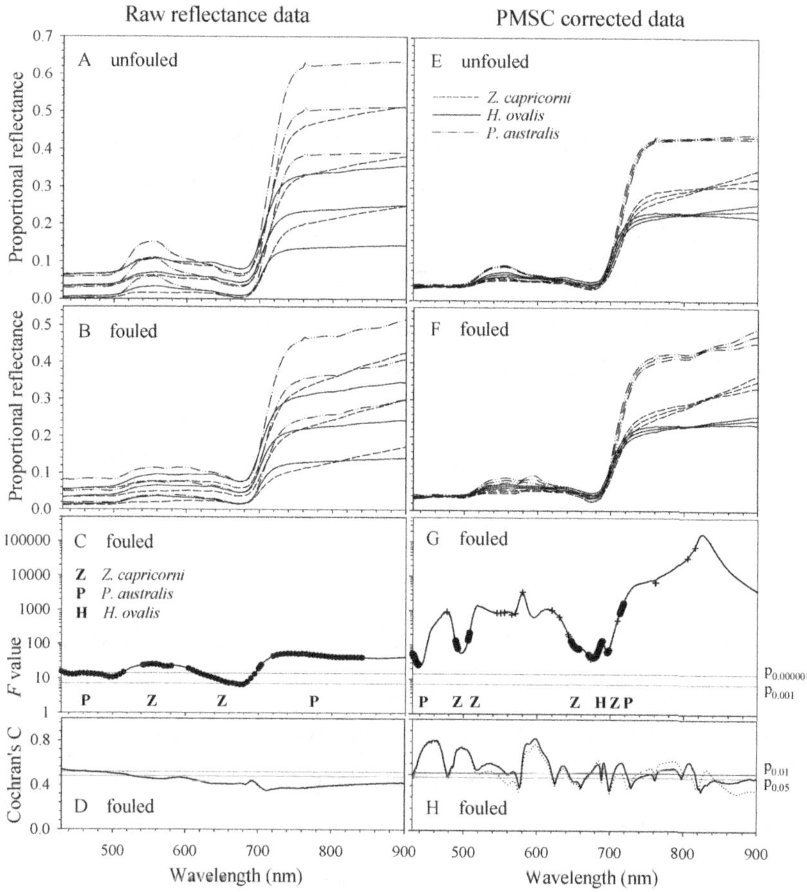

Figure 4.2. Mean ± SD spectral signatures of three seagrass species and wavelengths where significant reflectance differences between species occurred (Fyfe 2003).

For raw uncorrected reflectance data; (A) spectra of unfouled leaf samples, (B) spectra of fouled leaf samples, (C) results of ANOVA and Tukey test at each wavelength for fouled leaf samples, (D) results of Cochran's C test for homogeneity of variance of fouled leaf samples. For PMSC-corrected reflectance data; (E) spectra of unfouled leaf samples, (F) spectra of fouled leaf samples, (G) results of ANOVA and Tukey test at each wavelength for fouled leaf samples, (H) results of Cochran's C test for homogeneity of variance of fouled leaf samples. Wavelength regions where F values are represented as a solid line in (C) and (G) indicate that all species were significantly different from each other in pairwise tests. F values represented as circles are regions where only one species was significantly different from the others; the species is labelled below the circles. Crosses represent F values at select wavelengths after square root transformation of data. Solid lines in (D) and (H) indicate Cochran's C values for untransformed data, dotted lines are for square root transformed data. Critical p values are denoted by horizontal reference lines.

significant difference between species for the fouled seagrass samples (Figure 4.2G; $F_{(1), 2, 373} > 24$, $p << 0.001$) at all wavelengths. The fact that for many of the wavelengths Cochran's C test suggested significant heteroscedasticity (Figure 4.2H) does not greatly impact on the significance of the result. At those wavelengths where square root transformation of the data generated homogeneous variances, F values were reduced only very slightly (Figure 4.2G). Hence, Type I errors are unlikely to have had any measurable impact given the strong significance of the results. *A posteriori* multiple comparisons indicated a significant difference in the spectral reflectance of all three seagrass species over many of the visible and NIR wavelengths (Figure 4.2G). At wavelengths between 489-492 nm and 506-508 nm, however, only *Z. capricorni* can be discriminated by its lower reflectance.

Multivariate (nMDS) ordination of all PMSC-corrected seagrass samples, including leaves with and without epibionts (Figure 4.3), added support to the results of univariate analysis. There was a clear separation of the seagrass samples into species groups along the primary ordination axis (global R = 0.926, $p \leq 0.001$) with all species significantly different in pairwise tests at $p \leq 0.001$. Within each species cluster, there was a gradation of samples from fouled to unfouled along the second ordination axis (global R = 0.091, $p \leq 0.001$). Therefore, while leaf fouling has a significant influence on the spectral reflectance response within a seagrass species, it is apparent that the presence of epibionts does not impact on the spectral discrimination between species.

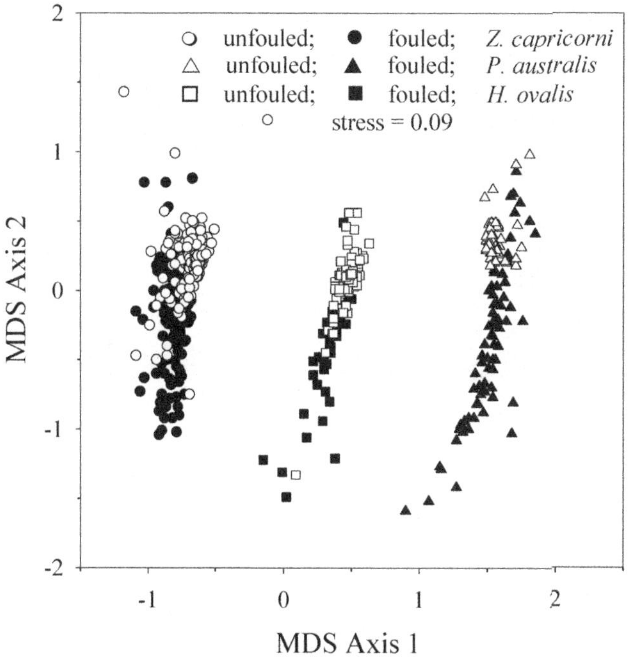

Figure 4.3. nMDS plot of the visible spectral reflectance response of fouled and unfouled leaf samples of three temperate seagrass species (Fyfe 2003).

4.3.2 Reflectance differences within species: spatial and temporal effects

Zostera capricorni

There was no significant difference between years in the spectral reflectance response of fouled *Z. capricorni* (Figure 4.4A; global $R = -0.003$, $p = 0.59$). The reflectance of this species was significantly affected by the season (global $R = 0.019$, $p = 0.027$), estuary (global $R = 0.125$, $p \leq 0.001$) and habitat (global $R = 0.302$, $p \leq 0.001$) in which leaves were sampled (Figure 4.4B-D). A clear pattern of spectral separation was apparent in the distribution of marine and brackish samples across the first ordination axis, with little overlap between these habitats

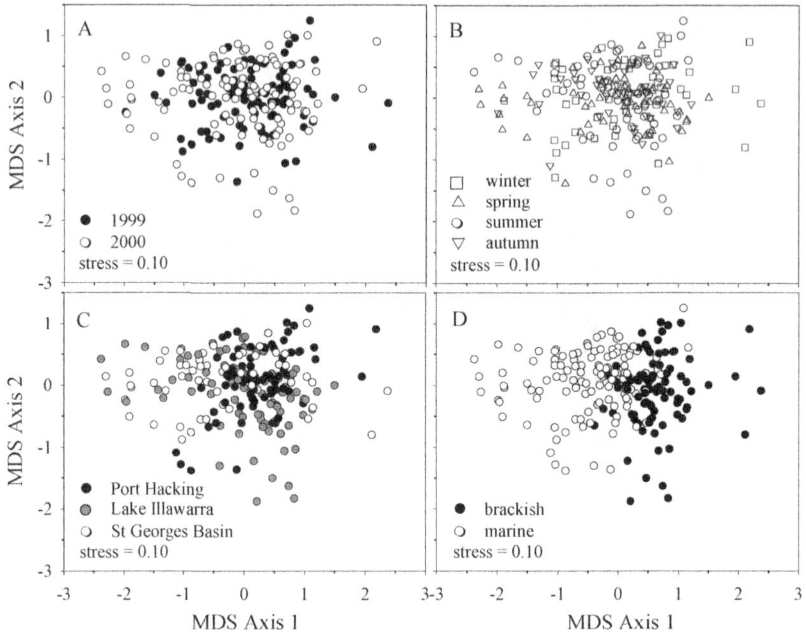

Figure 4.4. nMDS plots of the visible spectral reflectance response of fouled *Zostera capricorni* leaf samples associated with (A) year, (B) season, (C) estuary and (D) habitat of sample collection (Fyfe 2003).

(Figure 4.4D). On the other hand, the ANOSIM revealed significant differences ($p \leq 0.001$) between all estuaries in pairwise comparisons (Table 4.2) but the ordination plot comparing the reflectance of *Z. capricorni* leaves according to the estuary of sampling did not reveal a pattern of spectral separation across either nMDS axis (Figure 4.4C). Similarly, no pattern of separation was evident in the nMDS ordination plot according to the season of sampling (Figure 4.4B) and, in fact, only spring and summer samples showed a significant reflectance difference in pairwise tests ($p = 0.022$). Inspection of the third ordination axis of three-dimensional nMDS plots did not aid further in the separation of sample groups. It is important to note that ANOSIM is sensitive to heterogeneous

Table 4.2. Summary of ANOSIM and pairwise test results to determine within species differences in the spectral reflectance of seagrass leaves over the visible wavelengths (430-700 nm) associated with the year, season, estuary or habitat of sample collection (adapted from Fyfe 2003).

	YEAR	SEASON		ESTUARY		HABITAT
	Global *R* (pattern)	Global *R* (pattern)	Significant pairwise	Global *R* (pattern)	Significant pairwise	Global *R* (pattern)
Z. capricorni						
fouled	-0.003 ns (o'lap)	0.019* (o'lap)	spr-sum*	0.125*** (o'lap)	LI-PH*** LI-SG*** PH-SG***	0.302*** (seprtn)
unfouled	-0.008 ns (o'lap)	0.022** (o'lap)	win-spr* win-sum* spr-sum**	0.084*** (o'lap)	LI-PH*** LI-SG*** PH-SG***	0.249*** (seprtn)
P. australis						
fouled	-0.017 ns (o'lap)	0.076** (o'lap)	win-sum** aut-spr** aut-sum* spr-sum*	0.149*** (seprtn)	NA	NA
unfouled	0.129*** (seprtn)	0.044* (o'lap)	aut-spr* aut-sum**	0.071** (seprtn)	NA	NA
H. ovalis						
fouled	0.210*** (seprtn)	0.029 ns (o'lap)	aut-spr*	-0.083 ns (o'lap)	NA	0.173*** (o'lap)
unfouled	0.129*** (seprtn)	0.229*** (seprtn)	win-aut* win-spr** win-sum*** aut-spr** aut-sum*** spr-sum***	-0.028 ns (o'lap)	NA	0.117*** (o'lap)

Values are significant at the * $p < 0.05$, ** $p < 0.01$, *** $p < 0.001$ levels; ns = not significant. 'Year' factors are 1999 and 2000; 'Season' factors are winter (win), spring (spr), summer (sum) and autumn (aut); 'Estuary' factors are Lake Illawarra (LI), Port Hacking (PH) and St Georges Basin (SG); 'Habitat' factors are marine and brackish. 'Seprtn' indicates a pattern of spectral reflectance separation in the nMDS plot. Factors that 'o'lap' (i.e. overlap) are unlikely to be spectrally distinct.

multivariate dispersions among groups. Hence, a rejection of the null hypothesis that 'the groups do not differ' may be due to a difference in group dispersion rather than (or in combination with) a difference in central location of the groups (Anderson 1999). Multivariate dispersion appears to correspond to some measure of the overall variability in spectral reflectance of a group of seagrass leaf samples. This would suggest that groups that differ only on the basis of their multivariate dispersion would probably not be discriminated by image classification procedures. For example, the spectral signatures (mean $\pm$ SD) of fouled *Z. capricorni* leaf samples from marine and brackish habitats show small but discrete differences in the magnitude of their green reflectance and red absorption (Figure 4.5A) which could potentially be detected by remote sensing. On the other hand, the spectral signatures (mean $\pm$ SD) of fouled *Z. capricorni* leaves from different estuaries overlap to such an extent that spectral reflectance discrimination is unlikely (Figure 4.5B).

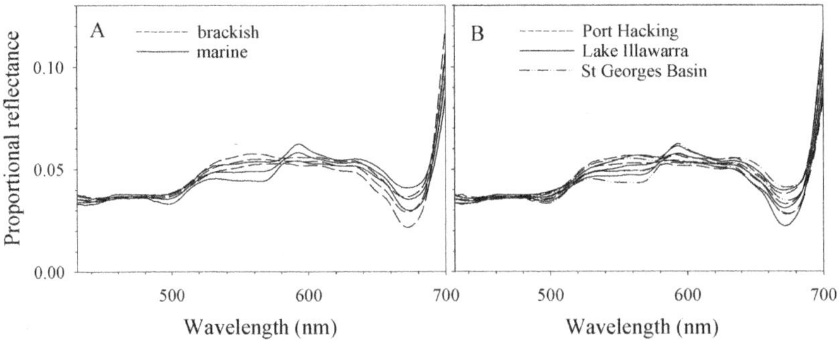

Figure 4.5. The mean $\pm$ SD spectral signatures of fouled *Zostera capricorni* leaf samples from (A) marine and brackish habitats and (B) three estuaries. Reflectance has been graphed across the visible wavelengths only (Fyfe 2003).

Given that fouling did not impact on the distinct spectral reflectance differences between seagrass species, it was not surprising that the presence or absence of epibionts generally had little influence on the reflectance differences within a species. Unfouled *Z. capricorni* leaf samples showed very similar patterns of spectral grouping according to year, season, estuary and habitat of sample collection to those of the fouled leaf samples (Table 4.2).

Posidonia australis

There was no significant grouping of fouled *P. australis* leaf samples by the year of sample collection based on spectral reflectance in the 430-700nm wavelength region (Figure 4.6A; global R = -0.017, $p = 0.906$). Significant seasonal groups were observed (global R = 0.076, $p = 0.003$), with the summer sample group different from all other seasons and the autumn group different from samples collected in spring (Table 4.2). However, there was no apparent separation of seasonal groups across either ordination axis in the nMDS plot (Figure 4.6B). Fouled *P. australis* were significantly grouped by estuary (global $R = 0.149$, $p \leq 0.001$). Some degree of separation in the reflectance of samples collected from Port Hacking and St Georges Basin could be observed across the second axis of the ordination plot (Figure 4.6C).

The patterns in reflectance associated with season and estuary for unfouled *P. australis* leaf samples were consistent with the results for fouled *P. australis* (Table 4.2). In contrast to fouled samples, unfouled *P. australis* leaves were significantly grouped by year (global $R = 0.129$, $p \leq 0.001$) although only a weak pattern of separation between year groups could be observed in the nMDS ordination plot.

Halophila ovalis

The spectral reflectance of fouled *H. ovalis* was significantly affected by the year (global $R = 0.210$, $p \leq 0.001$) and habitat (global $R = 0.173$, $p \leq 0.001$) from which the seagrass leaves were sampled. There was some degree of spectral reflectance separation between the samples from 1999

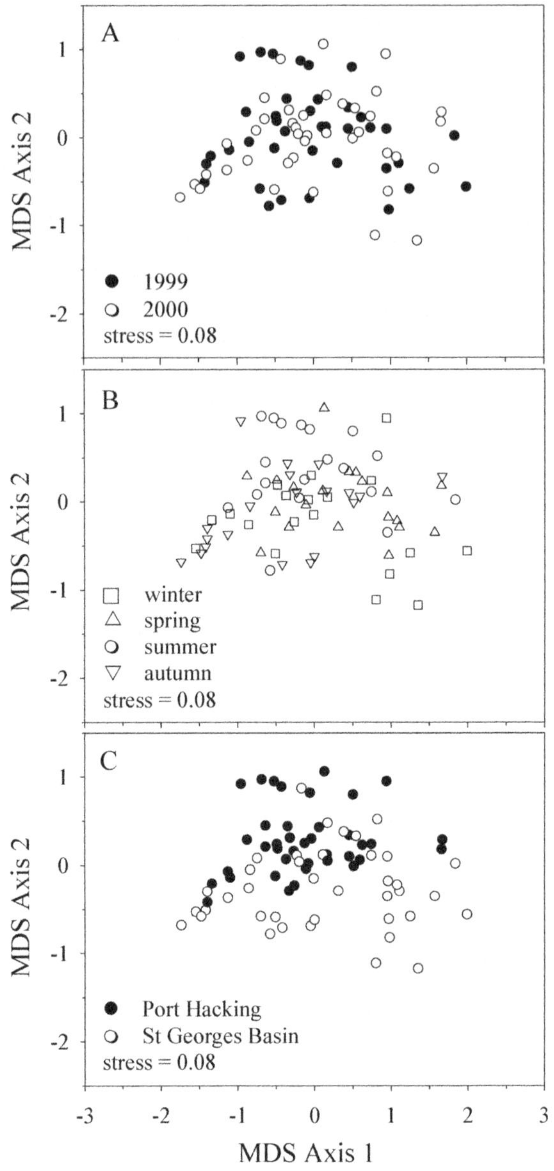

Figure 4.6. nMDS plots of the visible spectral reflectance response of fouled *Posidonia australis* leaf samples associated with (A) year, (B) season and (C) estuary of sample collection (Fyfe 2003).

and 2000 across the primary axis of the nMDS ordination (Figure 4.7A). In contrast, the marine and brackish samples were interspersed over the nMDS plot and could not be separated in any direction (Figure 4.7D). There was no significant grouping of fouled samples according to season (Figure 4.7B; global $R = 0.029$, $p = 0.068$) or estuary (Figure 4.7C; global $R = -0.083$, $p = 0.883$).

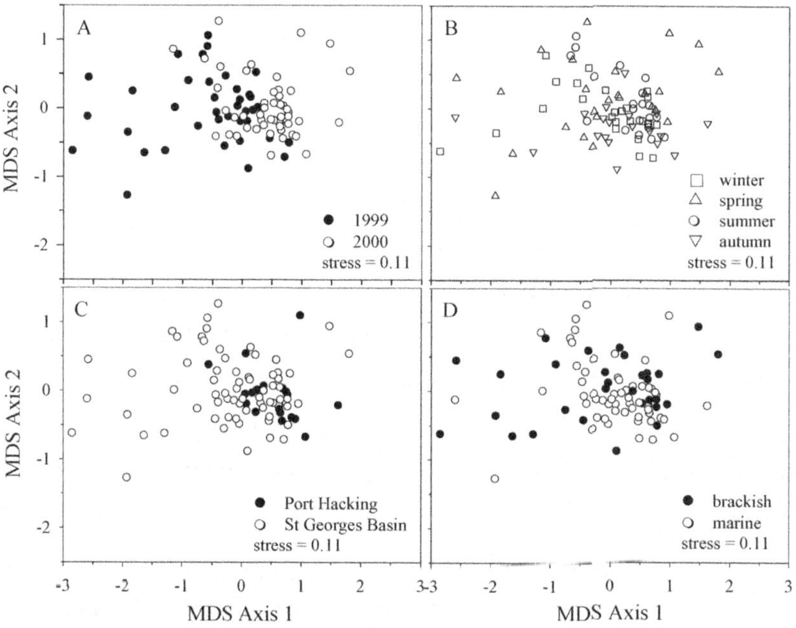

Figure 4.7. nMDS plots of the visible spectral reflectance response of fouled *Halophila ovalis* leaf samples associated with (A) year, (B) season, (C) estuary and (D) habitat of sample collection (Fyfe 2003).

The results of multivariate analysis of unfouled *H. ovalis* leaf samples were consistent with those for fouled leaves when investigating intraspecific reflectance differences associated with year, estuary and habitat of sampling (Table 4.2) but this was not the case for the season of sample collection. The reflectance of unfouled *H. ovalis* leaves was strongly affected by season (global $R = 0.229$, $p \leq 0.001$) with samples from all seasons significantly different from all others (Table 4.2). A transition in reflectance from summer to winter could be observed in the distribution of samples across the primary, and to some extent the secondary, ordination axes of the nMDS plot. Patterns in the distribution of spring and autumn samples were less easy to detect as these groups were variable and tended to intersperse with the summer and winter samples. As autumn reflectance lay more in the direction of the summer samples, this did allow for some separation from spring samples.

4.3.3 Wavelength selection for remote sensing of seagrass species

SIMPER analysis was used to select the wavelengths that contributed most to the dissimilarity in reflectance between pairs of species, or between significantly different pairs of year, season, estuary and habitat groups within each species. No single wavelength contributed more than 1% to the dissimilarity between species or 2% to the dissimilarity between factors within a species. The range of wavelengths offering discrimination between and within seagrass species proved to be similar for all species combinations irrespective of the presence of epibionts. Hence, individual SIMPER results were pooled to demonstrate the value of each wavelength in the detection of any spectral reflectance differences between (Figure 4.8A) and within species (Figure 4.8B).

Strong and consistent differences between seagrasses occurred across the green wavelengths 530-580 nm with additional discrimination in the regions 520-530 nm, 580-600 nm and 686-700 nm (Figure 4.8A). The wavelengths 580-606 nm were of particular importance in the discrimination of fouled *P. australis* while *Z. capricorni* and *H. ovalis*

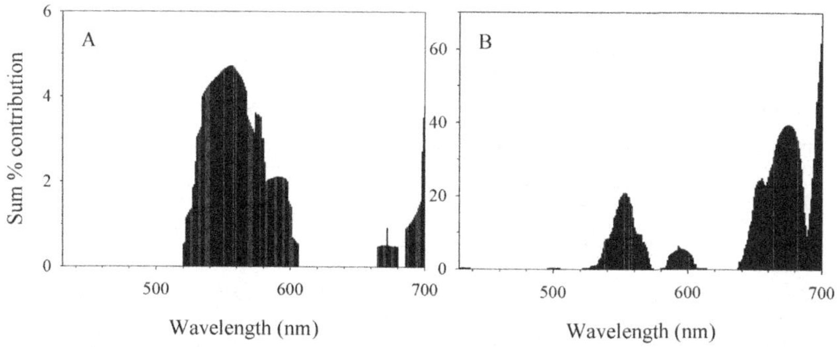

Figure 4.8. Sum percentage contribution made by each visible wavelength toward the significant dissimilarities (A) between seagrass species, and (B) between significantly different year, season, estuary and habitat groups within each seagrass species. Totals include only the percent contribution made by wavelengths contributing to the upper 50% of dissimilarity between groups in pairwise SIMPER analyses. Results for fouled and unfouled data were pooled (Fyfe 2003).

could be separated on the basis of their absorption of the red wavelengths 665-680 nm.

The red (637-700 nm) and green (522-574 nm) wavelengths were particularly effective at discriminating within species differences (Figure 4.8B). Maximum separation by year, season, estuary or habitat for any species occurred across the 549-556 nm, 649-686 nm and 692-700 nm wavelength regions. The wavelengths 430-438 nm, 496-505 nm and 580-614 nm contributed only to reflectance differences in fouled *P. australis* leaves. The region between 588-602 nm was, however, very important in the discrimination of estuary and seasonal differences in this species. Similarly, important reflectance differences within fouled *Z. capricorni* were observed across the wavelengths 550-570 nm.

4.4 Results: Biochemistry of seagrass leaves

4.4.1 Percent water content

Seagrass leaves generally contained around 75-82% water by weight. There was a significant difference in the percent water content of the samples tested (Figure 4.9; $F_{(2),11,57} = 3.514$, $p < 0.001$, power = 0.989). *A posteriori* comparison of means using Tukey- Kramer's HSD at the 5% level of significance indicated that unfouled *H. ovalis* leaves contained proportionately more water by weight than both the fouled *P. australis* and unfouled *Z. capricorni* from the brackish habitat in Lake Illawarra. In addition, the fouled *Z. capricorni* at this site contained proportionately more water than the fouled *P. australis* from Port Hacking (Figure 4.9A). However, no sample groups differed significantly in pairwise tests at the 1% significance level.

The percent water content of seagrass leaves sourced from marine habitats was not influenced by the species of seagrass, but was significantly affected by the presence of leaf epibionts (Table 4.3, Figure 4.9B). Fouling reduced the percent water content of *P. australis* and *H. ovalis* leaf samples to a much greater extent than it did for *Z. capricorni* leaves, possibly because the type and biomass of leaf epibionts varied among the seagrasses. For example, calcium-depositing epibionts such as spirorbid worms and coralline algae are likely to significantly increase the weight of dry matter attached to a leaf, and hence, reduce the proportion of leaf weight attributed to water.

The data for the *Z. capricorni* subset were found to be homoscedastic in the case of habitat*epibionts (pooling across estuaries) and nearly so for estuary*epibionts (pooling across habitats). There were no significant differences in the percentage water content of any of the *Z. capricorni* leaf samples tested (Table 4.4).

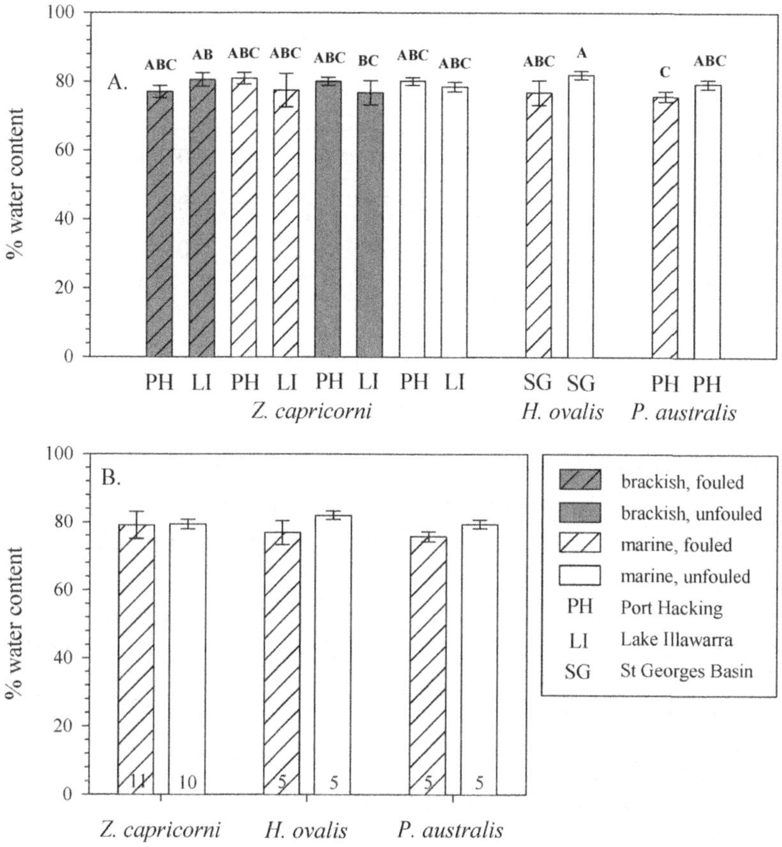

Figure 4.9. Mean ± SD percentage water content by weight for the leaves of three seagrass species for (A) all fouled and unfouled leaf samples tested from a range of different habitats and estuaries, and (B) leaf samples drawn from marine habitats only (pooling samples from different estuaries). Histograms joined by the same letter in (A) are not significantly different in Tukey-Kramer HSD pairwise comparisons at the 5% significance level. Sample sizes were not equal and have been included at the base of each histogram.

Table 4.3. Two-way analysis of variance to test for differences in the percent water content of seagrass leaves collected from marine habitats as a function of species (*Zostera capricorni*, *Posidonia australis*, *Halophila ovalis*) and the presence or absence of leaf epibionts.

Source	df	MS	*F*	*p*	power
Species	2	11.052	1.531	0.230	
Epibionts	1	83.799	11.610	0.002**	0.912
Species*Epibionts	2	23.099	3.200	0.053	
Error	35	7.218			

Values significant at the ** $p < 0.01$ level. Power was calculated at $\alpha = 0.05$.

Table 4.4. Two-way analyses of variance to test for differences in the percent water content of *Zostera capricorni* leaves as a function of (A) habitat type (marine, brackish) and leaf epibionts (fouled, unfouled), and (B) estuary (Port Hacking, Lake Illawarra) and leaf fouling.

Source	df	MS	*F*	p	power
A. Habitat	1	3.762	0.418	0.521	
Epibionts	1	7.277	0.809	0.373	
Habitat*Epibionts	1	13.665	1.518	0.224	
Error	45	9.001			
B. Estuary	1	18.174	2.157	0.149	
Epibionts	1	3.697	0.439	0.511	
Estuary*Epibionts	1	25.221	2.993	0.091	
Error	45	8.427			

Power was calculated at $\alpha = 0.05$.

The presence of significant differences in the percent water content of field samples suggested that sample groups might not be comparable based on their absolute pigment content per gram fresh weight. In most cases, this problem can be avoided by comparing fouled and unfouled leaf samples in separate analyses because this would act to remove the source of most of the variation. However, since some variability was also observed within fouled and unfouled groups, it was thought better to normalise the field data where possible by applying pigment ratios in the place of individual pigments. As field sampled *Z. capricorni* leaves did not vary significantly in their water content, it was assumed that the pigment concentrations calculated per gram fresh weight for *Z. capricorni* grown in controlled laboratory experiments (Chapter 5) would vary even less. Hence, the results of laboratory experiments should be comparable based on their absolute pigment concentrations.

4.4.2 Chlorophyll and carotenoid content

HPLC clearly separated the seagrass pigments, chlorophyll *a*, chlorophyll *b*, and the carotenoids neoxanthin, violaxanthin, antheraxanthin, lutein, zeaxanthin, α-carotene and ß-carotene in most samples (Figure 4.10A). The isomers of neoxanthin, violaxanthin, antheraxanthin, and ß-carotene that could obviously be identified as such by their absorption spectrum, time of elution, and comparison with chromatograms produced from pigment standards, were added together to calculate the concentrations of each of these pigments. Fouled seagrass samples also contained relatively large quantities of fucoxanthin, the dominant light-harvesting pigment of brown algae (e.g. Bjørnland and Liaaen-Jensen 1989; Rowan 1989). Despite the fact that a fucoxanthin standard was not run on the column, the identity of fucoxanthin was readily confirmed by the shape of its absorption spectrum and by its very early separation from the pigments normally found in green plants which are somewhat less polar (Figure 4.10B). A number of unidentified biochemicals were also observed as minor absorption features on the

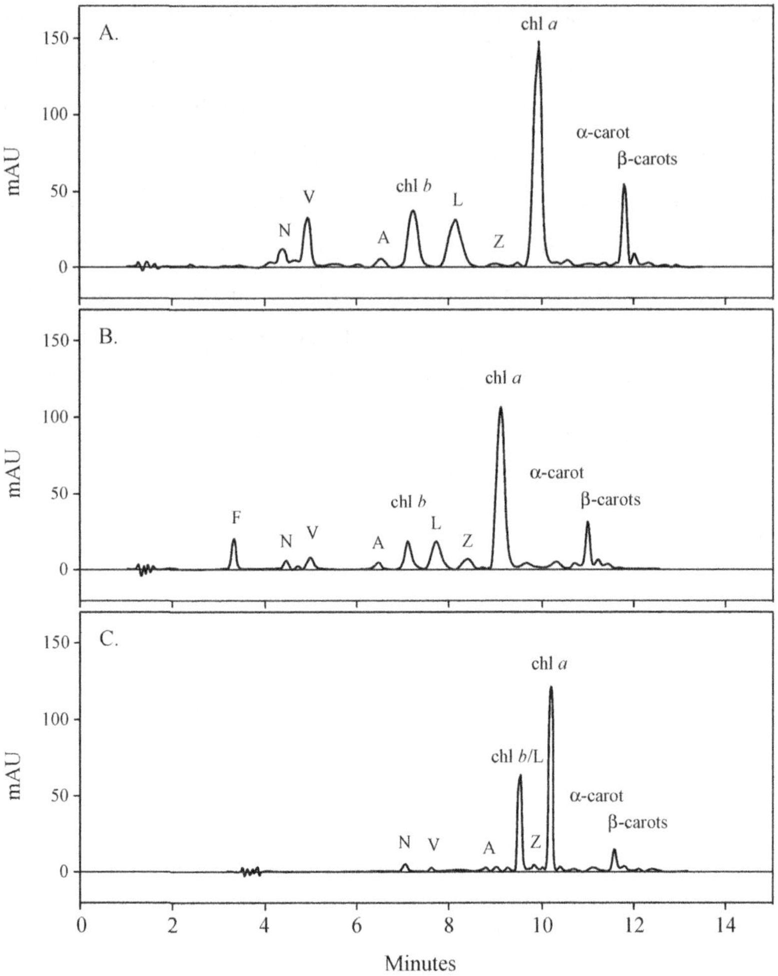

Figure 4.10. High performance liquid chromatography of carotenoids and chlorophylls from the leaves of *Zostera capricorni*; (A) a well separated chromatograph run from an unfouled *Z. capricorni* leaf sample, (B) a well separated run from a fouled leaf showing fucoxanthin eluting first in the series, and (C) a very poorly separated run from an unfouled leaf showing coelution of chlorophyll *b* and lutein. F, fucoxanthin; N, neoxanthin; V, violaxanthin; A, antheraxanthin; chl *b*, chlorophyll *b*; L, lutein; chl *a*, chlorophyll *a*; α-carot, α-carotene; β-carots, β-carotenes.

chromatograms but these were not included in any of the subsequent pigment data. It is likely that these features represent chlorophyll breakdown products such as pheophytin *a* and *b*, minor isomers or breakdown products of the seagrass carotenoids and some unidentified algal pigments.

Perfect separation of chlorophyll *b* and lutein was not achieved for all samples, however, but in these cases the order of elution and quality of separation of the remaining pigments was not affected (Figure 4.10C).

While chlorophyll *b* and lutein could be separated in some of these cases by scanning the absorption spectrum across the chromatogram, in other instances they could not. A preliminary investigation of the influence of the poor separation on the field sampled seagrass data sets confirmed that the values of chlorophyll *b* and lutein were skewed in comparison with unaffected data, but no other trends were apparent that might suggest problems with any of the other pigments. The concentrations of chlorophyll *b* and lutein calculated from affected HPLC runs were therefore considered too unreliable for individual use in subsequent pigment data analyses. Since both of these pigments play an important role in green reflectance and in the absorption of blue light across the broad 400-500 nm waveband, their combined value was included for analysis of the field data sets.

4.4.3 Pigment differences between species

There was a significant difference between all three seagrass species in the total photosynthetic pigment concentration per gram fresh weight of leaves without epibiont foulers (Table 4.5). The leaves of *P. australis* contained more pigment per gram than those of *H. ovalis*, with the lowest concentration found in *Z. capricorni* (Figure 4.11). The presence of leaf

Table 4.5. One-way analyses of variance to examine differences between the seagrass species, *Zostera capricorni* (*Z*), *Halophila ovalis* (*H*) and *Posidonia australis* (*P*) in their total and relative proportions of chlorophyll and carotenoid pigments for, A) unfouled and, B) fouled seagrass leaf samples.

Variable	df	MS	*F*	*p*	power	Significant pairwise
UNFOULED SEAGRASS LEAVES						
Total pigment content:						
T pigments#	2	1.511	15.358	<0.001***	0.999	*H-Z, H-P, P-Z*
Individual pigment content:						
chl *a* / T pigments#	2	0.034	29.283	<0.001***	1.000	*H-Z, P-Z*
chl *b* + lutein / T pigments#	2	0.013	18.040	<0.001***	1.000	*H-Z, P-Z*
neoxanthin / T pigments§	2	0.001	23.963	<0.001***	1.000	*H-Z, P-Z*
α-carotene / T pigments§	2	0.004	2.743	0.070		
β-carotene / T pigments§	2	0.000	2.720	0.071		
fucoxanthin (algal) / T pigments#	2	0.044	5.520	0.005**	0.842	*H-P, P-Z*
VAZ pool:						
V+A+Z / chl *a*	2	0.026	22.340	<0.001***	1.000	*H-P, H-Z*
A+Z / VAZ	2	0.145	7.090	0.001**	0.923	*H-P, H-Z*
Z / VAZ	2	0.173	5.570	0.005**	0.845	*H-Z*
FOULED SEAGRASS LEAVES						
Total pigment content:						
T pigments	2	860833	3.144	0.049*	0.588	*P-Z*
Individual pigment content:						
chl *a* / T pigments§	2	0.024	10.234	<0.001***	0.983	*H-Z*
chl *b* + lutein / T pigments	2	0.013	11.102	<0.001***	0.990	*H-Z, P-Z*
neoxanthin / T pigments#	2	0.002	16.631	<0.001***	1.000	*H-Z, P-Z*
α-carotene / T pigments	2	0.003	5.455	0.006**	0.835	*H-Z*
β-carotene / T pigments§	2	0.001	1.397	0.254		
fucoxanthin (algal) / T pigments	2	0.009	0.722	0.489		
VAZ pool:						
V+A+Z / chl *a*§	2	0.011	4.054	0.021*	0.796	*H-Z*
A+Z / VAZ	2	0.017	0.571	0.567		
Z / VAZ	2	0.260	0.998	0.373		

Cochran's C value significantly heteroscedastic; # at $0.01 < p < 0.05$, § at $p < 0.01$. Values significant at the *$p < 0.05$, ** $p < 0.01$, *** $p < 0.001$. Tukey-Kramer HSD comparison of pairs significant at $\alpha = 0.05$. Power was calculated at $\alpha = 0.05$.

epiphytes masked the differences between species to some extent, however fouled *P. australis* still displayed significantly greater pigment content than fouled *Z. capricorni* (Figure 4.11, Table 4.5).

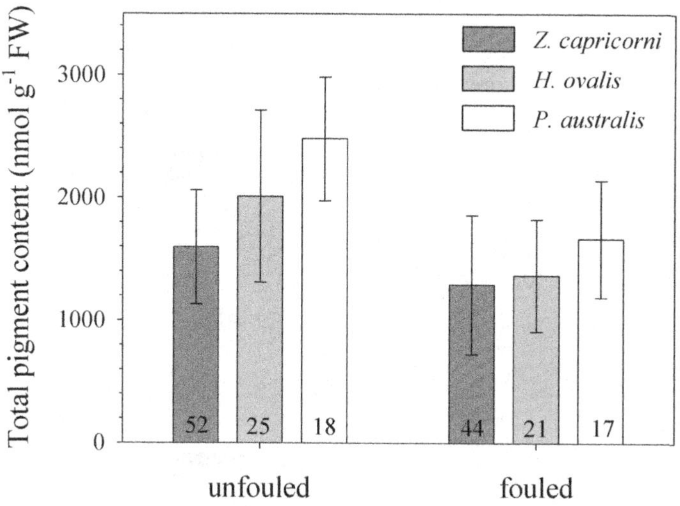

Figure 4.11. Mean ± SD total leaf (chlorophyll and carotenoid) pigment content for three temperate seagrass species with and without leaf epibionts. Sample sizes were not equal and have been included at the base of each histogram.

Unfouled *Z. capricorni* leaves contained a significantly greater proportion of chlorophyll *a*, and significantly less chlorophyll *b* + lutein and neoxanthin than unfouled leaves of the other seagrass species (Figure 4.12, Table 4.5). Zeaxanthin and antheraxanthin levels, and consequently the total available VAZ pool (V+A+Z), were greater for unfouled *H. ovalis* leaves than for unfouled *Z. capricorni* or *P. australis*. The proportion of carotenes in seagrass leaves did not vary significantly among unfouled leaf samples. Some level of fouling was indicated on 'unfouled' seagrass leaves by the presence of small quantities of fucoxanthin. Significantly less fucoxanthin was observed in unfouled *P. australis* samples (around 0.1% of total pigment content) than in the other two species, although fucoxanthin levels were < 1-2% of the total pigment content for all species.

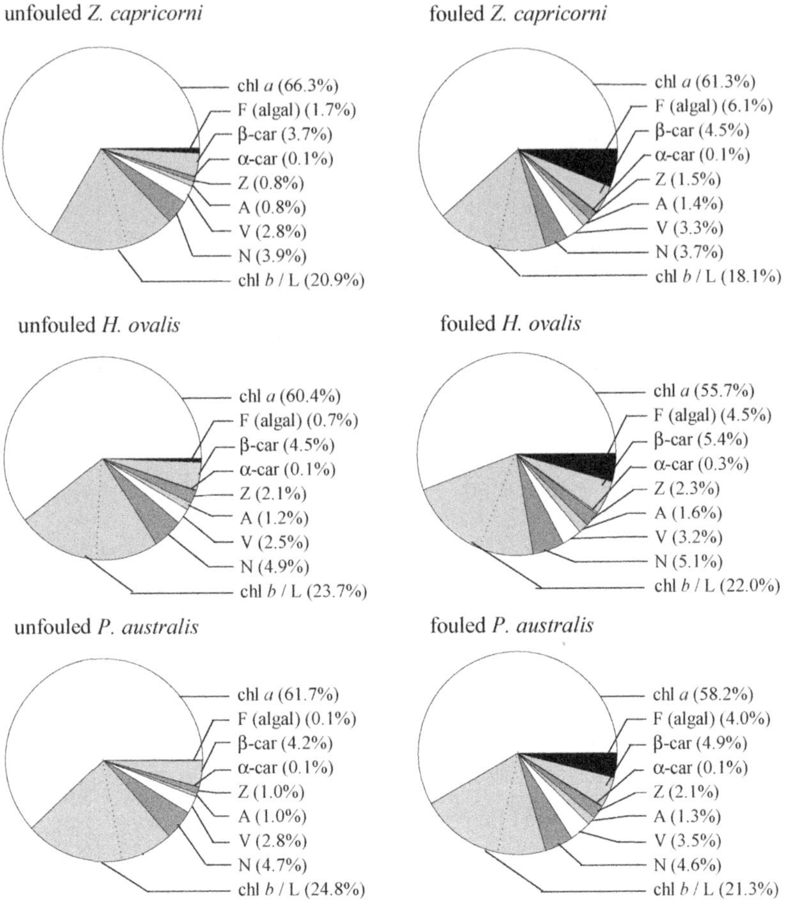

Figure 4.12. Comparison of the pigment (chlorophylls and carotenoids) composition of fouled and unfouled leaves of three temperate seagrass species. Pigments are; chl *a*, chlorophyll *a*; F, fucoxanthin; ß-car, ß-carotene; α-car, α-carotene; Z, zeaxanthin; A, antheraxanthin; V, violaxanthin; N, neoxanthin; chl *b* / L, combined chlorophyll *b* and lutein. Since chlorophyll *b* and lutein were not reliably separated from each other by HPLC for all samples, their approximate proportions are shown dotted.

Seagrass species did not differ significantly in the proportion of fucoxanthin to total pigments for fouled leaf samples, although

Z. capricorni leaves did appear to contain slightly higher proportions of the brown algal pigment (around 6% of the total pigment content). Fouled *Z. capricorni* leaves contained significantly more chlorophyll *a*, significantly less α-carotene, and had a proportionally smaller total VAZ pool than fouled *H. ovalis* (Figure 4.12, Table 4.5). In addition, the fouled leaves of *Z. capricorni* contained a lower proportion of chlorophyll *b* + lutein and neoxanthin than either fouled *H. ovalis* or fouled *P. australis* leaves.

The nMDS plot based on similarities in the chlorophyll and carotenoid composition of samples displayed relatively tight species groupings for the unfouled leaf samples, particularly in the case of *P. australis* and *Z. capricorni* (Figure 4.13). Fouled samples were more variable, but in general, seagrass species separated across the second MDS axis while the first MDS axis was indicative of the level of algal fouling of samples. The species were significantly grouped in a two-way ANOSIM (global $R = 0.222$, $p \leq 0.001$), as were the fouled and unfouled leaf samples (global $R = 0.261$, $p \leq 0.001$). *A posteriori* pairwise permutation tests indicated that each of the species differed significantly from the other two species (*H-Z*; $p \leq 0.001$, *P-Z*; $p = 0.002$, *H-P*; $p = 0.013$).

The results of SIMPER analysis suggested that the similarities within species groups and differences between species were predominantly related to the concentration of chlorophyll *a* in the leaf tissue. Chlorophyll *a* contributed 61-68% of the similarity within a species and 25-33% of the dissimilarity between species. The combined value of chlorophyll *b* + lutein was responsible for 20-23% of within species similarity and 20-23% of dissimilarity between species. Other pigments that made an important contribution toward the dissimilarity between the species included; fucoxanthin (16-18%), β-carotene (7-8%), zeaxanthin (7-10%), neoxanthin

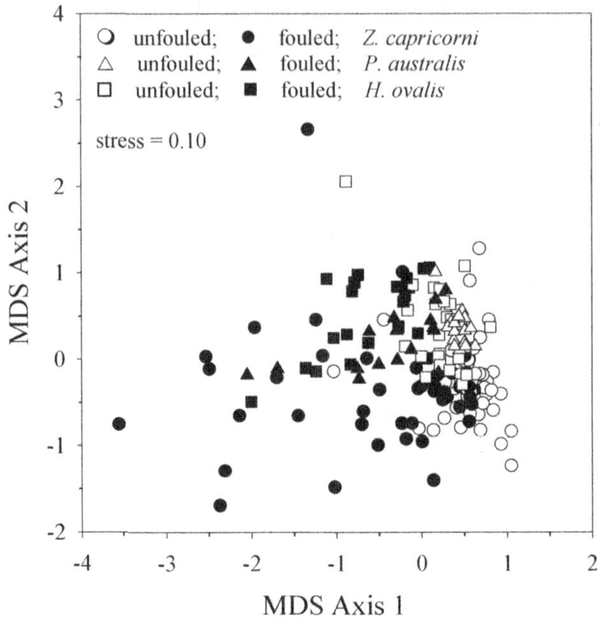

Figure 4.13. nMDS plot of the (chlorophyll and carotenoid) pigment composition of fouled and unfouled leaf samples of three temperate seagrass species.

(~6%) and violaxanthin (7-9%). Likewise, the similarities observed between samples in the fouled and unfouled groups, and the differences between these groups, were predominantly related to the concentration of chlorophyll *a* and chlorophyll *b* + lutein in the leaves. Chlorophyll *a* contributed 64-66% and chlorophyll *b* + lutein 20-22% toward within-group similarity and 29% and 20% respectively toward between-group dissimilarity. The dissimilarity between samples with and without leaf epibionts was also related to the proportion of fucoxanthin (contributed 21% toward dissimilarity), β-carotene (8%), violaxanthin (7%) and zeaxanthin (6%) in the sample.

4.4.4 Pigment differences within species: spatial and temporal effects

Zostera capricorni

The pigment composition of fouled *Z. capricorni* leaves varied with the season (*global* R = 0.137, p = 0.003) and estuary (global R = 0.237, $p \leq 0.001$) but not the habitat (global R = 0.047, p = 0.097) of sample collection (Figure 4.14A-C, Table 4.6). There was no clear pattern of separation between samples collected from warmer water conditions in January and cooler water in October (Figure 4.14A). Rather, the season groups differed in their multivariate dispersion with summer samples more variable across nMDS axis 2. Samples collected from Lake Illawarra did not differ significantly from those collected at St Georges Basin although separation between the two groups was evident across nMDS axis 2 (Figure 4.14B). The pigment content of leaf samples from Port Hacking differed significantly from both other estuary groups, partly because of the tighter multivariate dispersion of the Port Hacking group, but also because the Port Hacking samples separated clearly across nMDS axis 1.

There were significant differences in the pigment content of unfouled *Z. capricorni* leaves sampled from different seasons (global R = 0.047, p = 0.032), estuaries (global R = 0.080, p = 0.002) and habitats (global R = 0.075, p = 0.008) (Table 4.6, Figure 4.14D-F). Only habitat groups, however, showed the pattern of clear separation across the axes of nMDS plots that would suggest that the pigment concentrations of marine and brackish samples were indeed different (Figure 4.14F).

Posidonia australis

The pigment makeup of fouled *P. australis* leaves was strongly influenced by the estuary (global R = 0.385, $p \leq 0.001$) but not the season (global R = 0.069, p = 0.159) from which the leaves were taken (Table 4.6, Figure 4.15A-B). Samples from Port Hacking and St Georges Basin formed

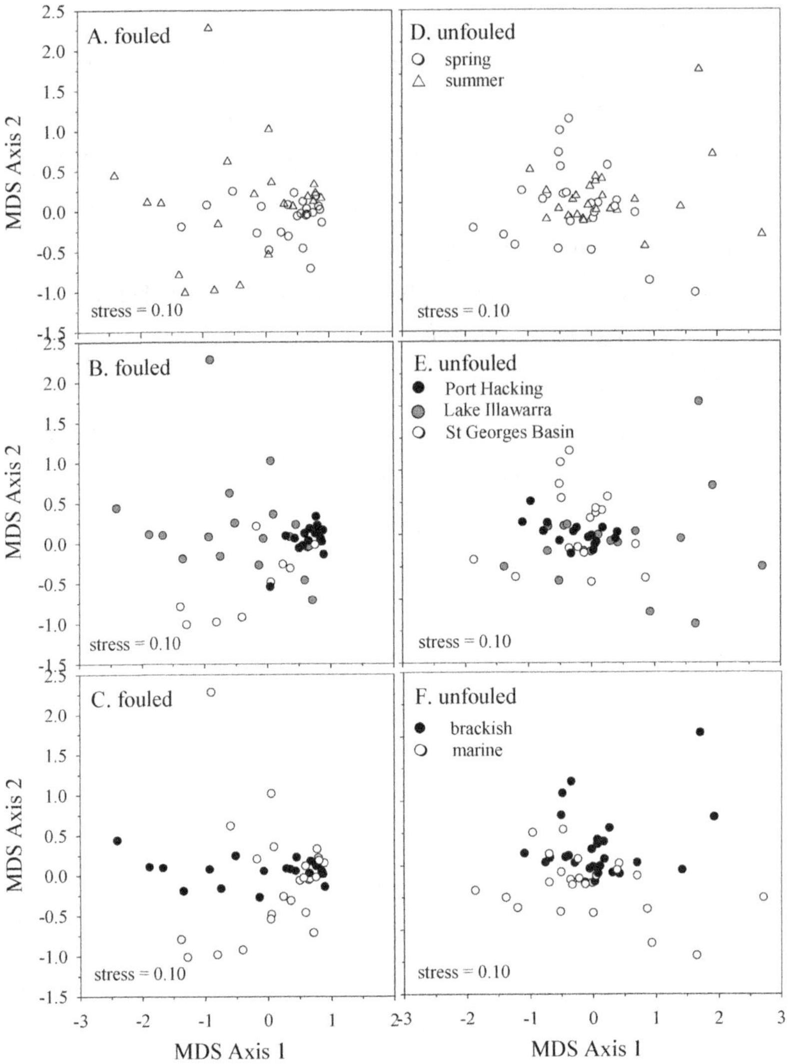

Figure 4.14. nMDS plot of the association between Zostera capricorni leaves based on their (chlorophyll and carotenoid) pigment composition; fouled leaf samples grouped by (A) season, (B) estuary, and (C) habitat, and unfouled leaf samples grouped by (D) season, (E) estuary, and (F) habitat of sample collection.

Table 4.6. Summary of ANOSIM and pairwise test results to determine within species differences in the pigment content (chlorophylls and carotenoids) of seagrass leaves associated with the season, estuary or habitat of sample collection.

	SEASON	ESTUARY		HABITAT
	Global *R* (pattern)	Global *R* (pattern)	Significant pairwise	Global *R* (pattern)
Z. capricorni				
fouled	0.137** (overlap)	0.237*** (separation)	LI-PH*** PH-SG***	0.047 ns (overlap)
unfouled	0.047* (overlap)	0.080** (overlap)	LI-SG* PH-SG**	0.075** (separation)
P. australis				
fouled	0.069 ns (overlap)	0.385*** (separation)	NA	NA
unfouled	0.110 ns (overlap)	0.121 ns (overlap)	NA	NA
H. ovalis				
fouled	0.278** (separation)	0.223* (separation)	NA	-0.114 ns (separation)
unfouled	0.246** (separation)	0.054 ns (overlap)	NA	0.014 ns (overlap)

Values are significant at the * $p < 0.05$, ** $p < 0.01$, *** $p < 0.001$ levels; ns = not significant. 'Season' factors are spring (October) and summer (January); 'Estuary' factors are Lake Illawarra (LI), Port Hacking (PH) and St Georges Basin (SG); 'Habitat' factors are marine and brackish. 'Separation' indicates a pattern of spectral reflectance separation in the nMDS plot. Factors that 'overlap' are unlikely to be spectrally distinct.

distinct groups across nMDS axis 1 based on differences in their pigment content (Figure 4.15B). Neither season (global $R = 0.110$, $p = 0.083$) nor estuary (global $R = 0.121$, $p = 0.066$) had a significant affect on the pigment composition of unfouled *P. australis* samples (Figure 4.15C-D).

Halophila ovalis

For fouled samples of *H. ovalis*, there were significant differences in the pigment content of leaves associated with the season (global $R = 0.278$, $p = 0.007$) and estuary (global $R = 0.223$, $p = 0.019$), but not the habitat

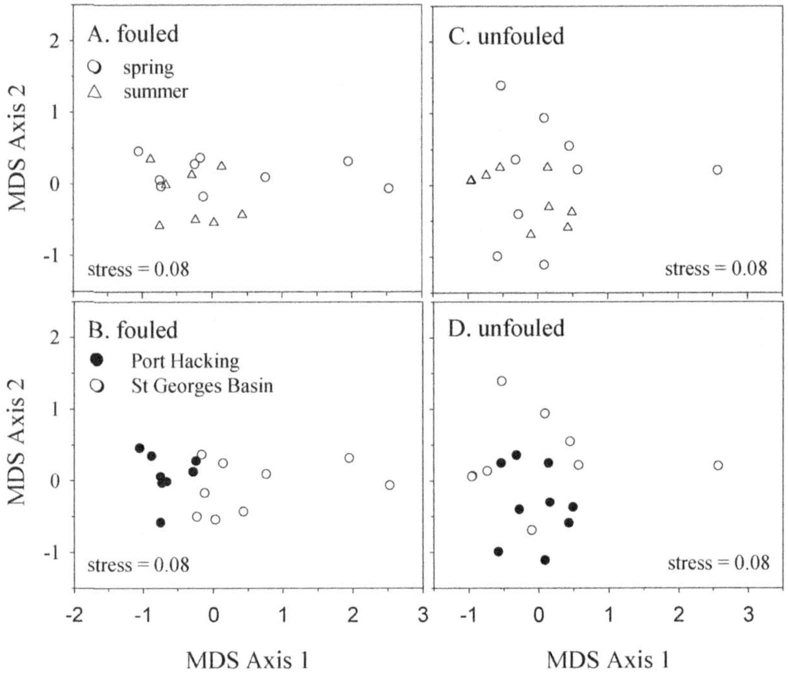

Figure 4.15. nMDS plot of the association between *Posidonia australis* leaves based on their (chlorophyll and carotenoid) pigment composition; fouled leaf samples grouped by (A) season, and (B) estuary, and unfouled leaf samples grouped by (C) season, and (D) estuary of sample collection.

(global R = -0.114, p = 0.800) of sample collection (Table 4.6, Figure 4.16A-C). January (warm water) samples separated from October samples (cooler water) across the primary ordination axis based on their pigment composition (Figure 4.16A). The Port Hacking sample group was distinct from St Georges Basin group across nMDS axis 2 (Figure 4.16B). Marine and brackish habitat groups also appear to display a pattern of separation in their pigment content although the differences were not significant (Figure 4.16C). There were no significant differences in the pigment composition of unfouled *H. ovalis* leaves collected from different estuaries (global R = 0.054, p = 0.272) or habitats (global R = 0.014, p = 0.393) (Figure

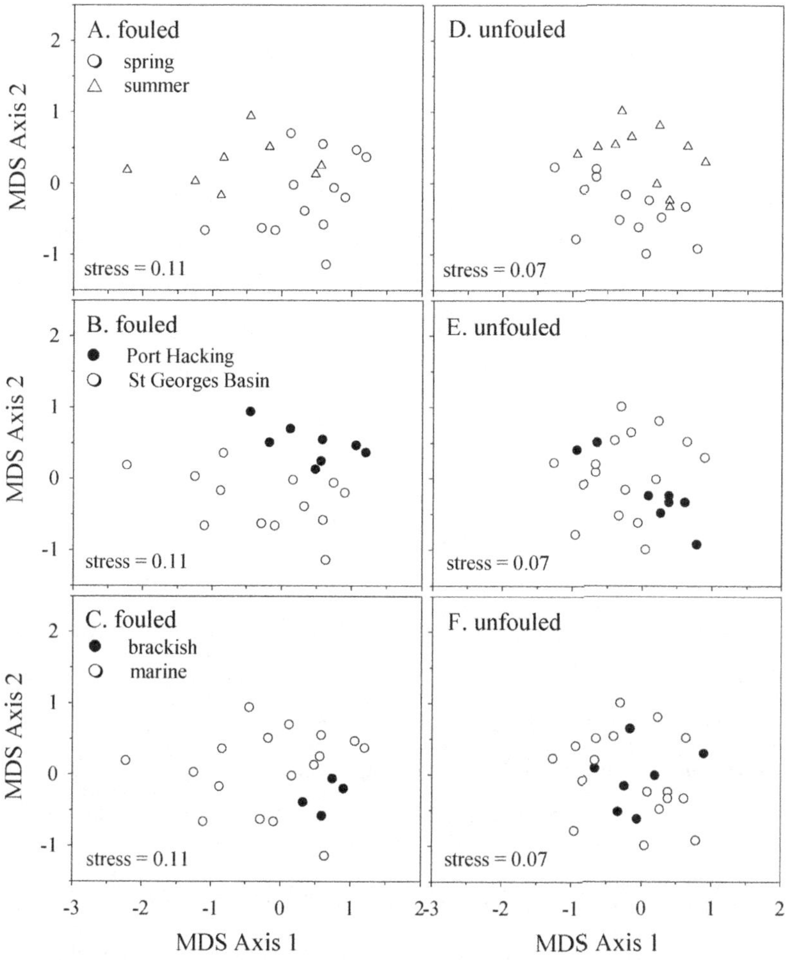

Figure 4.16. nMDS plot of the association between *Halophila ovalis* leaves based on their (chlorophyll and carotenoid) pigment composition; fouled leaf samples grouped by (A) season, (B) estuary, and (C) habitat, and unfouled leaf samples grouped by (D) season, (E) estuary, and (F) habitat of sample collection.

4.16E-F). Seasonal changes in pigment composition were apparent (global $R = 0.246$, $p = 0.006$) with the spring samples (October) clearly separating from the summer (January) samples across nMDS axis 2 (Fig 4.16D).

Pigments contributing to within species differences

The results of SIMPER analyses indicated that the similarities observed between seagrass leaf samples from any particular estuary, season or habitat group were primarily due to the concentration of chlorophyll *a* and chlorophyll *b* + lutein in the leaves. These pigments cumulatively contributed 80-90% towards within-group similarity in all cases. This result was not surprising since the three pigments comprised 78-87 % of the total photosynthetic pigment content observed in the leaves of these seagrass species (Figure 4.12).

The concentration of chlorophyll *a* and chlorophyll *b* + lutein in the leaves also contributed to between-group dissimilarities observed within the seagrass species. The proportion of chlorophyll *a* in the leaf tissue contributed the most (i.e. 25-30% of between-group dissimilarity) toward the separation of habitat groups in unfouled *Z. capricorni* (Figure 4.14F), estuary groups in fouled *Z. capricorni* (Figure 4.14B) and season groups in unfouled *H. ovalis* (Figure 4.16D). The contribution of chlorophyll *b* + lutein was of next highest importance in promoting significant dissimilarities between unfouled leaves of *Z. capricorni* and *H. ovalis* from different habitats and estuaries (21-25% contribution), with important contributions from ß-carotene (10-13%), and from zeaxanthin (16%) and violaxanthin (11%) for *H. ovalis*. Between-habitat variability in 'unfouled' *Z. capricorni* was also related to the proportion of fucoxanthin in samples, suggesting varying levels of algal contamination, probably by microscopic diatoms.

Less surprising was the contribution of fucoxanthin toward significant between-group differences in fouled seagrass leaves. Fucoxanthin was of prime importance in the separation of estuary groups in fouled *P. australis* (Figure 4.15B), contributing to 27% of the dissimilarity between samples sourced from different estuaries. Fucoxanthin contributed secondarily to the dissimilarities observed between samples from different estuaries for fouled *H. ovalis* (18%) and *Z. capricorni* (20-31%), and between samples from different seasons for fouled *H. ovalis* (22%). Seasonal and estuary differences in the pigment content of fouled *H. ovalis* (Figure 4.16A,B) could be attributed mostly to the levels of chlorophyll *b* + lutein in the leaf samples (24-27% contribution). These pigments also made important contributions to dissimilarities between estuaries for fouled *P. australis* (24%) and *Z. capricorni* (13-18%). In addition, ß-carotene (12-18%) and violaxanthin (11%) contributed to the differences between fouled *Z. capricorni* from different estuaries. The proportion of violaxanthin in samples was also important (11%) in separating estuary groups for fouled *P. australis* leaf samples, while zeaxanthin contributed to the dissimilarities observed between estuaries (11%) and seasons (13%) in fouled *H. ovalis* leaves. Dissimilarities associated with differences in the xanthophyll cycle pigments may not be consistent, however, because the relative concentrations of these pigments can change very rapidly with irradiance conditions (see Section 5. 1.3).

4.4.5 Anthocyanin content of seagrass leaves

It was immediately apparent from the colour of the anthocyanin extracts taken from the three species of seagrass, that each species was utilizing a slightly different combination of anthocyanins in their leaves. The extract of *Z. capricorni* was red or bronze, *P. australis* had a blue coloured extract while that of *H. ovalis* was distinctly yellow in colour, possibly suggesting lower anthocyanin concentrations (Plate 4.1).

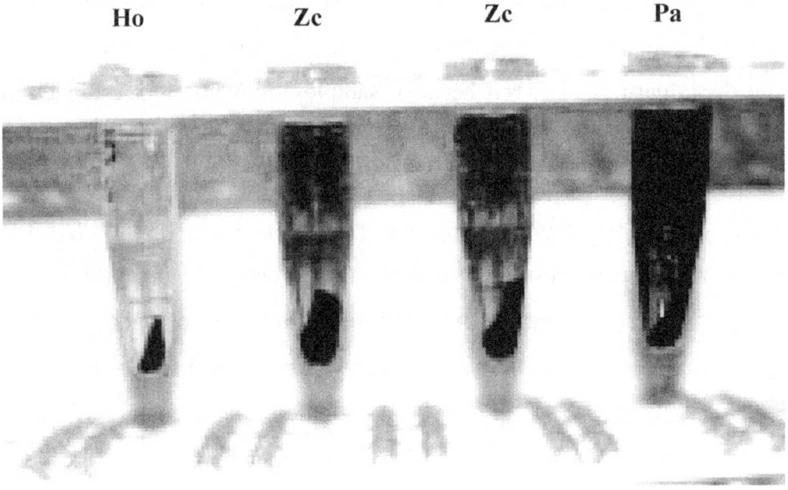

Plate 4.1. Anthocyanin extracts from the leaves of three seagrass species; *Halophila ovalis* (Ho; typically yellow in colour), *Zostera capricorni* (Zc; typically reddish) and *Posidonia australis* (Pa; blue colour).

Scanning with the spectrophotometer across a wavelength range of 400-700 nm at 0.5 nm intervals for a single unfouled and fouled sample of each species showed four absorption peaks common to all three species (Table 4.7). Maximum absorbance for all species occurring at the peak centred at 531 nm. The blue colour of the *P. australis* extract was attributed to a lack of absorbance at 416 nm; in the wavelength region where a distinct absorption peak could be observed in the case of fouled and unfouled *H. ovalis* and *Z. capricorni*. The colour difference between the extracts of *Z. capricorni* (red) and *H. ovalis* (yellow) were consistent despite only small differences in the positions of their absorbance peaks. Leaf epibionts appeared to contribute a novel water-soluble pigment absorption peak (471 nm) to the spectrum of fouled *P. australis*, but epibionts did not have much effect on the type or quantity of anthocyanins found in the leaves of the other two species. No attempt was made to separate or identify the specific anthocyanins observed in the seagrass leaves.

Table 4.7. Absorbance maxima (peaks) and minima (troughs) associated with the anthocyanins and other water-soluble pigments extracted from fouled and unfouled leaves of three temperate seagrass species.

Zostera capricorni		***Posidonia australis***		***Halophila ovalis***	
FOULED					
peaks	troughs	peaks	troughs	peaks	troughs
416.0	-	-	465.0	416.0	-
-	477.5	471.0	-	-	-
532.0	-	**531.5**	507.0	**531.0**	509.5
566.5	550.5	567.0	549.5	567.0	552.0
600.0	584.0	601.0	583.5	602.0	583.0
654.0	623.0	654.0	622.5	654.5	620.5
UNFOULED					
peaks	troughs	peaks	troughs	peaks	troughs
416.0	-	-	-	416.0	-
-	477.0	-	-	-	-
532.0	-	**531.0**	510.0	**530.5**	508.0
567.5	549.5	567.0	550.5	566.5	553.0
600.5	583.5	601.5	583.0	601.0	583.5
654.0	622.5	654.5	622.5	654.0	622.0

4.4.6 Differences between species in anthocyanin content

There was a significant difference between species in the anthocyanin content of both unfouled ($F_{(1), 2, 86} = 13.460$, $p < 0.0001$) and fouled seagrass leaves ($F_{(1), 2, 71} = 5.649$, $p = 0.005$). Fouled *Z. capricorni* leaves had significantly more anthocyanin per gram fresh weight than did

P. australis, while unfouled *Z. capricorni* leaves contained significantly higher concentrations of anthocyanins than both *P. australis* and *H. ovalis* (Figure 4.17). Although unfouled *P. australis* leaves contained somewhat higher levels of anthocyanins than unfouled *H. ovalis*, the differences between the species were not significant, either with or without epiphyte fouling on the leaves.

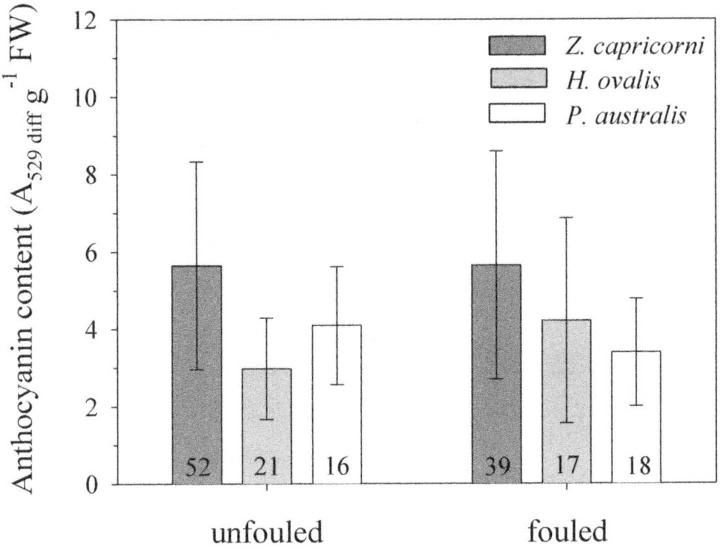

Figure 4.17. Mean $\pm$ SD anthocyanin pigment content for three temperate seagrass species with and without leaf epibionts. Sample sizes were not equal and have been included at the base of each histogram.

4.4.7 Within-species differences in anthocyanin content: spatial and temporal effects

Zostera capricorni

The anthocyanin concentration of unfouled *Z. capricorni* leaves did not vary significantly with season, estuary or habitat of sample collection (Table 4.8). Although differences between individual sample groups may

Table 4.8. Three-way analysis of variance to test for differences in the anthocyanin content of unfouled *Zostera capricorni* leaves as a function of season (spring, winter), habitat type (marine, brackish) and estuary (Port Hacking, Lake Illawarra, St Georges Basin).

Source	df	MS	*F*	*p*	power
Habitat	1	0.740	0.130	0.721	
Season	1	0.165	0.029	0.866	
Estuary	2	1.472	0.258	0.774	
Estuary*Season	2	30.924	5.428	0.008**	0.818
Estuary*Habitat	2	21.187	3.719	0.033*	0.649
Season*Habitat	1	19.645	3.448	0.071	
Season*Habitat*Estuary	2	5.684	0.998	0.378	
Error	40	5.697			

Values significant at the * $p < 0.05$, ** $p < 0.01$. Power was calculated at $\alpha = 0.05$.

be observed in Figure 4.18A, the results of the three-way ANOVA showed strong interaction between the factors 'estuary' and 'season' and an interaction between 'estuary' and 'habitat'.

Since fouled *Z. capricorni* were not available at all sites on both sampling dates, two-way ANOVA tests were performed on subsets of the available fouled *Z. capricorni* data to examine the influence of estuary, habitat and season on anthocyanin concentration (Figure 4.18B). The season and estuary from which leaf samples were collected had no effect on the anthocyanin content of fouled *Z. capricorni* leaves sourced from marine habitats (Table 4.9A). While the season may have had some influence on the anthocyanin concentration of fouled *Z. capricorni* leaves from Port Hacking, a strong interaction between the factors 'season' and 'habitat' would suggest that the result may not be consistent and should be treated with caution (Table 4.9B). There was no difference in the anthocyanin content of fouled *Z. capricorni* leaves collected from marine or brackish habitats in Port Hacking.

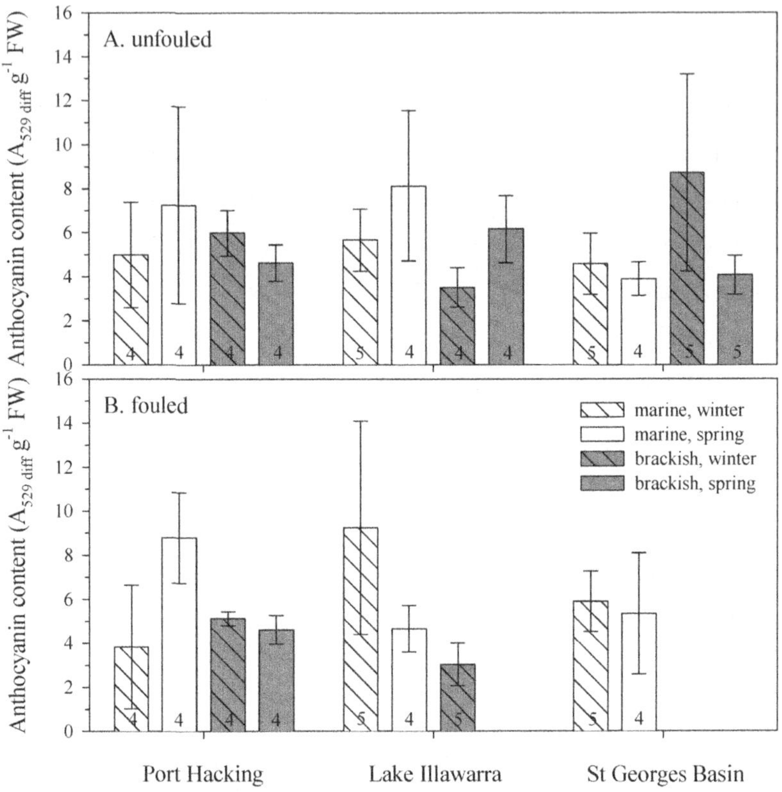

Figure 4.18. Mean $\pm$ SD anthocyanin content of (A) unfouled, and (B) fouled *Zostera capricorni* leaf samples collected from marine and brackish habitats in Port Hacking, Lake Illawarra and St Georges Basin during winter (July 2000) and spring (October 1999). Sample sizes have been included at the base of each histogram.

Posidonia australis

There was a significant difference in the anthocyanin concentration of both fouled and unfouled *P. australis* leaves sampled from different estuaries (Figure 4.19; Table 4.10). While the season of sample collection

Table 4.9. Two-way analyses of variance to test for differences in the anthocyanin content of fouled *Zostera capricorni* leaves (A) sourced from marine habitats as a function of season (winter, spring) and estuary (Port Hacking, Lake Illawarra, St Georges Basin), and (B) sourced from Port Hacking as a function of season and habitat type (marine, brackish).

Source	df	MS	*F*	*p*	power
A. Season	1	0.026	0.003	0.956	
Estuary	2	3.930	0.480	0.623	
Season*Estuary	2	48.203	5.885	0.010*	0.819
Error	20	8.191			
B. Season	1	19.747	6.258	0.023*	0.633
Habitat§	1	8.296	2.629	0.131	
Season*Habitat	1	29.984	9.502	0.010**	0.808
Error	12	3.156			

§Cochran's C value significantly heteroscedastic at $p < 0.01$. Values significant at the * $p < 0.05$, ** $p < 0.01$. Power was calculated at $\alpha = 0.05$.

significantly affected the anthocyanin content of unfouled *P. australis* leaves, the anthocyanins in fouled leaves did not display a seasonal trend.

Halophila ovalis

The sporadic appearance of *H. ovalis* at several of the sampling sites restricted statistical comparison of leaf anthocyanin concentration to those samples sourced from St Georges Basin where the *H. ovalis* meadows grew more consistently. The effects of 'season' and 'habitat' on the anthocyanin content of unfouled *H. ovalis* leaf samples were compared in a two-way ANOVA design (Table 4.11A). Unfouled *H. ovalis* leaves sourced from brackish water in St Georges Basin contained significantly more anthocyanin per gram fresh weight than those from marine locations (Figure 4.20), however, the power of the test was relatively low. The

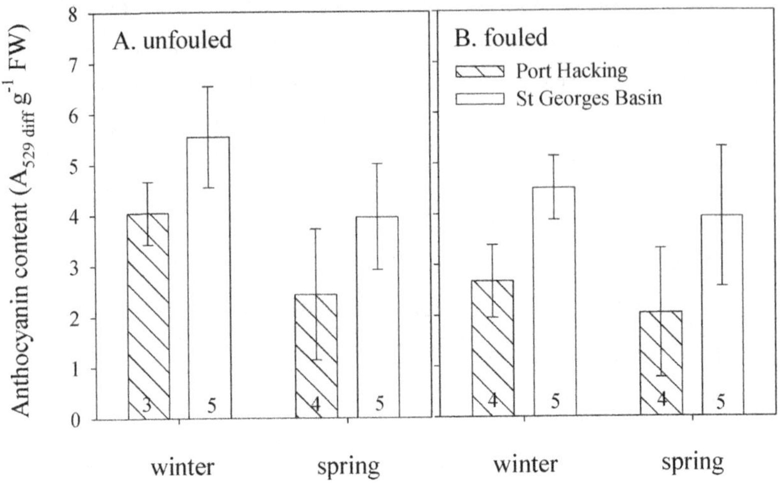

Figure 4.19. Mean ± SD anthocyanin content of (A) unfouled, and (B) fouled *Posidonia australis* leaf samples collected from marine habitats in Port Hacking and St Georges Basin during winter (July 2000) and spring (October 1999). Sample sizes have been included at the base of each histogram.

Table 4.10. Two-way analyses of variance to test for differences in the anthocyanin content of (A) unfouled and (B) fouled *Posidonia australis* leaves as a function of season (winter, spring) and estuary (Port Hacking, St Georges Basin).

Source	df	MS	*F*	*p*	power
A. Season	1	9.901	9.155	0.011*	0.793
Estuary	1	8.888	8.219	0.014*	0.749
Season*Estuary	1	0.001	0.001	0.980	
Error	12	1.081			
B. Season	1	1.619	1.457	0.247	
Estuary	1	15.518	13.965	0.002**	0.935
Season*Estuary	1	0.005	0.004	0.948	
Error	14	15.557			

Values significant at the * $p < 0.05$, ** $p < 0.01$. Power was calculated at $\alpha = 0.05$.

Table 4.11. Analyses of variance to test for differences in the anthocyanin content of *Halophila ovalis* leaves sourced from St Georges Basin; (A) two-way test of unfouled samples as a function of season (winter, spring) and habitat (marine, brackish), (B) one-way test of fouled samples sourced from marine waters as a function of season, and (C) one-way test of fouled samples collected in winter as a function of habitat.

Source	df	MS	*F*	p	power
A. Season	1	2.311	2.712	0.120	
Habitat	1	4.846	5.687	0.031*	0.607
Season*Habitat	1	1.132	1.328	0.267	
Error	15	0.852			
B. Season	1	14.714	5.401	0.053	
C. Habitat	1	0.550	0.166	0.694	

Values significant at the * $p < 0.05$. Power was calculated at $\alpha = 0.05$.

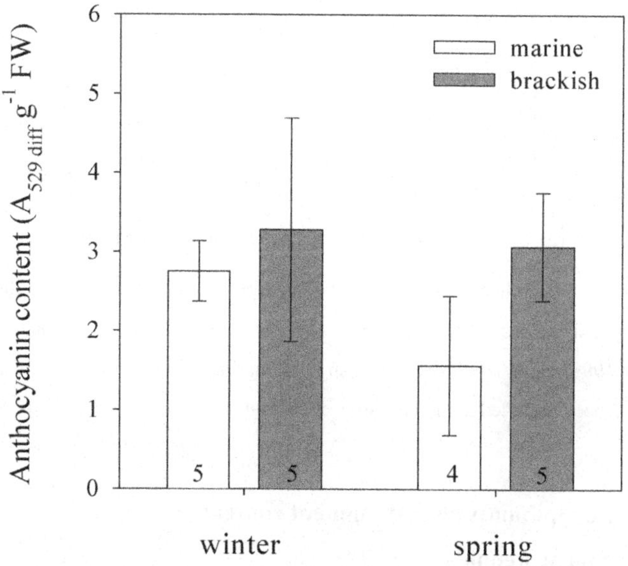

Figure 4.20. Mean ± SD anthocyanin content of unfouled *Halophila ovalis* leaf samples collected from brackish and marine habitats in St Georges Basin during winter (July 2000) and spring (October 1999). Sample sizes have been included at the base of each histogram.

season of sampling did not have a significant influence on the anthocyanin concentration of unfouled *H. ovalis* leaves.

There was no significant difference in the anthocyanin content of fouled *H. ovalis* leaf samples collected during different seasons (Table 4.11B), although there was an apparent trend towards higher anthocyanin concentrations in winter (Figure 4.21A). The habitat from which fouled *H. ovalis* leaves were collected had no effect on their anthocyanin concentration (Figure 4.21B).

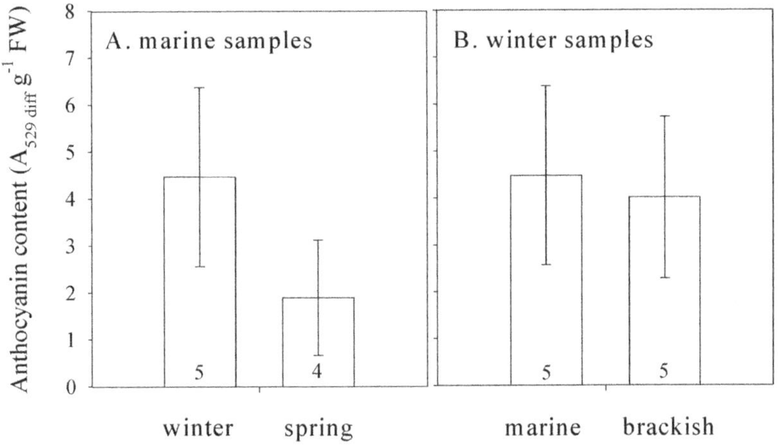

Figure 4.21. Mean ± SD anthocyanin content of fouled *Halophila ovalis* leaf samples collected from St Georges Basin (A) from marine habitats during winter (July 2000) and spring (October 1999), and (B) from brackish and marine habitats during winter. Sample sizes have been included at the base of each histogram.

4.4.8 Influence of epibionts on leaf pigment content

Chlorophylls and carotenoids

Scraping the seagrass leaf blade with a razor did not completely remove all epiphyte tissue from the leaf, as is evidenced by the persistence of small amounts of algal fucoxanthin in scraped samples of *Z. capricorni* (Figure

4.22). Since the scraped leaf segments contained pigments in quantities intermediate to those measured in the fouled and unfouled segments (Figure 4.22), they could not be considered to truly represent the pigment composition of a fouled seagrass leaf with all of its epibionts removed. The significant increase in the lutein concentration and significant decrease in the fucoxanthin concentration of scraped leaf segments compared to fouled segments (Table 4.12) was consistent with the removal of a reasonable amount of epiphyte material from the leaf. However, because of residual algal contamination, these results could not be used to demonstrate whether chromatic adaptation occurs in *Z. capricorni* leaves after they have been shaded by the growth of a layer of epibionts.

Microscopic algae (possibly diatoms) were also present in very small amounts on the surface of 'unfouled' *Z. capricorni* leaves (Figure 4.22), but it was assumed that their presence did not affect the pigment composition of the sample to any measurable extent since the quantity of algal pigment was so small. There was no significant difference between unfouled and fouled *Z. capricorni* leaves in the total chlorophyll and carotenoid content measured per gram fresh weight of leaf material (Table 4.12). Similarly, when fouling organisms occurred on *Z. capricorni* leaves they did not have a particularly notable impact on the relative proportions of chlorophylls and carotenoids observed in the leaves, although differences did occur (Table 4.12). As might be expected, fouled leaf samples contained significantly more fucoxanthin than unfouled samples (Figure 4.22), indicating the presence of the algal classes Baccilliariophyta (diatoms), Phaeophyta (juvenile brown algae) and possibly Dinophyta (dinoflagellates) (Table 4.1). Unfouled leaves contained proportionally more chlorophyll *b* and lutein than fouled leaves, whereas fouled *Z. capricorni* contained proportionally more zeaxanthin and α-carotene than unfouled leaves (Table 4.12). Chlorophyll *b* is the secondary

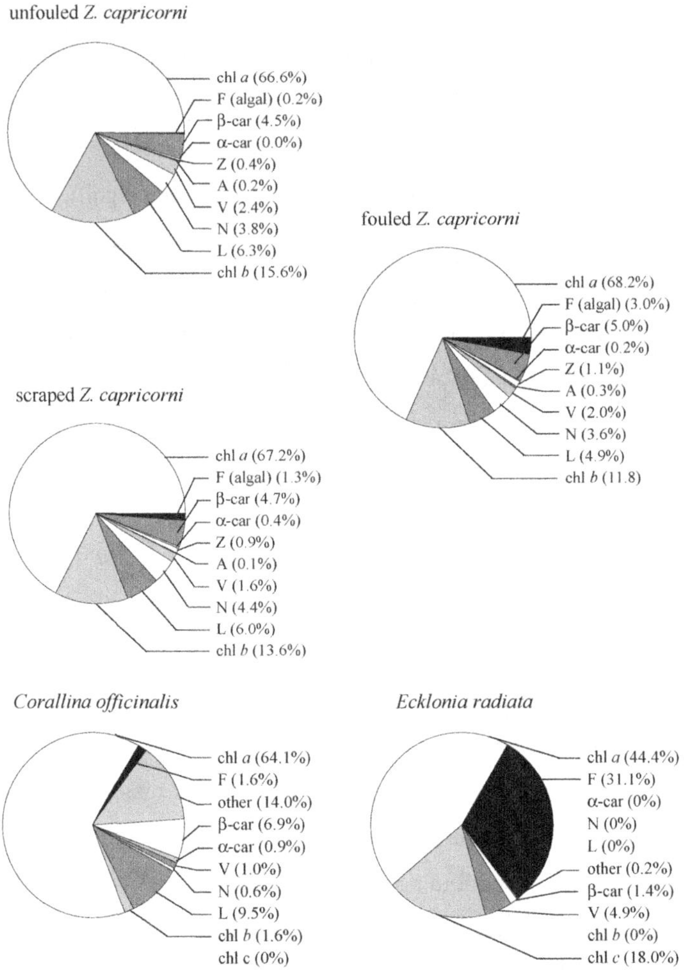

Figure 4.22. Comparison of the chlorophyll and carotenoid composition of unfouled leaves, fouled leaves and fouled leaves scraped of epibionts for *Zostera capricorni* samples collected in the field (n = 5). Also included for comparison are two marine algal species often associated with seagrass in SE Australia; coralline algae *Corallina officinalis*, a red alga (Rhodophyta) that occurs as an epiphyte on seagrass leaves (this sample contains traces of microscopic brown and green algae), and (B) a typical brown macroalga (Phaeophyta), the kelp *Ecklonia radiata*. Pigments separated by HPLC are; chl *a*, chlorophyll *a*; chl *b*, chlorophyll *b*; chl *c*, chlorophyll c; F, fucoxanthin; ß-car, ß-carotene; α-car, α-carotene; Z, zeaxanthin; A, antheraxanthin; V, violaxanthin; N, neoxanthin; L, lutein.

Table 4.12. Differences in the total pigment content per gram fresh weight, and in the proportional concentrations of chlorophylls and carotenoids, between fouled and unfouled *Zostera capricorni* leaves collected in the field. Fouled and unfouled leaf samples were compared using 2-sample t-tests (for unequal variance where appropriate). Paired t-tests were used to test for differences between fouled leaf segments and corresponding segments scraped of their epibionts prior to pigment analysis.

	FOULED vs UNFOULED			**FOULED vs SCRAPED**		
Variable	df	t	*p*	df	t	*p*
Total pigment content:						
T pigments gFW^{-1}	§5.1	1.326	0.241	4	0.893	0.423
Individual pigment content:						
chl *a* / T pigments	#6.7	-1.977	0.090	4	-1.192	0.299
chl *b* / T pigments	8	4.866	0.001**	4	1.761	0.153
lutein / T pigments	8	4.037	0.004**	4	5.860	0.004**
neoxanthin / T pigments	#7.9	0.282	0.785	4	1.123	0.324
violaxanthin / T pigments	8	1.100	0.303	4	-0.949	0.396
antheraxanthin / T pigments	#8.0	-0.990	0.351	4	-2.091	0.105
zeaxanthin / T pigments	8	-5.363	0.001***	4	-2.033	0.112
α-carotene / T pigments	8	-3.735	0.006**	4	0.774	0.482
β-carotene / T pigments	§4.9	-1.720	0.147	4	-1.142	0.317
fucoxanthin (algal) / T pigments	8	-5.323	0.001***	4	-3.673	0.021

Cochran's C value significantly heteroscedastic; # at $0.01 < p < 0.05$, § at $p < 0.01$. Values significant at *$p < 0.05$, ** $p < 0.01$, *** $p < 0.001$.

chlorophyll peculiar to higher plants and green algae (Chlorophyta) while lutein is the predominant light-harvesting carotenoid of these groups (and the Rhodophyta) but is not found in other algal classes (Table 4.1).

While the pigments zeaxanthin and α-carotene are also characteristic of the green plants and red algae, their concentrations in leaves are often linked to plant stress and are far more variable within any particular species. It is also important to note that 'unfouled' leaves represent newly expanded young leaves that become progressively more fouled over a period of days or weeks as the seagrass leaves mature and adapt to the ambient environment. Hence, 'fouled' leaves also represent mature leaves under most estuarine conditions.

Anthocyanins

There was no difference in the anthocyanin content per gram fresh leaf weight of unfouled and fouled *Z. capricorni* samples (Figure 4.23; $t = -0.497$, $df = 8$, $p = 0.633$), although fouled leaves were somewhat more variable in their anthocyanin content than unfouled leaves. In contrast, leaf segments scraped of epibionts just prior to pigment extraction had a significantly higher anthocyanin concentration than their fouled counterparts (Figure 4.23; $t = 23.105$, $df = 4$, $p < 0.0001$).

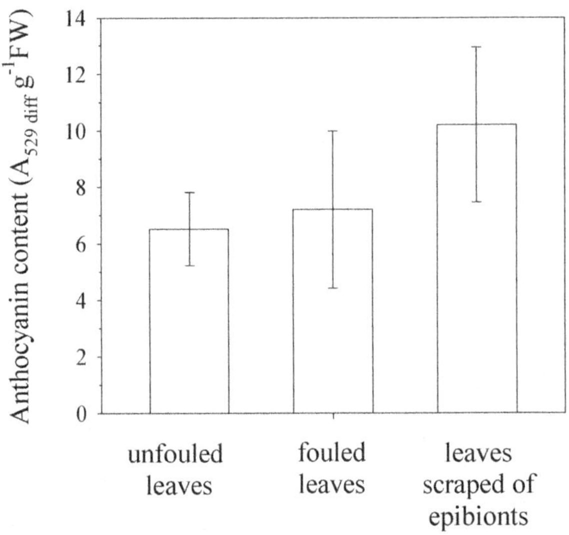

Figure 4.23. Mean ± SD anthocyanin content per gram fresh weight of *Zostera capricorni* leaves without epibionts, fouled with epibionts and scraped of leaf epibionts (n = 5).

4.5 Discussion

4.5.1 Spectral discrimination of seagrass species

The three common seagrasses of the south eastern Australia were clearly spectrally distinct over wide regions of the visible wavelengths regardless of the presence of leaf epibionts, and despite small but significant within species variability related to the year, season, estuary or habitat of sample collection.

The optimal wavelengths for the discrimination of seagrass species (530-580 nm) and other important regions of spectral reflectance separation (520-530 nm, 580-600 nm) conveniently lie within the range of wavelengths that are least attenuated by coastal waters (Figure 4.24).

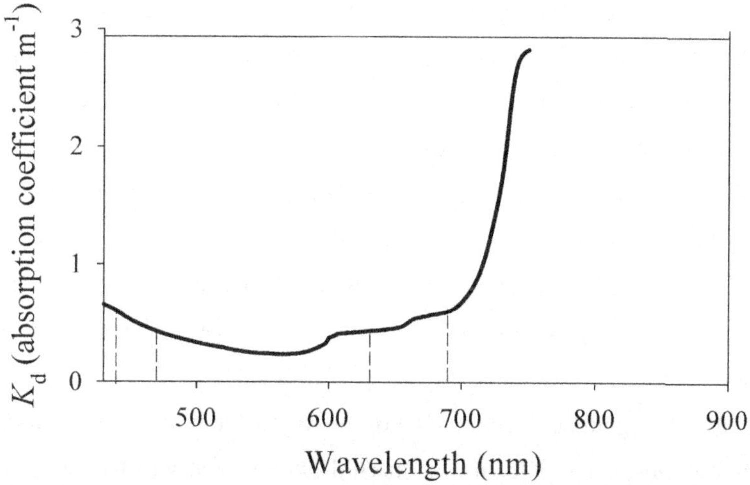

Figure 4.24. The wavelength dependant absorption of light (K_d) by a typical south eastern Australian estuarine water column of 2.1 m depth (A.G. Dekker, pers. comm). Inner drop lines mark the wavelengths of optimal penetration of light through the water column, outer drop lines mark regions of significant but less optimal penetration (Fyfe 2003).

The red wavelengths also penetrate shallow water sufficiently to allow detection of the chlorophyll absorption features of different species (665-680 nm). In south eastern Australia, seagrasses rarely grow deeper than three metres and the meadows, although small, are usually dense and monospecific. Therefore, it should be possible to map the seagrasses of this region to species level in all but the most turbid of estuaries.

4.5.2 Advantages of PMSC correction

While significant differences between species were observed in raw spectral data, differences between all three species were much more pronounced after PMSC correction of the sample spectra. Indeed, intraspecific variability was surprisingly small after the variations caused by illumination conditions and sample geometry were removed. An advantage of the PMSC procedure is that it provides a clearer picture of the inherent differences between species, and of the level of variability within species, that can be related specifically to leaf physiology, biochemistry and morphology. The high variability of uncorrected spectral data is primarily caused by external environmental conditions during spectral reflectance measurement. Hence, PMSC offers a means of generating spectral libraries from field-based spectral reflectance measurements that are equivalent to those collected under controlled laboratory conditions. In addition, because PMSC-corrected spectral signatures retain the actual shape and magnitude of reflectance of the raw spectral reflectance curves, they are most suitable for use in spectral libraries that will be applied as reference data for modelling or image classification. PMSC-corrected spectra can be resampled or binned to match the resolution of any narrow or broadband image data and it can therefore be used for spectral matching classification of non-continuous imagery as well as continuous hyperspectral imagery. In contrast, while derivative analysis has proven valuable for identifying spectral features characteristic of different species of corals, macroalgae (Hochberg and Atkinson 2000), wetland plants (Wang *et al.* 1998), eucalypts (Kumar and Skidmore 1998) and phytoplankton (Richardson *et al.* 1994; Aguirre-Gómez 2001), derivatives can only be used in the classification of continuous hyperspectral imagery, e.g AVIRIS (Richardson *et al.* 1994) or the Senstron Airborne Imaging Spectrometer (SAIS; Wang *et al.* 1998).

Hyperspectral remote sensing procedures to determine plant growth, health and productivity by estimating pigment content may also benefit from the use of this simple transformation. Datt (1998b) found that simple linear multiplicative scatter correction (MSC) improved estimation of the chlorophyll content of eucalypt leaves using single wavelength reflectance, 1st derivatives and simple reflectance ratios. The PMSC method is generally superior to MSC, particularly where scatter variation is large in comparison to chemical variation, and where the spectra have relatively broad and strongly non-selective peaks (Isaksson and Kowalski 1993). These conditions are typical of the spectral signatures recorded from plants, particularly those collected in the field where illumination conditions are continually changing over time and space.

4.5.3 Determinants of reflectance

Seagrass leaf morphology

Maximum separation in the spectral reflectance of the three seagrass species occurred at NIR wavelengths (720-900 nm) where differences between species at the leaf level will primarily be explained by differences in the internal structure of the leaves (Gausman *et al.* 1969; Woolley 1971; Gausman *et al.* 1975; Gausman *et al.* 1984; Vogelmann 1989; Guyot 1990). The number and size of intercellular air spaces in leaf mesophyll tissue predominantly controls light scattering within a leaf (Gausman *et al.* 1975; Gausman *et al.* 1984; Vogelmann 1989), and therefore, the amount of NIR and other weakly absorbed wavelengths including green light reflected from that leaf (Jacquemoud and Baret 1990). There are no intercellular spaces in the mesophyll of seagrass leaves, but rather, a regular system of air canals called lacunae that perform a similar function. While the lacunae of *Posidonia* spp. are smaller than the lacunae of *Zostera* and *Halophila* spp., they are more numerous and occur in several layers (see Figure 2.1), providing increased cell wall-air interfaces for light refraction and probably a greater overall internal air volume. *Zostera* spp. have much larger lacunae but they occur in a single layer with the canals regularly spaced across the leaf. Hence, *P. australis* reflects more strongly in the NIR compared to *Z. capricorni* and *H. ovalis* (Figure 4.2). The significantly lower NIR reflectance of *Halophila* spp. can be attributed to its single layer of small lacunae that only occur in the mid-vein region of its thin leaves.

The degree to which each seagrass species scatters and reflects both NIR and weakly absorbed visible wavelengths will also be influenced to a lesser extent by other factors of leaf internal and external morphology and surface quality (e.g. Shul'gin *et al.* 1960; Gates *et al.* 1965; Woolley 1971; Gausman and Allen 1973; Vogelmann 1989; Guyot 1990). For example, Knapp and Carter (1998) found strong relationships between leaf thickness and NIR reflectance for a number of plant species from a range of different habitats. There are apparent differences between *P. australis*, *Z. capricorni* and *H. ovalis* in leaf thickness as well as leaf surface texture, the size, shape and arrangement of epidermal and mesophyll cells, cell organelles and cell wall structure (see Section 2.3; Figure 2.1). While the epidermal cells of most seagrasses are cubic in shape, those of *P. australis* are elongated axially while *Zostera* spp. cells are more rounded and this would have some effect on internal scattering. The large, smooth surfaced blades of *P. australis* would certainly increase the incidence of specular reflectance from its leaves.

It is interesting to note how similar the spectral response of seagrasses is to that of terrestrial angiosperms, since the optical properties of seagrass leaves would be unlike those described by Vogelmann *et al.* (1996) for typical terrestrial C_3 plants. The epidermal cells are the site of photosynthesis in seagrasses and therefore, they probably do not act to focus light into the underlying leaf layers as they do for a typical terrestrial plant. The epidermal cells lack pigmentation in most terrestrial plants and photosynthesis is carried out by the mesophyll instead. In these plants, up to 90% of the available blue and red light wavelengths are absorbed by the first 1-2 layers of photosynthetic cells (Vogelmann 1989) and the effect of internal scattering may be to spread the more weakly absorbed green wavelengths diffusely around the leaf and facilitate their uptake by the chloroplasts (Vogelmann *et al.* 1996). Green light has been shown to drive considerable photosynthesis within the mesophyll of terrestrial plants (McCree 1971). In many estuarine waters, green wavelengths dominate the PAR available to a seagrass (Kirk 1994). In addition, light is scattered by water molecules and by organic and inorganic particles in a water column, so an underwater light field irradiating a seagrass meadow will tend to be more diffuse than directional (Kirk 1994), except where wave focussing is sending directional beams of intense light into the canopy. Since diffuse light is absorbed more efficiently by chloroplasts than collimated light, this could explain why seagrasses and only a few terrestrial deep shade plants have their chloroplasts concentrated in the epidermis.

Seagrass leaf pigment content

While there are important differences between seagrasses and terrestrial angiosperms (and even other aquatic plants) in their leaf anatomy and metabolism, the biochemistry of all angiosperms is based on the same complement of pigments whose proportions vary by only small amounts. The spectral signatures of seagrasses consequently display the absorption features characteristic of the spectral response of all green plants. Remote sensing discrimination of seagrass species must therefore rely on differences in the magnitude of these absorption features, i.e. the depth and width of absorption troughs, or the height and shape of reflectance peaks. Species differences in NIR reflectance can be very useful for mapping intertidal seagrasses exposed at low tide or for detecting floating mats of leaves and algae but these wavelengths are rapidly attenuated by water (Figure 4.24). Of greater importance are differences observed in the visible wavelengths that can be used to remotely sense benthic vegetation through a water column.

There were consistent differences in the visible spectral reflectance of the three seagrass species that can, in part, be attributed to consistent differences in the total and relative concentrations of photosynthetic and accessory pigments in their leaves. The seagrass species differed significantly in their capacity for blue light absorption, but the differences were small and may not be of a magnitude that can be detected using remote sensing instruments. The broad region of blue wavelength absorption is the sum of the absorption features of both chlorophylls *a* and *b* and a range of carotenoids that extend absorption to shorter wavelengths of the visible spectrum (Gausman 1982). Although each of the chlorophylls and carotenoids absorb maximally in slightly different blue wavelengths (e.g. see Foppen 1971), the influence of individual pigments on reflectance in the blue region cannot be observed in the spectral signatures of seagrasses, a fact noted previously for terrestrial plants by Thomas and Gausman (1977). For example, *Z. capricorni* contained relatively more chlorophyll *a* and less neoxanthin than both *H. ovalis* and

P. australis, which did not differ significantly in their proportional concentrations of these pigments. These pigment differences were not apparent in the spectral response of the seagrasses since the reflectance of fouled *P. australis* was significantly higher than that of both other species at 430 nm (where chlorophyll *a* absorbs maximally) and at 437 nm (a neoxanthin absorption maxima), while *Z. capricorni* and *H. ovalis* did not differ in their reflectance of these wavelengths. Essentially all of the chlorophyll and most of the carotenoids in chloroplasts occur on the thylakoid membranes complexed to a protein in various combinations of several chlorophyll molecules plus one or more carotenoid molecules per polypeptide (Kirk 1994). It is therefore not surprising that individual pigments have little impact on absorption within specific wavelengths. Molecular interactions that occur in the pigment–protein complexes modify the absorption spectra of each of the pigments to varying degrees from those observed for the pure pigments in organic solvent. In general, chlorophylls and carotenoids in pigment-protein complexes absorb at longer wavelengths and across a broader range than they would in isolation (Kirk 1994). This can also be seen in the broad reflectance trough associated with *in vivo* chlorophyll absorption of red wavelengths by seagrass leaves which is centred near 675 nm despite the fact that chlorophyll *a* absorbs maximally at around 660 nm and chlorophyll *b* at 640 nm in organic solutions.

The significantly higher proportion of chlorophyll *a* in *Z. capricorni* leaves corresponded with a deeper trough of red absorption and generally lower reflectance over the blue wavelengths compared with *P. australis* and *H. ovalis* leaves. A higher concentration of any predominant pigment, including chlorophyll, will reduce reflectance across all visible wavelengths, as well as deepening and broadening the absorption features associated with that pigment (Curran *et al.* 1991). Hence, *Z. capricorni* could be spectrally discriminated based on the higher proportion of chlorophyll *a* to total pigments in its leaves. On the other hand, *P. australis* contained significantly more total pigment per gram fresh weight of leaf than *Z. capricorni*, with *H. ovalis* intermediate between the two, which should have resulted in more absorption and lower reflectance of visible wavelengths. In fact, the spectral signature of *P. australis* displayed significantly higher reflectance across most wavelengths. It is likely that the higher overall reflectivity of *P. australis* can be attributed to leaf morphology rather than pigment content, in particular, to the thicker leaves and distinctive lacunal structure of this seagrass. The elevated visible reflectance and distinctive green peak of *P. australis* might also be explained by a higher total carotenoid:chlorophyll ratio though problems with HPLC separation did not allow such a comparison to be made. However, the ratio of VAZ cycle carotenoids to chlorophyll *a* was greatest in the leaves of *H. ovalis*, with no apparent difference between *P. australis* and *Z. capricorni*.

The magnitude of green and near red reflectance from a leaf is not only based on the lack of absorption of these wavelengths by photosynthetic pigments, but on the total and relative concentrations of chlorophylls, carotenoids and accessory pigments in the leaf. A decrease in chlorophyll concentration relative to carotenoid content, or a sufficient increase in the concentration of carotenoids or accessory pigments, may mask the green colour of chlorophyll and increase reflectance in wavelengths dependant on the colour of the dominant pigment (Curran *et al.* 1991). For example, in healthy, mature *H. ovalis* leaves the green colour of chlorophyll masks the yellow colour apparent in both the immature and senescent leaves. While the chlorophyll content of *Z. capricorni* leaves was greater than that of the other seagrass species, the leaves also contained significantly higher concentrations of anthocyanin pigments. Anthocyanins mask the reflectance of chlorophyll at around 560 nm by absorbing maximally in the green and produce pronounced reflectance in the red end of the spectrum (Gausman 1982; Boyer *et al.* 1988). Minor absorption features of anthocyanins are probably also responsible for the shoulders (bumps) observed in the reflectance curves of the seagrasses between 560-660 nm. In south eastern Australia, the red colours of the anthocyanins are most obvious in the immature leaves of *Z. capricorni* but are retained as a dark bronze colour in the adult leaves. It is likely that *Z. capricorni* requires the photoprotection offered by anthocyanins (Chalker-Scott 1999) because it grows at shallower depths than the other two seagrasses and the leaves are more likely to be exposed at the surface at low tide (Dawson and Dennison 1996). The spectral reflectance curve of *Z. capricorni* completely lacks a noticeable green peak and reflects significantly less light than *P. australis* and *H. ovalis* across all visible wavelengths because of the combined effect of relatively high concentrations of both chlorophyll and anthocyanin pigments. *Posidonia australis* does not display the enhanced red reflectance observed for the other seagrasses, even though anthocyanin content of the leaves of this species was not significantly different from that of *H. ovalis*. Since *P. australis* leaves are rarely exposed at the surface of the water, they are unlikely to require the same level of photoprotection as *Z. capricorni* but anthocyanins are known to function in other important roles such as UV (ultraviolet) light filtering and defense against predators or pathogens (Chalker-Scott 1999). In general, the spectral signatures of *H. ovalis* and particularly *Z. capricorni* appear to be relatively flat compared to those of healthy green terrestrial plants that reflect around 20% of incident light at the green peak and around 50% of NIR light. The spectral signature of *P. australis* is more typical of green plants, though somewhat depressed in the green and NIR reflectance. This must to a large extent be associated with the very different leaf anatomy of seagrasses and the impact this has on scattering, e.g. the epidermal chloroplasts, lacunae replacing large intercellular spaces and stomata, lack of palisade mesophyll, etc., as well as the presence of green light-absorbing anthocyanins and possibly other secondary compounds. Mosses display similarly flat spectral signatures which have been linked to their very different leaf morphology, including leaf surface chloroplasts (Lovelock and Robinson 2002). The leaves of dicotyledons in general reflect more strongly than monocotyledon leaves of the same thickness because they have better developed spongy mesophyll (reviewed in Guyot 1990).

A comparison of the nMDS plots of the visible spectral reflectance response of seagrass leaves (Figure 4.3) and the chlorophyll and carotenoid pigment composition of seagrass leaves (Figure 4.13) clearly illustrates the point that leaf level factors other than chlorophyll and carotenoid concentrations are responsible for the spectral differences observed between seagrass species. While the level of similarity and scale of separation between fouled and unfouled samples is the same for both graphs, the high degree of spectral separation between species is not observed in the pigment data. Species separate into groups along the primary ordination axis when analyzed by their spectral reflectance, but when analyzed by their pigment content, they form far less distinct species groupings across the secondary ordination axis. Hence, the leaves of different seagrass species are far more similar to each other in their pigment make up than in their spectral reflectance. The absorbance and reflectance of visible light by plant leaves is determined by leaf morphology, the internal optics of leaf tissue, the formation of pigment-protein complexes plus the arrangement of pigments within chloroplasts and of chloroplasts within cells, as well as by the absorption characteristics of the leaf pigments (Buschmann and Nagel 1991).

Influence of epibionts

As mature leaves at the top of a seagrass canopy are rarely observed in temperate Australian estuaries without a moderate to heavy growth of epibionts, the spectral reflectance of fouled leaf samples can be considered most applicable to the development of spectral libraries for remote sensing purposes. Epiphytes may be beneficial to seagrasses if they reduce desiccation damage by trapping water or increase the levels of nutrients available to the seagrass (Penhale and Smith 1977). However, fouling is most often detrimental to the seagrasses because shading by epiphytic algae interferes with photosynthesis (e.g. Sand-Jensen 1977; Fong *et al.* 2000). Epibionts are most abundant at the tops of the oldest leaves while younger leaves at the centre of the seagrass plant are more likely to be bare (May *et al.* 1978). Even on the surface of newly emerged leaves there is a 'phyllosphere' containing numerous bacteria and diatoms (McComb *et al.* 1981), as was evidenced by the presence of fucoxanthin in 'unfouled' seagrass samples. Differences observed in the spectral signatures of seagrass samples with and without epibionts are therefore, not only due to the contribution that algal epiphytes make to the spectral reflectance, but can also be related to differences in the age and vigour of the leaves.

Fouled and unfouled seagrass leaves did display significantly different patterns of reflectance but the clear spectral reflectance differences between seagrass species were not diminished by the presence of epibionts. Fouling did not change the comparative magnitude of reflectance for any of the species over the visible - NIR wavelengths, although it did have an influence on the shape of their spectral reflectance curves since epiphytes contributed their own absorption and reflectance features to the response. Epibionts tended to reduce seagrass reflectance at the green peak while increasing variability in the reflectance of green wavelengths. The red chlorophyll absorption trough was deepened but the most obvious effect of fouling was to increase reflectance in the far green – near red (575-630 nm) region.

The spectral reflectance responses of the seagrass species were not affected equally by the presence of epibionts. Fouling changed the spectral signature of *H. ovalis* less than it did for *Z. capricorni*, and much less than it did for *P. australis* where the epibionts produced a distinctive reflectance peak centred around 590 nm. Although leaf epiphytes are not host specific, certain algal taxa may be associated with particular seagrass species because of leaf size, shape and turnover time (Borowitzka and Lethbridge 1989). *Posidonia australis* supports a higher diversity of epiphytes, not only because of the higher likelihood of the occurrence of species on a larger host plant, but because the large robust leaves can support larger sized epiphyte species (May *et al.* 1978). In addition, *P. australis* supports a higher biomass of epiphytic algae because its leaves are stronger and retained by the plant for a longer time. The coralline alga *Corallina officinalis* was a ubiquitous epiphyte of *P. australis* leaves, often covering 80-90% of the leaf surface of mature leaves. This red alga appeared to be most responsible for the unusual near red reflectance peak observed in the spectral signature of fouled *P. australis* because of the biliproteins that distinguish the Rhodophyta (Rowan 1989). Biliproteins absorb predominantly green wavelengths and reflect mostly red (in the case of phycoerythrins) or blue light (in the case of phycocyanins) (Kirk 1994). In addition, significant quantities of brown algae (Phaeophyta), diatoms (Bacilliariophyta) and other red algal species would have contributed to the strength of this reflectance peak. Brown algae and diatoms are both distinguished by having fucoxanthin as the major light-harvesting carotenoid. This pigment absorbs in green wavelengths *in vivo* and reflects across a broad range of far green – near red (orange) wavelengths (Rowan 1989). There were also small amounts of unidentified carotenoids in the seagrass samples that provided evidence for the presence of other algal groups, e.g. dinoflagellates (peridinin) and blue-green algae (echinenone, myxoxanthophyll) (Table 4.1), but the quantities of these were presumed too small to have any impact on reflectance.

Zostera capricorni and *H. ovalis* support lower diversities of epibiont species than *P. australis* because of their smaller blade size and faster leaf turnover rates (Walker *et al.* 1999). The cover and composition of algal species on the leaves of *Z. capricorni* was also more variable since this seagrass grows across a wider range of environmental situations. Coralline algae and other red and brown algal macrophytes were abundant on *Z. capricorni* leaves in marine habitats. Under brackish conditions, epibiont cover varied from almost no fouling to a heavy load of predominantly diatoms with infrequent occurrences of large quantities of green algae (Chlorophyta). Diatoms were present in measurable quantities on nearly all *Z. capricorni* and *H. ovalis* leaves, including the 'unfouled' younger leaves. In general, *H. ovalis* carried lower densities of fewer and smaller epibionts, and were usually only fouled by diatoms. Absorption of the green wavelengths by fucoxanthin may have contributed to the uncharacteristically (for a plant) flat visible reflectance curve observed for *Z. capricorni* and *H. ovalis*, considering that a strong green peak was associated with much lower levels of fucoxanthin in unfouled *P. australis*. The overall influence of fouling on the spectral reflectance of *Z. capricorni* and *H. ovalis* was similar to that for *P. australis*, though less pronounced; i.e. enhanced near red reflectance and reduced green reflectance. The distinct reflectance peak centred at around 590 nm, and presumed too be associated with high levels of red algal fouling on *P. australis* leaves, was missing or far less defined in the spectra of fouled *Z. capricorni* and *H. ovalis* leaves.

4.5.4 Intraspecific variability in reflectance

Within species differences were observed in the spectral signatures and pigment composition of seagrass leaves collected from different years, seasons, estuaries and habitats in south eastern Australia. Both interspecific and intraspecific differences in the pigment content of seagrass samples could primarily be related to their chlorophyll concentration. However, only the intraspecific differences in visible spectral response could be principally related to chlorophyll absorption of red wavelengths (Figure 4.8). Presumably, the chlorophyll and carotenoid composition of samples had more influence on the spectral reflectance differences occurring within a species because the kinds of changes that could occur in leaf internal structure for a single species would be comparatively limited. On the other hand, visible spectral reflectance differences between seagrass species were principally a function of their green reflectance (Figure 4.8), suggesting that structural differences between seagrass leaves play a more important role in the spectral discrimination of species than does their pigment composition.

Interannual changes were apparent in the spectra of *H. ovalis* and unfouled *P. australis*, while distinct seasonal changes could only be observed in the spectral reflectance of unfouled *H. ovalis*. There are many reasons as to why spectral reflectance differences might be observed in seagrasses over consecutive years. For example, average weather patterns and the frequency and intensity of storms will affect longer term patterns in freshwater, sediment and pollutant inflow to estuaries. *Zostera capricorni* is an extremely adaptable seagrass that can tolerate extreme change and rapidly fluctuating environmental conditions, while *P. australis* is more sensitive to environmental change (Kirkman 1997). There is insufficient information available to speculate on this point, however. Seasonal changes in solar irradiance and water temperature did not influence the pigment content or spectral signatures of the perennial seagrass species in this subtropical climatic region of south eastern NSW. However, significant seasonal changes were apparent in the pigment content of both fouled and unfouled *H. ovalis*, although these changes only corresponded with a significant spectral reflectance response for the unfouled leaves. *Halophila ovalis* often (though not always) grows as an annual plant (Kirkman 1997) and therefore its leaves probably respond to a light or temperature induced cycle of growth and senescence in which chlorophyll breaks down and carotenoids persist for longer into the colder season (e.g. Tageeva *et al.* 1961; Boyer *et al.* 1988).

The estuary from which samples were collected clearly influenced the spectral signatures of *P. australis* although only the fouled leaves showed a similar change in pigment content. The concentration of fucoxanthin in the sample had most influence on the pigment differences observed between estuaries for the leaves of this seagrass. This can be explained by differences in the epiphyte populations inhabiting the meadows at each estuary. *Posidonia australis* leaves from St Georges Basin appeared to support a higher biomass of juvenile brown macroalgae and diatoms than those from Port Hacking, although the degree of fouling by coralline algae was probably about the same at both estuaries. Epibiont cover or biomass was not quantified in this study, however.

Since spectral reflectance differences also occurred in the unfouled leaves of *P. australis* sampled from different estuaries, without a concurrent change in chlorophyll and carotenoid content, then factors other than different epiphyte populations might be implicated. Small changes in leaf thickness and/or internal structure may induce changes in leaf reflectance, and such changes can occur as the leaf morphology of seagrass plants adapts to the ambient environmental conditions in an estuary. For example,

P. australis growing in the inlet to St Georges Basin are subjected to a stronger tidal current than those at Port Hacking, so their leaves possibly develop more fibrous support tissue. In addition, *P. australis* from St Georges Basin had higher concentrations of anthocyanins than those from the deeper and more turbid Port Hacking site. Anthocyanins are often transient and may be induced by a variety of environmental stresses including exposure to high levels of UVB and visible light (Mancinelli 1985), cold temperatures or stress (Chalker-Scott 1999). Since the anthocyanin content of *P. australis* also responded to seasonal influences (for unfouled but not fouled samples) but no seasonal reflectance differences were observed, then anthocyanins probably had only a small influence on the spectral response.

For the adaptable seagrass *Z. capricorni*, leaf samples from marine and brackish sites proved to be spectrally distinct. Although this pattern could not be explained by pigment composition when leaves were fouled, the pigment content of unfouled leaves was significantly associated with habitat. Brackish estuary backwater habitats are consistently more turbid and more nutrient enriched than estuary inlets that are regularly flushed with oligotrophic ocean water. Total chlorophyll content may be higher in seagrasses growing in relatively eutrophic areas compared to those at oligotrophic sites (Alcoverro *et al.* 2001). It has also been shown that seagrass plants exposed to low light intensities have higher total chlorophyll content (Alcoverro *et al.* 2001) and lower chlorophyll *a*/*b* ratios (Longstaff and Dennison 1999). The relatively strong spectral reflectance divergence in the leaves of *Z. capricorni* from marine and brackish habitats is more probably a function of their leaf morphology. Several botanists (e.g. Robertson 1984) recognise two ecological forms of *Z. capricorni* (syn. *Z. muelleri*); the 'estuarine form' has shorter, narrower and thinner leaves containing less fibrous support tissue than the 'marine form' that is found in open bays and estuary inlets. These forms are an indication of the phenological plasticity of this species since they do not differ genetically.

Spatial and temporal variations in light and nutrient availability, water temperatures, salinity levels and the degree of water movement around plants influence growth, photosynthesis (Hillman *et al.* 1989) and therefore the spectral reflectance response of seagrasses. The ability of a seagrass species to chromatically adapt and maintain photosynthetic vigour under a range of different environmental conditions will affect the direction and degree of spectral reflectance change, and the level of spectral variability observed in the reflectance of that species. Environmental conditions will also determine the biomass and species composition of epibionts that contribute to the spectral reflectance of seagrass leaves. Many algal epiphyte species cannot tolerate the extreme fluctuations in salinity and water temperatures (King 1981) that commonly occur in the shallow backwaters of estuaries. May *et al.* (1978) noted that higher pollution levels reduced the diversity of algal epiphytes on *P. australis* and *Z. capricorni* and suggested certain 'indicator' species of clean and less than pristine environments. One easily identified "indicator" species of clean water, *Corallina officinalis,* is known to occur less frequently on *P. australis* and *Z. capricorni* in polluted or turbid water while the green algae *Chaetomorpha linum* and *Enteromorpha intestinalis* show a concurrent increase (May *et al.* 1978). In general, macroalgal epiphytes tend to die quickly under shade and are replaced by encrusting invertebrates such as bryozoans and spirorbid polychaetes (Kirkman 1989).

The sensitivity of epibionts to environmental conditions may contribute to spectral reflectance differences within a species but they do not allow for any further spectral separation that is not already provided by differences within the plant leaves themselves. For *P. australis* and *H. ovalis*, the intraspecific reflectance differences associated with year and season (respectively) are masked by their presence.

4.5.5 Waveband selection and application

Although it appears that the three seagrass species could be discriminated based on their green reflectance alone, the signals sensed from an aeroplane or satellite will contain different information from that collected by ground spectroradiometry. This will mainly be due to scattering and absorption of light in the atmosphere and water column, reflectance of light from the water surface, mixing of signals in image pixels, the density and geometry of the seagrass canopy and background effect of the substrate. Hence, there are further challenges for researchers in scaling the spectral reflectance response of field measured leaf samples up to real mapping situations (e.g. Lubin *et al.* 2001). It would be an advantage to utilise as many of the distinct spectral reflectance features that characterise species as is practical when selecting remote sensing imagery.

An appropriate band set for remote sensing should target the range of absorption and reflectance features that characterise each species, whether fouled or unfouled. For discrimination of the seagrass species of south eastern Australian estuaries, a band set could include narrow (5-15 nm) bands centred around;

- the major peaks of reflectance and troughs of absorption for the seagrass photosynthetic and accessory pigments (550, 620 and 675 nm). An alternative is to place two bands near the shoulders of wide peaks or troughs to gain more information (e.g. 530 and 570 nm for the green peak).

- the major peaks of reflectance and troughs of absorption for characteristic epiphyte photosynthetic and accessory pigments (590 and 640 nm)
- at least one spectrally insensitive region where separation between species is poor to use as a reference wavelength where possible (e.g. 440 or 495 nm).
- one NIR waveband to assist in discrimination of plant matter at the surface of the water, e.g. floating seagrass and algal mats.

The actual number and width of the bands may not be critical (Thomson *et al.* 1998) but bands should not be wider than the peak or trough they represent and must not overlap with the spectral information provided by other features. Successful mapping of intertidal macroalgae has been performed using as few as five wavebands (Zacharias *et al.* 1992). If the spectral response of the plant species expected in a remote sensing mapping project have not already been characterised (by ground radiometry, for example) it may be better to collect image data across a larger number of narrow bands (Bajjouk *et al.* 1996, Kumar and Skidmore 1998). Feature reduction techniques can then be applied during processing, if necessary.

The specific selection and placement of narrow wavebands require a programmable imaging spectrometer such as the CASI, or a sensor with a suitably narrow spectral resolution and appropriate band centres, bandwidths and interband centre distances. Rollings *et al.* (1993) recommended using the CASI after pilot studies indicated its suitability for the remote sensing of benthic cover, water quality and water depth. As this airborne hyperspectral sensor can also provide image data of high spatial resolution, it is well suited for mapping the often small and patchy seagrass meadows that occur in the estuaries of New South Wales. Fixed-band sensors such as HyMap, AVIRIS and Hyperion (with bandwidths of 15 nm,

9 nm and 10 nm respectively) will perhaps be less suitable unless their band positioning coincides with the relevant visible spectral reflectance features, despite the large number of bands they offer.

This compilation of data forms a functional spectral library for application in future remote sensing classifications and for radiative transfer modelling of atmospheric and water column effects for comparison with the range of signatures actually observed in airborne and satellite image data sets. Since sampling was carried out at sites where seagrasses have been known to have persisted for at least 40-50 years (West *et al.* 1985; Meehan 2001), this data also provides a baseline characterization of 'healthy' seagrass that will be used to identify the spectral reflectance and pigment changes that occur in stressed seagrasses (see Chapter 5).

Chapter 5

Spectral and Physiological Response of Eelgrass to Light Stress

S.K. Fyfe & S.A. Robinson

5.1 Introduction

5.1.1 Stress and acclimation in seagrasses

Seagrass plants will be stressed by physical and biotic environmental factors that are outside any particular species' range of tolerances (e.g. reviews of Hillman *et al.* 1989; Shephard *et al.* 1989; den Hartog 1996) and such stressors can cause changes to a plant's physiological processes (reviewed in Hale and Orcutt, 1987). Short term exposure to stress will induce changes in biochemical processes, while stress over longer periods may result in morphological modifications. Death will result if the stress becomes too extreme or persistent for the plant to be able to adapt or continually repair the damage. Over evolutionary time, however, stressful conditions may select for structural and physiological adaptations that will optimise the growth of a species in that environment (Keely 1991).

Seagrass plants have the ability to tolerate stress to a certain degree by acclimatizing to environmental conditions through changes in their morphology or metabolism, including modifications to their photosynthetic pigment content and stoichiometry. For example, chromatic acclimation of pigments occurs as the plant adjusts its photosynthetic apparatus to maximise its ability to harvest light under the particular conditions of light quantity and light quality. Chromatic acclimation can occur over the course of a day in aquatic plants grown under changing conditions of water depth or clarity (e.g. Falkowski and LaRoche 1991; Dawes 1998).

Acclimation of biochemistry and metabolism results in optimization of photosynthesis and growth and in some cases may increase tolerance to, and the ability to recover from, extreme exposures. Some temperate species are renowned for their ability to adapt to the large environmental fluctuations experienced in temperate estuaries (Dennison and Alberte 1985; Abal *et al.* 1994; Goodman *et al.* 1995). For example, *Zostera noltii* and *Z. marina*, have been observed to chromatically adapt within one hour to a change in irradiance regime (Jiménez *et al.* 1987). Such potential for photosynthetic acclimation is genetically determined and is characteristic of many types of plants adapted to living in variable environments (Björkman 1981; Falkowski and LaRoche 1991). Tropical seagrass species typically grow in more stable environments and therefore demonstrate far less potential for phenotypic plasticity (e.g. *Halodule wrightii* and *Syringodium filiforme*; Dunton 1994; Dunton and Tomasko 1994; Major and Dunton 2000). The ability of a seagrass species to photoadjust and the speed with which it can acclimate to the new environment will determine its ability to survive under stressful conditions, e.g. sedimentation of an estuary associated with dredging or development of the catchment.

Seagrasses have been shown to suffer physiological stress when subjected to a variety of less than optimal environmental conditions (Table 5.1). Light, however, is the primary environmental factor influencing photosynthesis, growth and the depth distribution of seagrasses. Under natural conditions, the depth to which a seagrass meadow extends will be limited by light availability while the shallow boundary of a seagrass meadow will be controlled by exposure to sun (i.e. too much light), wind, wave action and sediment movement (Dennison and Kirkman 1996). The productivity of seagrasses in eutrophic temperate regions is nearly always light-limited while tropical waters tend to be oligotrophic and the seagrasses may instead be limited by nutrient availability (e.g. Dennison

Table 5.1. Environmental stress factors shown to induce physiological stress responses in seagrasses.

STRESS FACTOR	REFERENCES
low and excess light conditions	Dennison & Alberte 1985 Dennison 1987 Dawes & Tomasko 1988 Duarte 1991 Dennison *et al.* 1993 Abal & Dennison 1996 Vermaat *et al.* 1996 Moore & Wetzel 2000
nutrient deficiencies	Short 1987 Udy & Dennison 1997b Touchette & Burkholder 2000a Alcoverro *et al.* 2001
toxic concentrations of nitrates, micro-nutrients and other pollutants	Burkholder *et al.* 1992 Moore & Wetzel 2000 Touchette & Burkholder 2000a Ralph & Burchett 1998a; 1998b Prange & Dennison 2000 Macinnis-Ng & Ralph 2002
a lack of water movement limiting inorganic carbon diffusion across the boundary layer and uptake of nutrients from the water column	Fonseca & Kenworthy 1987 Larkum *et al.* 1989 Moriarty & Boon 1989
dessication when exposed by tides	Pérez-Llorens *et al.* 1994
high and low water temperatures	Bulthuis 1987 Masini *et al.* 1995 Masini & Manning 1997
changes in salinity and pH especially when low salinity limits inorganic carbon uptake	Beer 1996 Invers *et al.* 1997 Ralph 1998 Kamermans *et al.* 1999 Hellblom & Björk 1999

1987; Duarte 1991; Alcoverro *et al.* 2001) or high light exposure. More importantly, light reduction is nearly always implicated in the dieback of seagrass meadows impacted on by human activities, whether the shading is a result of increased turbidity of the water column, or through excessive epiphyte growth in eutrophied waters.

5.1.2 Light and photosynthetic rate

The rate of photosynthesis achieved by a plant cell depends on its capacity to capture light energy and on the efficiency by which the photosynthetic apparatus uses the captured energy to fix CO_2. Light capture is dependent on the pigment makeup of the light-harvesting antennae in photosystems I and II, since the various chlorophylls and carotenoids have unique absorption spectra and absorb maximally at slightly different wavelengths. The efficiency with which light energy is converted into chemical energy, i.e. the quantum yield, ø, is the number of CO_2 molecules fixed per quantum of light absorbed by the plant. Light use efficiency therefore depends on the rate at which the electron transfer chain components in the thylakoid membranes and the CO_2 fixing enzymes in the stroma can utilise the excitation energy collected by light-harvesting pigments arriving at the reaction centres.

Photosynthesis-irradiance (P-I) curves, in which rates of photosynthesis or production are plotted as a function of light intensity, are a useful way to describe the efficiency with which light energy is used for photosynthesis (e.g. Vermaat *et al.* 1996). Photosynthetic rate, *P*, increases linearly in the first part of the curve since all of the photons absorbed by chlorophyll are used in photochemistry (measured using fluorescence as photochemical quenching, qP), i.e. to drive photosynthesis (Figure 5.1). With increasing irradiance, there is a gradual onset of saturation because energy is being delivered to the reaction centres faster than the electron transfer

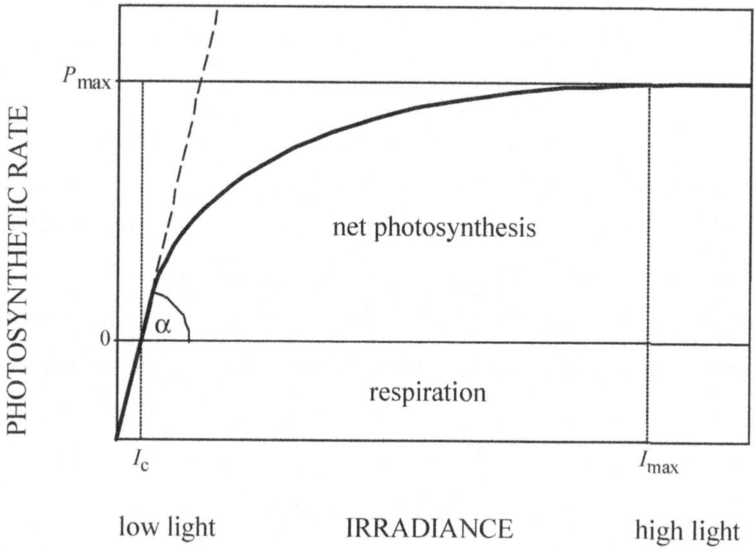

Figure 5.1. Theoretical photosynthesis-irradiance (*P-I*) curve where P_{max} is the maximum photosynthetic rate, I_c is the compensation irradiance, I_{max} is the maximum photosynthetic irradiance and α is the photosynthetic efficiency at low light intensities (maximum quantum yield). Note that irradiance refers to quanta absorbed by the plant rather than incident quanta.

components and/or CO_2 fixation enzymes can make use of it. At this stage, the P-I line curves over towards an asymptotic maximum photosynthetic rate, P_{max} and excess energy must be dissipated by the processes of non-photochemical quenching (NPQ). The initial slope of the P-I curve at low light intensities describes the maximum quantum yield that a particular plant can obtain and therefore provides a relative measure of optimal photosynthetic efficiency. In healthy plants this value rarely exceeds 0.06 moles CO_2 fixed per mole of quanta (Kirk 1994). The 'light compensation point' (I_c) is the level of irradiance at which photosynthetic oxygen liberation equals respiratory oxygen consumption in photosynthetic tissues. Net photosynthesis, and therefore growth, can only occur if light intensity exceeds the compensation point.

Photosynthesis-irradiance relationships have been used to investigate the light requirements for the maintenance of healthy growth in a wide range of tropical and temperate seagrasses (see review of Touchette and Burkholder 2000b). Seagrasses have P-I curves similar in shape to other plants (Vermaat *et al.* 1996) but species can vary considerably in their compensation points, light levels required to saturate photosynthesis (e.g. Drew 1979; Masini *et al.* 1995; Agawin *et al.* 2001), maximum photosynthetic rates and the maximal quantum yields (e.g. Ralph *et al.* 1998a). In contrast, some studies have shown little variation in the I_c of different species collected from the same location (e.g. Masini *et al.* 1995) and within-species seasonal variation in I_c can be of comparable magnitude to between-species variation (Vermaat *et al.* 1996).

In any case, the photosynthetic response of seagrasses is highly dependent on recent light-history (Kirk 1994; Goodman *et al.* 1995; Vermaat *et al.* 1996; Ralph *et al.* 1998). Sun-grown terrestrial plants typically have higher photosynthetic capacity at a saturating light intensity than shade plants (see review in Björkman, 1981) and the same is true for seagrasses where P_{max} decreases with depth (e.g. Drew 1979; Dennison and Alberte 1982; Goodman *et al.* 1995) or varies with season (Agawin *et al.* 2001). For example, *Posidonia sinuosa* collected from close to the depth limits for this species had lower P_{max}, and higher ø than plants collected from shallow water (Masini and Manning 1997).

Photosynthetic rate and I_c may also be influenced by the nutritional status of the seagrass (Touchette and Burkholder 2000a; Alcoverro *et al.* 2001), water temperature, the CO_2 concentration of the aquatic medium (Masini *et al.* 1995; Masini and Manning 1997) and other seasonal factors (Agawin *et al.* 2001). The strong positive relationship between light-saturated photosynthetic rate and nitrogen (N) content of leaves is widely recognised since leaf N is essentially allocated to chlorophyll and other photosynthetic components. Temperature is the primary environmental factor believed to influence photosynthetic and respiration rates, and hence the compensation point, in seagrasses (Bulthuis 1987; Perez and Romero 1992; Masini *et al.* 1995; Masini and Manning 1997). In fact, growth rates of seagrasses have been correlated with water temperature in shallow waters where light is not limiting (e.g. Kirkman and Reid 1979 cited in Kirk 1994).

5.1.3 Photodamage, photoinhibition and photoprotection

As irradiance increases it may exceed that required for light saturation at P_{max} and the plant will be absorbing quanta faster than the reaction centres can use the excitation energy for photochemistry. The excess energy in excited singlet state chlorophyll *a* molecules must be dissipated or it will result in photo-oxidative damage. Only a very small amount can be re-emitted as chlorophyll fluorescence, so the excited chlorophyll must decay via the triplet state or through the thermal dissipation processes of NPQ.

Photo-oxidative damage (photodamage) occurs when excess irradiance causes the production of singlet oxygen and oxygen-free radical species which are potentially extremely destructive to photosynthetic membranes (reviewed in Björkman 1981; Demmig-Adams and Adams 1992a). Photodamage results in damage to the antennae or reaction centre pigments and proteins (chlorophyll bleaching), damage to membranes and possibly tissue death (reviewed by Osmond 1994; Carpentier 1997; Falkowski and Raven 1997; Robinson 1999). Some excess excitation energy may be dissipated safely via decay of the chlorophyll triplet state, but this process can result in transfer of energy to ground state oxygen to generate reactive singlet oxygen. While carotenoids are known to be able to quench singlet oxygen directly, de-excitation through either qP or NPQ is preferred because it minimises the production of singlet oxygen in the PSII antennae (Müller *et al.* 2001).

The processes of non-photochemical quenching offer photoprotection by dissipating excessive light energy as heat (Krause and Weis 1991). There are several components of NPQ, but ΔpH-dependent non-photochemical quenching (qE) is the major and most rapid (seconds to minutes) and reversible process whereas the photoinhibitory quenching (qI) component of NPQ may take days to repair (Osmond 1994; Owens 1994; Müller *et al.* 2001). Photoinhibition via qI depresses photosynthetic rates by damaging the reaction centre of PSII in response to excess light (Critchley, 1981; Critchley, 1988). This primarily affects the light reactions by impairing both electron transport and photophosphorylation. Alternately, photoinhibition through qE regulates the absorption of light energy by PSII antenna to limit the amount reaching the reaction centres of PSII and is therefore often referred to as light-dependent down-regulation of the quantum yield of photosynthesis (e.g. Demmig-Adams 1990; see reviews of Osmond 1994; Robinson 1999). qE plays a central role in overall regulation of photosynthesis since the down-regulation of PSII avoids accumulation of photoinhibitory damage (Pearcy 1994).

One mechanism for qE energy dissipation is provided by the interconversion of the xanthophyll cycle carotenoids in response to excess light. Violaxanthin is de-epoxidised, first to the intermediate mono-epoxide form, antheraxanthin, and finally to the epoxide-free form, zeaxanthin (Demmig *et al.* 1987). The de-epoxidation reaction is rapid and leads to a proportional increase in zeaxanthin in the total xanthophyll pool.

The reverse reaction is somewhat slower but most of the pool will be converted back to violaxanthin after exposure to low light for a period of time, usually overnight (Demmig *et al.* 1987; Demmig-Adams 1990; Adams and Demmig-Adams, 1992; Adams *et al.*, 1995; Schiefthaler 1999). The process is mainly regulated by the extent of the pH gradient (ΔpH) across the thylakoid membrane which represents a balance between the rate of electron transport and the rate of energy utilization measured by ATP consumption (reviewed in Owens 1994; Björkman and Demmig-Adams 1995). This ΔpH is generated by proton migration during electron transport. As the lumen is acidified by the build up of a proton gradient, it promotes the enzymatic de-epoxidation of violaxanthin. It also induces conformational changes in the proteins of LHCII that further facilitate the dissipation of excess energy (Schiefthaler 1999). Zeaxanthin and the ΔpH together convert the light harvesting complexes into quenching complexes that draw energy away from chlorophylls thus preventing over -reduction of the photosynthetic electron transport chain (Horton *et al.* 2000). The requirement for both zeaxanthin and ΔpH means this quenching is rapidly reversible and is responsive to rapid changes in light environment, since the ΔpH is fine tuned to these changes. The other component that regulates qE in plants is the PSII protein PsbS (for review see Murchie and Niyogi 2011).

The dynamic down-regulation of photosynthesis results in a substantial inhibition of electron flow through PSII and therefore temporarily lowers ø, although P_{max} remains unchanged. This photoprotection offered by the xanthophyll cycle prevents a more sustained depression in PSII activity that results when photodamage occurs. Photoinhibition will cause a decline in both quantum yield and P_{max} due to closure of reaction centres but it will only occur when qE is inadequate to dissipate the excess energy (Demmig Adams and Adams 1992a, Murchie and Niyogi 2011). Additional stresses such as inorganic carbon or nutrient limitations and temperature extremes will exacerbate the onset and impact of photoinhibition because the maximum potential rate of photosynthesis in the plant has already been reduced (see reviews of Björkman 1981; Björkman and Demmig-Adams 1995; Murchie and Niyogi 2011). Furthermore, the longer the exposure to high light, the more significant the inhibitory effects and the longer the recovery will take once the plant is restored to lower light intensities (Belay, 1981).

5.1.4 Response of seagrasses to light stress

The daily quantum flux available for photosynthesis by plants is highly variable diurnally, seasonally and spatially, even within an individual plant canopy. Seagrasses grow in particularly dynamic light fields (Ralph *et al.* 2002) since tides, waves, water depth and changing water quality can rapidly alter the intensity and spectral distribution of the irradiance (Kirk 1994). A plant must maximise efficient light capture and use when light conditions are limiting, but regulate light harvesting to avoid photo-oxidative damage to the photosynthetic reaction centres when cells are exposed to excess light above the level of saturation (Björkman and Demmig-Adams 1995). The response of seagrasses to the ambient light climate appear to occur at all levels of plant structure and function; e.g. pigment content, photosynthetic performance, anatomy, canopy height and density, growth, biomass, stable isotope and tissue composition (Abal *et al.* 1994).

Low Light stress

A plant stressed by low light levels will eventually consume its carbohydrate stores and waste away if it cannot fix enough CO_2 by photosynthesis to balance the amount consumed in cellular respiration. The fraction of the total lifetime spent at low light is a key factor in determining whether a seagrass will survive light deprivation (Dennison and Alberte 1985; Williams and Dennison 1990; Masini *et al.* 1995; Moore and Wetzel 2000). Seagrass species differ in their long-term light requirements and tolerances to transient periods of low light stress (Longstaff and Dennison 1999). Persistence under pulsed light deprivation appears to be related to rhizome size and the ability to store carbohydrates, especially in the roots (e.g. Czerny and Dunton 1995; Kraemer and Alberte 1995; Longstaff *et al.* 1999), as well as the ability of the roots to tolerate anoxia (Zimmerman *et al.* 1991). Anoxia will occur when above ground photosynthetic biomass is lost from heavily shaded plants so that O_2 is no longer transported to the roots in a sufficient quantity for aerobic respiration to continue (Smith *et al.* 1988). The ability of a seagrass species to survive longer term light deprivation, however, depends on its capacity to photoacclimate to the new environmental conditions.

Seagrasses are generally considered to be shade-adapted plants (Ralph and Burchett, 1995) but in fact, seagrasses have high minimum light requirements for survival compared to other plants (Dennison *et al.* 1993; Abal *et al.* 1994). The light requirements of seagrasses differ between species but range from 4-29% of surface light (Duarte 1991). This high requirement is thought to be related to the large proportion of seagrass biomass often located in anoxic sediments (Gallegos and Kenworthy 1996) since relatively high rates of photosynthesis are required to supply sufficient O_2 to the roots to support aerobic respiration (Smith *et al.* 1984). The maximal depth at which submersed plants can survive will therefore increase with increasing light penetration. Since light penetration through a water body is a function of vertical attenuation (K_d) in the water column and latitude (Dennison 1987; Kirk 1994), Duarte (1991) was able to establish a generalised colonisation depth for seagrasses of $z_{col} = 1.86/K_d$. This equation has been shown to be an accurate predictor of maximum seagrass colonisation depth across a range of situations and it stresses the importance of light availability in the distribution of seagrasses (Vermaat *et al.* 1996).

Shade-adapted terrestrial plants tolerate low light conditions by a variety of physiological and morphological mechanisms. For example shade leaves tend to be larger, thinner, have larger chloroplasts with large grana stacks, higher total chlorophyll content, lower chlorophyll *a*:*b* ratios, increased PSII:PSI ratio, and lower rates of electron transport and CO_2 fixation (e.g. Nishio *et al.* 1994; Björkman and Demmig-Adams 1995; Schiefthaler *et al.* 1999). Shade leaves also tend to display lower rates of dark respiration and therefore have lower compensation points which allow them to maximise light harvesting and maintain a positive carbon balance (Björkman 1981). Shade-adapted terrestrial plants arrange their leaves to minimise self-shading in the canopy but allocate more carbon to photosynthetic structures so their root:shoot ratios tend to be lower than those of sun plants (Björkman and Demmig-Adams 1995).

A number of laboratory and field-based studies have assessed the physiological responses of various seagrass species to light deficits. No decrease in the photochemical efficiency of photosystem II (F_v:F_m) has been recorded in seagrass leaves subjected to reduced light levels because the plant makes efficient use of all light quanta absorbed (Dawson and Dennison 1996; Longstaff *et al.* 1999). The photosynthetic rate will, however, be limited by available light energy (e.g. Pirc 1986; Longstaff *et al.* 1999; Alcoverro *et al.* 2001) so less carbon will be fixed by the plant during the course of a day. Prolonged exposure can induce physiological photoacclimation to the lower light regime and decreases in P_{max} and I_c have been observed for seagrasses (Alcoverro *et al.* 2001). For example, the minimum light requirement for survival of the adaptable *Heterozostera tasmanica* varies from 5-20% of surface light depending on the light climate in which it was grown (Dennison *et al.* 1993). Not all species use this strategy to adjust to *in situ* light reduction, however (Dennison and Alberte 1982; Neely 2000). Since decreases in the compensation point are associated with a drop in respiration rates, growth rates will be slower in plants showing I_c acclimation (e.g. *Z. marina;* Dennison and Alberte 1982).

In general, seagrasses stressed by low light conditions are characterised by reduced productivity (Dennison and Alberte 1982; 1985; Abal *et al.* 1994; Gordon *et al.* 1994; Grice *et al.* 1996; Longstaff *et al.* 1999; Longstaff and Dennison 1999) and consequently reduced standing crop, biomass (reviewed by Duarte 1991; Abal *et al.* 1994; Longstaff and Dennison 1999; Dixon 2000; Neely 2000) and shoot densities (e.g. Neverauskas 1988; Abal *et al.* 1994; van Lent *et al.* 1995; Longstaff and Dennison 1999; Dixon 2000; Neely 2000; Agawin *et al.* 2001; Alcoverro *et al.* 2001). Shoot thinning appears to be an important, energy-efficient photoadaptive response by which seagrasses can reduce self-shading and increase light harvesting efficiency (Masini *et al.* 1995; Enriquez *et al.* 2002). Leaves tend to be lost before reductions in below-ground biomass become apparent (van Lent *et al.* 1995; Longstaff *et al.* 1999) and this generally results in lower above- to below-ground biomass ratios for shaded seagrass plants in the short term although the ratios may be affected by other factors as well (Dixon 2000). Other morphological changes include lengthening of the leaves and stems to raise the seagrass closer to the water surface and therefore to the light source. This strategy is particularly effective for seagrass growing in shallow turbid water since light is attenuated exponentially (Vermaat *et al.* 1996).

An increase in the total chlorophyll concentration of the leaves may increase the efficiency of light capture for low-light stressed seagrass plants

(Drew 1979; Dennison and Alberte 1982; 1985; Abal *et al.* 1994; Major and Dunton 2002). This photoadaptive response appears to be quite sensitive for some species (Williams and Dennison 1990; Masini and Manning 1997; Longstaff *et al.* 1999) but cannot be generalised across all seagrasses. For a number of seagrasses, total chlorophyll content does not change significantly with low light stress induced by shading, or increasing depth of collection of the seagrass material (Masini *et al.* 1995; van Lent *et al.* 1995; Longstaff and Dennison 1999; Neely 2000). Significant increases occurred only occasionally in other species (Dennison and Alberte 1982; 1985). At extremely low light levels, total chlorophyll concentration will drop via the process of etiolation (Longstaff *et al.* 1999). However, reductions in total chlorophyll concentration were also measured in five seagrass species growing at 50% ambient irradiance (Dawson and Dennison 1996), well above the light levels required to stimulate chlorophyll synthesis. Where shade adaptation does result in an increase in chlorophyll content, it is largely due to an increase in the number of photosynthetic units per chloroplast, though their average size may also increase (Kirk 1994). Abal *et al.* (1994) found that, despite an increase in total leaf chlorophyll content, the number of chloroplasts per cell was much lower in both the epidermal and mesophyll cells of seagrasses grown under low light conditions.

Light deprived seagrasses often display decreased chlorophyll *a:b* ratios (Dennison and Alberte 1982; 1985; Abal *et al.* 1994; Longstaff *et al.* 1999; Longstaff and Dennison 1999) although shading may not influence this ratio at all for some species (Major and Dunton 2002) or for plants already growing under relatively low irradiance levels at depth (Williams and Dennison 1990). If nitrogen is limiting in oligotrophic waters, then chlorophyll *a:b* ratios have actually been observed to increase in the leaves of seagrasses after shading (Williams and Dennison 1990).

Concomitant with increases in the chlorophyll content of seagrasses subjected to reduced irradiance levels are increases in the nitrogen content of leaves, rhizomes and roots (van Lent *et al.* 1995; Moore and Wetzel 2000). Seagrass plants suffering from light deprivation sometimes experience a decrease in the carbon content of leaves (Longstaff and Dennison 1999) and stems (van Lent *et al.* 1995) or a decline in sugar concentrations in the rhizomes (Longstaff *et al.* 1999) as the plants carbohydrate stores are depleted by continued respiration and growth in the absence of net photosynthesis.

High light stress

Excess light is a common stress factor that reduces the productivity of seagrasses exposed at or near the water surface (Ralph *et al.* 2002) and seagrasses respond to high light stress in much the same way as terrestrial higher plants (Dawson and Dennison 1996). For many seagrasses, P increases daily with irradiance from dawn to a mid-morning high but P_{max} shows a midday depression associated with photoinhibition, particularly in situations where maximum irradiance corresponds with low tide (Ralph *et al.* 1998; Beer and Björk 2000). Photoinhibition has been recorded in both tropical and temperate seagrass species, most often at light intensities greater than 1000 μmol photon m^{-2} s^{-1}, and with maximal photoinhibition occurring in the 3 hours after midday (e.g. Ralph and Burchett 1995; Flanigan and Critchley 1996; Dawson and Dennison 1996; Beer and Björk 2000; Ralph *et al.* 2002). Only under severe conditions of high light stress will chlorophyll bleaching be observed (Lovelock *et al.* 1994) but this has been observed to occur when seagrasses are exposed for longer periods than is usual, e.g. when extreme low tides coincide with midday irradiation and high temperatures in summer (e.g. Fyfe pers. obs., Seddon *et al.* 2000).

In high light environments plants may protect against photodamage and maintain photosynthesis using a range of structural, anatomical and biochemical modifications. For example, photoacclimation may produce changes in leaf size and orientation, leaf surface reflectance, rearrangement of, or a change in the number or size of leaf chloroplasts, and higher rates of electron transport and CO_2 fixation (e.g. Nishio *et al.* 1994; Björkman and Demmig-Adams 1995; Robinson 1999; Schiefthaler 1999). Biochemical adaptations may include changes in the amount or proportions of chlorophyll and carotenoid per leaf area, or an increase in the concentration of other photoprotective pigments such as anthocyanins which act either by quenching excess energy directly or as antioxidants. Of the molecular and physiological changes that can occur within the cells in response to high light, the most effective photoprotective mechanism is thermal energy dissipation through NPQ (Biswal 1997).

Ralph *et al.* (1998) observed that for three seagrass species photochemical processes dominated under low light while NPQ dominated in high light conditions. The modulation of NPQ was concomitant with large changes in the violaxanthin (V), zeaxanthin (Z) and antheraxanthin (A) content of *Z. marina* (Ralph *et al.* 2002) and *Z. capricorni* leaves (Flanigan and Critchley 1996). Similarly, Dawson and Dennison (1996) observed decreases in the proportion of the total xanthophyll cycle present in the photoprotective form, zeaxanthin (Z/V+A+Z) for five seagrass species when shaded to 50% of ambient light, though the response was only significant for *C. serrulata* and *Syringodium isoetifolium*. Hence, down-regulation of photosynthesis through the xanthophyll cycle can be seen to provide a photoprotective response for a number of seagrasses. Some species appear to have less capacity for NPQ modulation in response to fluctuating irradiance levels, e.g. photoinhibition took 2 hours to develop in *H. ovalis* when exposed to light at 1000 $\mu E\ m^{-2}\ s^{-1}$ (Ralph and Burchett,

1995) while *Z. marina* responded almost immediately to increased light levels (Ralph *et al.* 2002). However, these results may simply indicate that higher light levels are required to saturate photosynthesis in *H. ovalis* and therefore the light energy is not in excess, but is being utilised in photochemistry.

The current level of photoprotection in a leaf may be identified by measuring the de-epoxidation state (EPS = (Z+0.5A)/V+A+Z) of the xanthophyll cycle pigments (VAZ) as defined by Thayer and Björkman (1990). EPS can be used as a biochemical monitor of high light exposure in plants, including seagrasses (e.g. Ralph *et al.* 2002). The pool size of VAZ pigments, on the other hand, indicates the level of longer term photoacclimation to light stress in a plant (Björkman and Demmig-Adams 1995).

Photorespiration, the fixation of O_2 by the key photosynthetic enzymes Ribulose bisphosphate carboylase/oxygenase (Rubisco), is also favoured by high temperatures, high light and limited water movement around the plant (Touchette and Burkholder, 2000b), i.e. the conditions to which seagrasses may be exposed to daily at low tide in many estuaries. It has been observed to occur in the seagrasses *Halophila stipulacea, Cymodocea nodosa* (Beer *et al.* 1998) and *H. wrightii* (Beer and Björk 2000). In submerged aquatic plants with carbon concentrating mechanisms like the majority of seagrasses, photorespiration is rare and plants can maintain high photosynthetic rates under elevated oxygen levels (Falkowski and Raven, 1997). Photorespiration wastes carbon otherwise available for photosynthesis but it probably acts as an additional photoprotective mechanism as it limits damage to the photosynthetic apparatus during periods of high light intensity and low CO_2 availability (Osmond and Grace 1995; Heber *et al.* 1996; Touchette and Burkholder 2000b).

A number of researchers have examined the physiological responses of seagrasses to high light stress. While high light exposure leads to increases in photosynthetic rate and productivity (Abal *et al.* 1994; Grice *et al.* 1996; Agawin *et al.* 2001), there is typically a decrease in the photochemical efficiency (F_v:F_m) of high light stressed seagrasses (Ralph and Burchett 1995; Dawson and Dennison 1996; Agawin *et al.* 2001; Major and Dunton 2002; Ralph *et al.* 2002). Increased growth and productivity, plus increased shoot densities (e.g. Alcoverro *et al.* 2001), lead to high meadow biomass unless the stress becomes severe enough to cause significant levels of photodamage. Canopy heights will often decrease in high irradiance situations since the seagrass plants tend to display shorter, thicker leaves with a larger lacunal area (Abal *et al.* 1994; Grice *et al.* 1996). The plants often have lower above-ground to below-ground biomass ratios (Abal *et al.* 1994) because more carbon is allocated to the structures involved in nutrient uptake to support high rates of photosynthesis.

Seagrass plants grown in high light tend to have lower total chlorophyll contents than plants grown in moderate light (Abal *et al.* 1994; Agawin *et al.* 2001) although the decrease may not be significant (Dawson and Dennison 1996). No consistent changes have been observed in the chlorophyll *a*:*b* ratios of high light stressed seagrasses (Abal *et al.* 1994; Agawin *et al.* 2001). High PAR levels increased the number of chloroplasts per epidermal and mesodermal cell in *Z. capricorni* (Abal *et al.* 1994) but the reverse pattern was seen for five seagrass species exposed to increased levels of UV (Dawson and Dennison 1996). High UV is a factor that compounds the effects of strong sunlight in the field; hence seagrasses may display increased levels of anthocyanins and other UV blocking pigments when exposed to high irradiance levels (Trocine *et al.* 1982; Abal *et al.* 1994; Dawson and Dennison 1996).

Dawson and Dennison (1996), Flanigan and Critchley (1996) and Ralph *et al.* (2002) have measured changes in the xanthophyll cycle carotenoids in seagrass leaves exposed to different irradiance levels. The proportion of Z:VAZ in *C. serrulata* and *S. isoetifolium* leaves was significantly greater in high light treated plants compared to those grown in moderate light but the change was not significant for *Z. capricorni*, *H. ovalis* or *H. uninervis* (Dawson and Dennison 1996). Flanigan and Critchley (1996) and Ralph *et al.* (2002) noted that A occurred in higher concentrations and was more responsive than Z to changes in irradiance for *Z. capricorni* and *Z. marina* respectively. Ralph *et al.* (2002) suggested that NPQ was more highly correlated to A+Z:VAZ than Z:VAZ, at least for these species. In addition, the VAZ pool size of *Z. marina*, and hence its potential for NPQ, was observed to increase with increasing light levels during the course of a day (Ralph *et al.* 2002). Changes in carotenoids other than the VAZ cycle pigments may also be induced by high irradiance conditions and may have some role in photoprotection; lutein as a quencher of excess energy (Young and Britton 1990; García-Plazaola *et al.* 2007) and β-carotene as a scavenger of free radical species (Young *et al.* 1997). However, the response of other carotenoids to changing irradiance has not been examined for seagrasses.

High light stressed seagrasses show decreased leaf nitrogen content as a consequence of their lower chlorophyll concentration (Dawson and Dennison 1996; Abal *et al.* 1994; Grice *et al.* 1996). The carbon concentration of leaves may not change significantly, e.g. in the case of *Z. capricorni*, but there is a general trend for carbon concentrations to increase when seagrass species are grown under conditions of high light (Grice *et al.* 1996). As a consequence, higher carbon:nitrogen ratios (C:N) can be observed in high light grown seagrasses (Grice *et al.* 1996).

5.1.5 Spectral shifts associated with stress in plants

Healthy leaves of most higher plants appear green because photosynthetic pigments absorb mostly in the red and blue regions of the visible spectrum and reflect a large proportion of the green light incident upon them. In the remote sensing of vegetation canopies, chlorophyll content is commonly considered to be a surrogate for photosynthetic health because healthy canopies usually contain a larger biomass of leaves rich in chlorophyll (Chappelle *et al.* 1992; Stone *et al.* 2001). If stress leads to a degradation of the chlorophylls then the carotenoids will be exposed (Young 1993b), and reflectance in the green wavelengths between 520-600 nm will increase (Tageeva *et al.* 1961; Thomas and Gausman 1977; Gausman 1982; 1984; Brakke *et al.* 1989). A number of workers have recorded a strong correlation between green reflectance at around 550 nm and chlorophyll concentration (e.g. Thomas and Gausman 1977; Gausman *et al.* 1984; Saxena *et al.* 1985; Buschmann and Nagel 1993; Adams *et al.* 1999; Lovelock and Robinson 2002). For a diverse range of plant groups, green peak reflectance has offered more information on chlorophyll content than the main chlorophyll absorption feature at 675 nm (Carter 1993). This is because increasing leaf chlorophyll concentration causes a broadening rather than a deepening of this absorption trough (Björkman and Demmig-Adams 1995; Curran *et al.* 1998). The magnitude of red absorption does not usually change much unless significant chlorophyll breakdown occurs, as is usually the case for senescing leaves (Gausman *et al.* 1971; Gausman 1982).

The red edge occurs around 700 nm and is possibly the most prominent and unique feature in the spectral signature of healthy green vegetation. The red edge (RE) is the steep slope between strong red wavelength absorption by chlorophyll, and strong NIR reflectance caused predominantly by scattering in the leaf mesophyll and a lack of absorption of NIR wavelengths by pigments (Woolley 1971). Since the red chlorophyll absorption trough narrows as chlorophyll content decreases, a 'blue shift' of the RE to shorter wavelengths may be seen when vegetation is stressed (Rock *et al.* 1986; 1988; Ustin *et al.* 1990). The wavelength position of the RE has been used for many years in remote sensing to collect accurate quantitative information on the photosynthetic vigour of terrestrial vegetation because it is strongly correlated with total chlorophyll content (e.g. Gates *et al.* 1965; Horler *et al.* 1983; Rock *et al.* 1986; 1988; Boochs *et al.* 1990; Curran *et al.* 1990; 1998; Kupiec 1994). Reflectance in the red wavelengths can increase when chlorophyll loss is significant and NIR reflectance may simultaneously decrease (Malthus and Madeira 1993) or increase (Sinclair *et al.* 1973) due to degradation of internal cell structure (Gausman and Allen 1973). Hence, changes in the magnitude of the red edge near 700 nm can also provide useful information about the condition of vegetation in addition to the position of the RE (REP) (Carter and Knapp 2001).

No specific spectral features have been associated with any one type of stress, e.g. the deficiency of individual plant trace minerals (Ponzoni and Gonçalves 1999), although Carter *et al.* (1989) linked reflectance changes in blue wavelengths to increases in potassium and magnesium in the leaves of *Pinus taeda* (at 401 and 470 nm respectively). In general, the regions 535-640 nm and 685-700 nm are sensitive to a wide variety of plant stressors (Carter 1993; Luther and Carroll 1999). Attempts have been made to correlate specific wavelengths with the concentrations of certain pigments. For example, Chappelle *et al.* (1992) suggested that the best indicative wavelengths for determination of chlorophyll *a*, *b* and carotenoid content were 675, 650 and 500 nm respectively. However, green reflectance will increase as the chlorophyll content of stressed plants decreases and this change may or may not be accompanied by increases in carotenoid content, so the magnitude of the green peak may not be a dependable measure of carotenoid content. Thomas and Gausman (1977) and Saxena *et al.* (1985) tested a range of individual wavelengths (including the carotenoid absorption maxima around 450 nm and minima around 550 nm) but found none that were significantly correlated with total carotenoid content. Nevertheless, the inclusion of carotenoid concentration did improve the statistical significance of the correlation between the wavelength 550 nm and chlorophyll concentration (Thomas and Gausman 1977) and Peñuelas *et al.* (1993) could roughly gauge the carotenoid:chlorophyll ratio of wetland emergents based on differences observed at 430 nm referenced against chlorophyll absorption at 680 nm.

In addition to the blue shift of the RE, a 'green shift' can be observed when chlorophyll content decreases and/or carotenoid content increases. The shoulders of the green reflectance peak broaden as green reflectance increases and the blue and red chlorophyll absorption troughs narrow. Gamon *et al.* (1992) noted that there was some relationship between the green shift and photosynthetic radiation use efficiency. Bilger *et al.* (1989) first observed subtle changes in the absorbance between 505-515 nm that could be linked to the epoxidation state of the xanthophyll cycle pigments. Subsequently, the VAZ pool and EPS were strongly correlated to reflectance at the 531 nm wavelength centred at the green edge (Gamon *et al.* 1990).

Stressed leaves commonly display visible symptoms like chlorosis in response to nutrient deficiencies (e.g. Thomas and Oerther 1972; Carter *et al.* 1989; Curran *et al.* 1998; Ponzoni and Gonçalves 1999), certain diseases (e.g. Lorenzen and Jensen 1989), high light exposure (Gausman 1984), seasonal changes (Boyer *et al.* 1988; Peñuelas *et al.* 1995b; Adams *et al.* 1999) or simply senescence (Boyer *et al.* 1988). Photo-oxidative processes induced by toxic pollutants that cause pigment bleaching (Young and Britton 1990; Young 1993b) will have a similar influence on reflectance (e.g. Curran *et al.* 1998). Insect attack on eucalypts may result in a reddening of the leaves rather than yellowing because anthocyanins are synthesised in response to the damage and become more prominent as the insects remove chlorophyll from the leaf (Stone *et al.* 2001; Coops *et al.* 2004). Anthocyanins are also exposed in the senescing leaves of many deciduous plants (Boyer *et al.* 1988) and result in an increase in orange reflectance around 600-650 nm (Gausman 1982; Boyer *et al.* 1988) with a possible decrease in green peak reflectance between 500-550 nm (Gausman 1982) and a blue shift in REP near 750 nm (Stone *et al.* 2001).

Sometimes a loss of vigour causes a decrease in reflectance across the whole of the visible-NIR wavelengths (e.g. Luther and Carroll 1999). Spectral reflectance shifts in aging or diseased plants have been related to morphological or anatomical changes in the leaves and cells as well as pigment changes, e.g. changes in the size or shape of the intercellular air spaces (Tageeva *et al.* 1961; Gausman *et al.* 1969; 1971), dissolution of organelles or dessication of tissue (Boyer *et al.* 1988). Reflectance differences between sun and shade grown leaves can be attributed to differences in their size and mesophyll air volume as well as their total chlorophyll concentrations (Gausman 1984). Certain fungal infections cause a collapse in leaf cell structure and a flattening of the reflectance spectrum (e.g. Malthus and Madeira 1993). REP is not only controlled by chlorophyll concentration but is also affected by differences in the composition of the pigment-protein complexes and by internal damage in the leaves and cells (Lee *et al.* 1990). For example, air pollutants caused a blue shift in conifer leaves that was attributed to chlorophyll denaturation as well as cell collapse (Westman and Price 1988).

5.1.6 Potential for spectral detection of stress in seagrass

Human activities in and around estuaries will almost always affect seagrass health by reducing the quantity of light available for photosynthesis. This may occur through changes in the sedimentation regime (Vermaat *et al.* 1996) or through eutrophication of the waterway (Shepherd *et al.* 1989). Occasionally human activities impact on seagrass meadows by increasing the potential for photo-oxidative damage following high light exposure. In shallow waters or intertidal areas where seagrasses occur at or close to the water surface, the susceptibility of the plants to photoinhibition and photodamage will be dramatically increased by accumulated stress on the plants. For example, industrial developments that pollute the waterways with toxic chemicals, or change the salinity, temperature or depth of the water may leave these sensitive ecosystems susceptible to chlorophyll bleaching and dieback.

Declining seagrass health and impending dieback can be detected in the photophysiological responses of the plants before irreversible changes in meadow biomass or distribution occur (Longstaff and Dennison 1999; Major and Dunton 2000; 2002). Hence, the physiological state of the seagrass meadow provides a biologically based early warning of impending meadow dieback and water quality problems in an estuary. Managers could use this information to correct problems in an estuary before they lead to significant and irreversible seagrass loss and ecosystem decline. Remote sensing may provide a cost-effective, repeatable, synoptic means of monitoring the health of seagrass meadows. The availability of high spectral, radiometric and spatial resolution imagery opens up possibilities for monitoring vegetation health and/or fine scale change. The basic premise is that stress causes biochemical or morphological changes in the plants that are matched by characteristic spectral reflectance changes of a magnitude that can be remotely-sensed.

Seagrass species vary somewhat from each other, and from terrestrial angiosperms, in their reaction to different types of stress and in their ability to photoacclimate to changes in the environment. It is therefore important to determine the stress-induced pigment changes characteristic of a particular species and the spectral responses that typify that stress (see review of Peñuelas and Filella 1998).

Zostera capricorni is one of the most adaptable and tolerant seagrass species and the most common and abundant seagrass in NSW, yet its conservation is under constant threat from anthropogenic pressure. The aim of this laboratory-based study was to identify the pigment changes that occur in *Z. capricorni* leaves after exposure to low and high light stress and to investigate whether such pigment-based 'stress indicators' can be spectrally detected. The study applies a combination of spectroradiometry, pigment analysis, chlorophyll fluorescence and other techniques to provide baseline data that will assist in the future development of remote sensing methods for monitoring seagrass condition.

5.2 Study methods

5.2.1 Field sampling of photosynthetic efficiency

The photosynthetic efficiency (F_v:F_m) of *Z. capricorni* was sampled in the field at Costen's Point and Gray's Point, Port Hacking between 10 am and 2 pm in mid-summer (January 2003) so that information could be gathered about the performance of the species under the high irradiance conditions it naturally experiences in the field. Leaf material was collected haphazardly from the meadow at each site and placed in a black, plastic bag of seawater without removing the samples from the water. At each site, the leaves were dark adapted for approximately 20-30 minutes by suspending the bag in the water beside the boat to maintain the samples at ambient water temperature. The water was subsequently drained from the bag and the material divided into 20 handful sized samples which were measured using the PAM fluorometer (according to methods described in Section 3.3.2) without removing the leaf samples from the dark environment of the bag.

5.2.2 Diurnal cycle in the field

The variation in the photosynthetic efficiency of *Z. capricorni* was tracked in relation to varying environmental conditions over a daily cycle of sunlight and tides in the meadow at Gray's Point, Port Hacking on a fine, clear autumn day (May 2000). Five replicate leaf samples were collected haphazardly from within the meadow at approximately two hour intervals between sunrise (6.37 am) and sunset (5.05 pm). The samples were placed in seawater-filled sample jars (without removing them from the water) and were dark adapted for 15-20 minutes under black plastic. Samples were drained and measured for F_v:F_m using the PAM fluorometer (according to methods described in Section 3.3.2) without removing the leaf samples from the dark environment. Immediately prior to collecting leaf samples at each sampling period, the temperature and depth of the water column and the PAR at the water surface over the seagrass meadow was also recorded.

The significance of changes observed in photosynthetic efficiency between sampling times was assessed using two-tailed 2-sample t-tests for unequal variance after the data sets had been checked for normality and homogeneity of variance.

5.2.3 Estimating photosynthetic parameters in the laboratory

Sampling of electron transport rate (ETR) was performed to gain some insight into the light intensity at which *Z. capricorni* achieves maximum photosynthetic efficiency under laboratory conditions. *Zostera capricorni* plants were grown in tanks in the laboratory (as described in Section 3.2) under a range of artificial illumination levels for one month prior to sampling in November, 1999. Small leaf samples were cut from each tank and immediately placed into the leaf clip of the fluorometer above the tank for measurement of yield and ETR. Four measurements were collected at each tank to provide an overview of the performance of the plants under laboratory conditions;

1. one leaf sample from the uncovered side of the tank was measured under ambient irradiance levels while,
2. a second sample from the uncovered side was measured under shadecloth in the adjoining side of the tank, then,
3. one leaf sample from the shadecloth covered side of each tank was measured under shadecloth in the position from which it was collected, and
4. another sample from the shadecloth covered side was measured in the adjoining, unshaded side of the tank.

The results were graphed as a scatterplot of ETR versus measuring irradiance with samples grouped into categories according to the irradiance levels in which they were grown; low light (grown in PAR less than 100 μmol photon m^{-2} s^{-1}), control (grown in PAR between 100-400 μmol photon m^{-2} s^{-1}) and high light (grown in PAR greater than 400 μmol photon m^{-2} s^{-1}). Photosynthesis-irradiance curves were fitted to the data in each category using an exponential non-linear regression model (e.g. Masini and Manning 1997; Agawin *et al.* 2001); in this case the Exponential Rise to Max (simple exponent, two parameter) curve fitting function, $y = a(1-b^x)$, available in SigmaPlot 8.02 (SPSS Inc.). Maximum photosynthetic rate, P_{max}, was estimated from the horizontal portion of the curve generated for each category. The initial linear portion of each PI curve was determined by eye and a simple linear regression line was fit to selected points including the origin. Efficiency, α was determined from the slope of the regression line, i.e. $y = \alpha x + b$ where $b = 0$.

5.2.4 Manipulative laboratory experiments

Low light stress

Zostera capricorni plants were grown in 12 paired replicate sample tanks (according to the conditions described in Section 3.2) under control (PAR 168 $\pm$ 49 μmol photon m^{-2} s^{-1}) lighting conditions for one month prior to commencement of the experiment in February 2000. Shadecloth covers were installed over the outer half of each tank to induce low light 'treatments' (PAR 39 $\pm$ 12 μmol photon m^{-2} s^{-1}), while the 'control' half of each paired sample tank was left uncovered (see Section 3.2.4).

Sampling was carried out on the initial day of experimental treatment, every three days thereafter for two weeks and once again after three months. Tank water temperatures and salinity were checked just before sampling commenced. Leaf samples were cut from the treated and control halves of each tank at least 2 hours into the light period so that photosynthesis had a chance to stabilise to daily levels of illumination. The leaves were placed in sample jars filled with tank water and held in an esky at laboratory room temperature for a minimum of 20 minutes dark adaptation. Each sample was drained of water in a dark room and the chlorophyll fluorescence parameters (F_o, F_m and F_v:F_m) were measured (according to Section 3.3.2) immediately before collecting the spectral reflectance signature under controlled artificial lighting (according to the methods described in Section 3.3.1). Approximately 10 leaves were removed from the centre of the FOV of the spectroradiometer, wrapped in foil, snap frozen in liquid nitrogen then stored at -80 °C for subsequent analysis of photosynthetic pigment content by HPLC (according to Section 3.3.3), anthocyanin content by spectrophotometer (according to Section 3.3.4) and carbon and nitrogen content by mass spectrometer (according to Section 3.3.5).

Pigment and carbon/nitrogen analysis were carried out on a subset of the samples that were available. Five tanks were randomly selected for HPLC analysis of photosynthetic pigment concentrations. Pigments were quantified in paired samples taken from the same five tanks on the initial day of the experiment, on day 7, day 13 and day 86. Since clear HPLC separation of chlorophylls and carotenoids was achieved in both laboratory experiments, and because no significant algal fouling was apparent on the blades of laboratory grown seagrass, then comparisons of pigment content could be made on the basis of absolute pigment concentration per gram fresh leaf weight for all laboratory experiments (see Section 4.4.2). Preliminary experimentation had demonstrated that the leaves of *Z. capricorni* did not differ significantly in percent water content (Section 4.4.1). Four tanks were randomly selected from those not assigned to HPLC analysis and the paired samples from day 1, day 7 and day 13 were dried to constant weight at 60 °C. Half of each dried sample was utilised in the analysis of carbon and nitrogen content and half in the determination of anthocyanin content.

The tanks were dismantled after completion of the final sampling event and all remaining live seagrass was carefully washed and sorted into above-ground (leaves and shoots) and below-ground (roots and rhizomes) material. This material was dried at 60 °C to constant weight and weighed for measurement of the above- and below-ground dry biomass.

High light stress

Zostera capricorni plants were grown in 24 paired replicate sample tanks distributed over 2 benches (as described in Section 3.2). The plants were stabilised in the laboratory for one day prior to commencement of the experiment in October 1999 (see note in Section 3.2.2) when the tank halves were randomly assigned to high light 'treatment' (uncovered; PAR 661 $\pm$ 190 µmol photon m^{-2} s^{-1}) or 'control' (shadecloth covered; PAR 238 $\pm$ 127 µmol photon m^{-2} s^{-1}).

Sampling of chlorophyll fluorescence, spectral reflectance and the collection of leaf material for subsequent pigment analysis was carried out on the initial day of experimental treatment and every three days after until day 17 according to the procedures described above for the low light experiment (Section 5.2.2). Measurement of the chlorophyll fluorescence parameters and tank water temperatures were continued for a further 2½ months at approximately 1-2 week intervals. Eight replicate pairs of leaf samples were randomly selected from the 24 available tanks for HPLC analysis of pigment content. Samples from the same set of tanks were analysed for days 1, 7, 13 and 17 of the high light experiment. A further eight pairs of frozen samples were randomly selected from those remaining after HPLC analysis and were utilised in the determination of anthocyanin content (by fresh weight) for days 1 and 13 of the experiment only.

Data analysis

The difference between control and treated samples in the laboratory experiments was analysed for each of the seagrass parameters that were measured on each sampling date, using two-tailed paired t-tests. Prior to analysis, each data set was visually assessed for normality, although paired t-tests are known to be quite robust to departures from normality (Zar 1984).

HPLC analysis separated the chlorophylls *a*, *b* and a range of light-harvesting and/or photoprotective carotenoids as well as some minor products (see Section 4.4.2). T-tests were performed on individual pigments and on a range of pigment ratios to examine whether there were differences in the pigment makeup of control and high or low light treated samples. The 23 tested ratios and individual pigments describe pigment changes that have been previously reported in the literature as being associated with high and low light stress in plants.

The spectral reflectance data was PMSC corrected prior to graphing and statistical analysis (as described in Section 3.3.1), then two-tailed paired t-tests were performed at each wavelength between 430–750 nm.

5.3 Results: Photosynthetic parameters of *Zostera capricorni*

5.3.1 Photosynthetic efficiency in the field

The photosynthetic efficiency (F_v:F_m) recorded for 20 leaf samples of field grown *Z. capricorni* during high irradiance conditions in the middle of a hot summer's day at Gray's Point, Port Hacking ranged from 0.578 to 0.764 (mean $\pm$ SD = 0.708 $\pm$ 0.044). The F_v:F_m range of 20 *Z. capricorni* leaf samples collected from clearer and shallower water on the same day at Costen's Point, Port Hacking was 0.493 to 0.720 (mean $\pm$ SD = 0.622 $\pm$ 0.048).

5.3.2 Diurnal cycle of photosynthetic efficiency

Throughout the course of a day, the photosynthetic efficiency of *Z. capricorni* growing near the boat ramp at Gray's Point, Port Hacking, was observed to vary inversely with the ambient level of solar irradiance (Figure 5.2). The highest F_v:F_m of 0.806 $\pm$ 0.009 was recorded in the early morning when water surface PAR (48 µmol photon m^{-2} s^{-1}) was below the level associated with the onset of light saturation in these seagrass plants. The photosynthetic efficiency of samples became more variable over time and began to fall (though not significantly), even before a noticeable increase in solar PAR had been experienced. The irradiance at 8.30 am

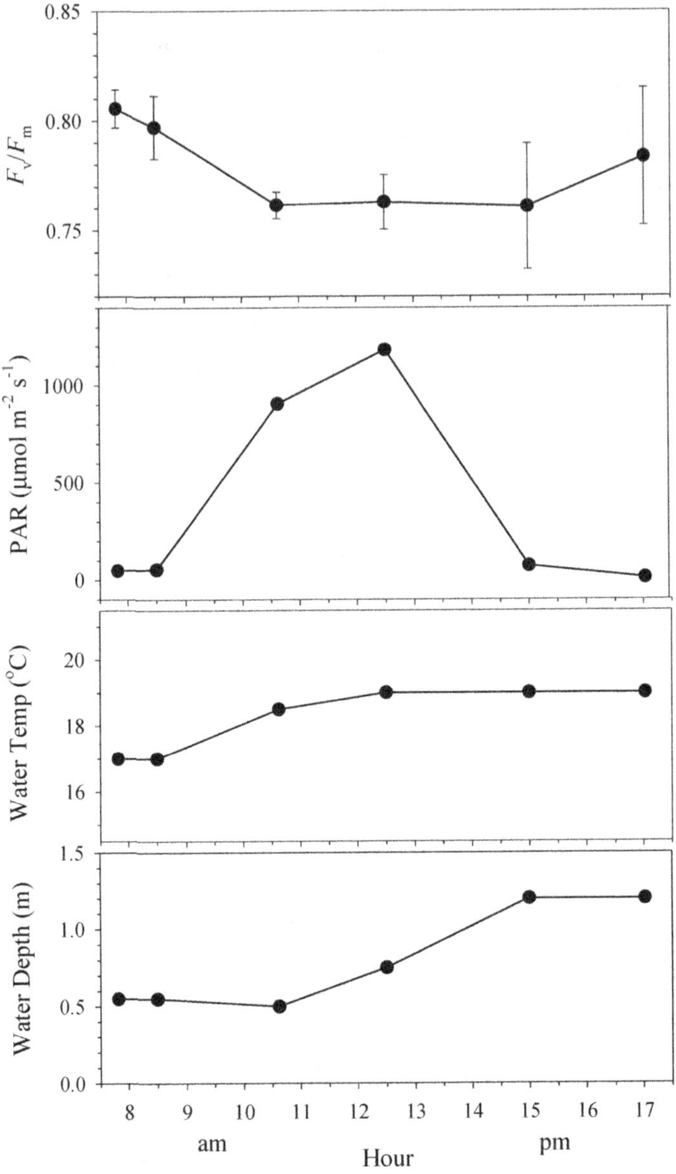

Figure 5.2. Example of a diurnal cycle of photosynthetic efficiency (F_v:F_m) for the seagrass *Zostera capricorni* in relation to daily changes in solar irradiance (PAR), and the tidal influences of water temperature and water depth. Plants were growing in brackish water at Gray's Point, Port Hacking.

was only 55 μmol photon m^{-2} s^{-1} yet the average F_v:F_m had fallen to 0.797 $\pm$ 0.014. Photoinhibition was obvious by 10.37 am when the lowest F_v:F_m readings with lowest variability (0.761 $\pm$ 0.006) were recorded from the leaves, concomitant with conditions of strong sunlight (PAR 905 μmol photon m^{-2} s^{-1}), low tide (10 cm depth of water remaining over seagrass canopy) and 1.5 °C warmer water temperatures. This was the only significant change observed in F_v:F_m between time periods (t = 4.207, df = 8, p = 0.003). Photosynthetic efficiency did not continue to fall as PAR increased to midday levels of 1181 μmol photon m^{-2} s^{-1}, possibly because the rising tide was synchronised with increasing sunlight, and 35 cm water covered the seagrass canopy by this time. The F_v:F_m of *Z. capricorni* leaves had become much more variable by 3 pm in the afternoon (PAR 73 μmol photon m^{-2} s^{-1}) but there was no noticeable recovery from photoinhibition until 5.02 pm (PAR 11 μmol photon m^{-2} s^{-1}), despite tidal submersion under 0.8 m water reducing PAR at the seagrass canopy to well below surface PAR. Due to high variability in afternoon measurements, the observed increase in F_v:F_m by 5.02 pm was not significantly higher than levels recorded during the midday depression.

5.3.3 Photosynthetic parameters of laboratory-grown *Z. capricorni*

Typical plant light response (P-I) curves were generated by regression of the scatterplot data of photosynthetic rate (ETR) vs. irradiance intensity (Figure 5.3). Increasing irradiance initially caused a linear rise in ETR in the light-limited parts of the P-I curves, after which each line curved over toward an asymptote at P_{max} due to the onset of saturation. Though regression of the data for each growing PAR category was highly significant, regression coefficients were not particularly high (notably r^2 = 0.360 for control samples) because of the high variability in the ETR of samples at high light levels. P decreased below P_{max} at irradiance levels

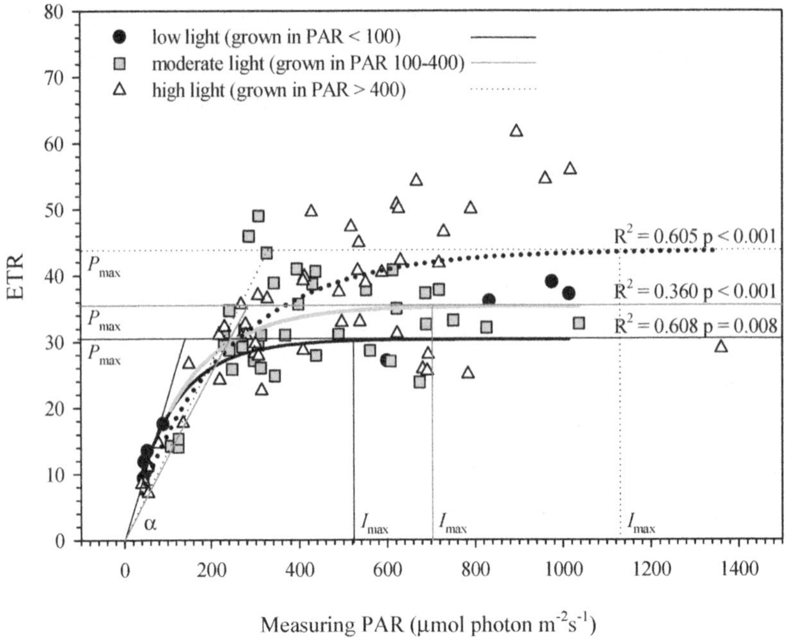

Figure 5.3. Photosynthesis-Irradiance curves based on electron transport rates (ETR) observed for the seagrass *Zostera capricorni* grown in the laboratory for 1 month under low, moderate or high levels of irradiance (PAR). P-I curves were fitted to the data for each category using an exponential non-linear regression model. Derived photosynthetic parameters are; P_{max} the maximum photosynthetic rate, α the maximum quantum yield and I_{max} the maximum photosynthetic irradiance (i.e. the minimum irradiance required to support P_{max}).

above 500 μmol photon $m^{-2}s^{-1}$ for some, but not all, of the moderate and high light samples. The downward turn in the P-I curve due to photoinhibition was most clearly observed in the in the moderate light group. Despite the variability in leaf sample response, *Z. capricorni* displayed clear photoacclimation to 1 month of growth under high, moderate or low PAR levels. Plants grown at high light levels achieved the highest maximum photosynthetic rates (ETR = 44), while low light grown plants displayed lower P_{max} (ETR = 30). The irradiance level required to support P_{max} was lower for shade grown plants (I_{max}= 520 µmol photon $m^{-2}s^{-1}$) than for moderate light (I_{max} = 700 µmol photon $m^{-2}s^{-1}$) and high light grown plants (I_{max} = 1130 µmol photon $m^{-2}s^{-1}$). Respiration rates were not measured in this experiment, so compensation points (I_c) could not be calculated.

The maximum quantum yield achieved at low light intensities, α, also varied between groups due to acclimation of the photosynthetic process. Photosynthetic efficiency was relatively lower for high light (α = 0.133) and moderate light categories (α = 0.125) than for shade grown *Z. capricorni* (α = 0.215). In this experiment P_{max} was based on ETR rather than gross photosynthetic O_2 evolution and in accordance with the theoretical stoichiometric relationship; 4 mol electrons produces 1 mol evolved O_2 (Beer *et al.* 1998). In addition, ETR was calculated from chlorophyll fluorescence in this experiment using an assumed green leaf absorbance of 0.83. In reality, the absorbance for dark green-bronze coloured *Z. capricorni* leaves could be as low as 0.44 reported for the very similar species, *Z. marina* (Beer *et al.* 1998). Beer *et al.* (1998) calculated that the molar ratio of ETR to the rate of O_2 evolution was 0.5 for *Z. marina.* Using this relationship to allow comparison with other studies, α can be estimated to be 0.107 for low light, 0.063 for moderate light and 0.066 for high light grown *Z. capricorni.*

5.4 Results: Manipulative laboratory experiments

5.4.1 Response of *Z. capricorni* to low light stress

There was no difference in the photosynthetic efficiency, or in the initial or maximal fluorescence, of *Z. capricorni* leaves grown at control or low irradiance levels over a three month period (Table 5.2; Figure 5.4A-C). The F_v:F_m for both control and shaded samples was generally maintained at between 0.750 – 0.800 throughout the experiment although the data were

Table 5.2. Paired t-tests to examine the affect of shading on the photosynthetic efficiency, initial and maximal chlorophyll fluorescence of *Zostera capricorni* leaves. Leaves were analysed on the initial day of the experiment and on the subsequent days as specified after continual exposure to low light stress.

day	F_v:F_m			F_o			F_m		
	df	t	p	df	t	p	df	t	p
initial	11	0.068	0.949	11	0.037	0.971	11	0.337	0.742
day 4	11	0.327	0.750	11	0.290	0.777	11	0.027	0.979
day 7	11	0.514	0.618	11	0.784	0.449	11	1.891	0.085
day 10	11	0.006	0.994	11	0.114	0.911	11	0.640	0.535
day 13	11	0.786	0.449	11	0.217	0.832	11	0.383	0.709
day 86	9	0.552	0.594	9	0.048	0.963	9	0.070	0.946

quite variable on days 7 and 13, containing outliers as low as 0.299. The average values were close to the maximum recorded in the field for this species, i.e. 0.801-0.821 recorded just after sunrise at Gray's Point in autumn (Section 5.3.2), and were higher than those recorded for photoinhibited *Z. capricorni* measured at Port Hacking on a hot summer afternoon (Section 5.3.1). The F_o and F_m values recorded for both treatments had both approximately halved after 86 days of the experiment but this had no apparent effect on photosynthetic efficiency. Water temperature in the tanks was maintained at a comfortable level for photosynthesis (Figure 5.4D) and although fluorescence parameters rose and fell slightly in synchrony with small changes in water temperature, this did not appear to influence photosynthetic efficiency to any noticeable extent.

Although photosynthetic efficiency was not affected by shading, there was a highly significant difference in the biomass of *Z. capricorni* from the control and shaded sides of the tanks after three months of growth in the laboratory (Table 5.3). Shaded samples displayed almost complete dieback

(dieback was complete in five of the 12 samples) while the dry biomass of control samples was around two orders of magnitude higher (Figure 5.5A).

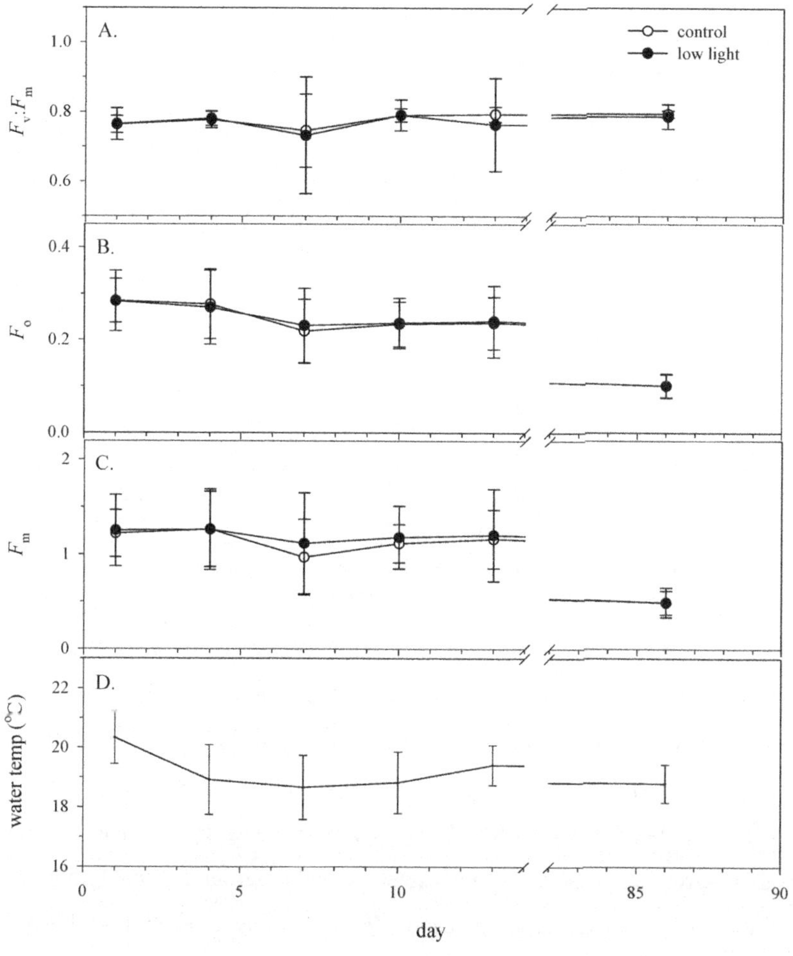

Figure 5.4. Fluorescence parameters of *Zostera capricorni* leaves affected by low light stress in a laboratory shading experiment; (A) photosynthetic efficiency (F_v:F_m), (B) initial fluorescence (F_o), (C) maximal fluorescence (F_m) and (D) tank water temperatures. Control and shade treated plants (shadecloth covered) were paired in tanks. (n = 12 for days 1 - 13, n = 10 for day 86). All data are means ± SD.

Table 5.3. The influence of shading on some growth parameters of *Zostera capricorni*. A paired t-test was not performed on the initial day of the experiment since only a representative subset of the leaf samples were analysed for initial leaf carbon and nitrogen content, however, paired samples were tested on subsequent days as specified after continual exposure to low light stress. Plant biomass (gDW per sample) was measured destructively upon completion of the experiment after 3 months.

Variable	day	**CONTROL**	**LOW LIGHT**	df	t	*p*
whole dry weight	86	10.32 $\pm$ 7.09	0.16 $\pm$ 0.19	11	5.051	<0.001***
gDW: above-ground	86	4.47 $\pm$ 4.84	0.04 $\pm$ 0.05	11	5.128	<0.001***
gDW: below-ground	86	6.79 $\pm$ 7.37	0.13 $\pm$ 0.14	11	4.898	<0.001***
% carbon	1	34.35 $\pm$ 0.90				
	7	34.65 $\pm$ 1.64	34.90 $\pm$ 0.40	3	0.257	0.814
	13	35.48 $\pm$ 0.98	33.85 $\pm$ 0.54	3	4.745	0.018*
% nitrogen	1	2.05 $\pm$ 0.33				
	7	2.25 $\pm$ 0.37	2.10 $\pm$ 0.41	3	1.567	0.215
	13	2.13 $\pm$ 0.21	2.30 $\pm$ 0.41	3	1.400	0.256
carbon:nitrogen	1	17.05 $\pm$ 2.47				
	7	15.61 $\pm$ 1.71	17.30 $\pm$ 4.29	3	1.530	0.224
	13	16.83 $\pm$ 1.96	15.04 $\pm$ 2.47	3	3.172	0.050*

Values significant at the $^*p < 0.05$, $^{**}p<0.01$, $^{***}p<0.001$

Eelgrass grown under control light conditions retained high shoot densities and displayed a very dense, healthy leaf canopy. The biomass of laboratory grown control samples (above-ground 30.95 $\pm$ 2.75 gDW m^{-2}, below-ground 47.08 $\pm$ 4.42 gDW m^{-2}) was lower than the average values that have been reported for *Z. capricorni* growing in the field (average above-ground 191.4 gDW m^{-2}, average below-ground 176.0 gDW m^{-2}; reviewed in Duarte and Chiscano 1990; standing crop at Port Hacking 50-70 gDW m^{-2}; Kirkman *et al.* 1982). However, the seagrass probably achieved close to the maximum possible biomass for the water volume available in the laboratory, since space was severely restricted by the 230 mm depth of the tanks.

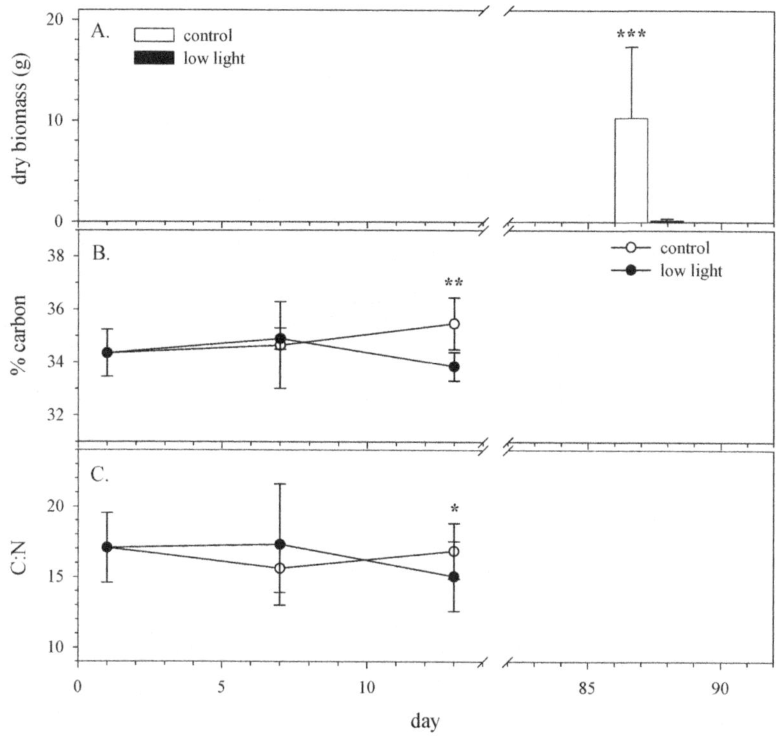

Figure 5.5. Some growth parameters of *Zostera capricorni* leaves affected by low light stress in a laboratory shading experiment; (A) total dry plant biomass (n = 12), (B) carbon content (n = 4) and (C) carbon:nitrogen ratio (n = 4). Control and shadecloth treated plants were paired in tanks. All data are means ± SD. $*p < 0.05$, $**p < 0.01$, $***p < 0.001$.

The nitrogen content of control and shaded leaves did not differ significantly in the first two weeks of the experiment, however a significant decrease in the carbon content of shaded samples was responsible for the significant decrease in carbon:nitrogen observed by day 13 (Table 5.3, Figure 5.5B-C). The concentrations of antheraxanthin and zeaxanthin had significantly decreased in shaded samples by day 13 of the experiment (Figure 5.6 A-B; Table 5.4). In addition, the ratio of chlorophyll *a*:*b* increased in all samples across the first two weeks of the experiment and

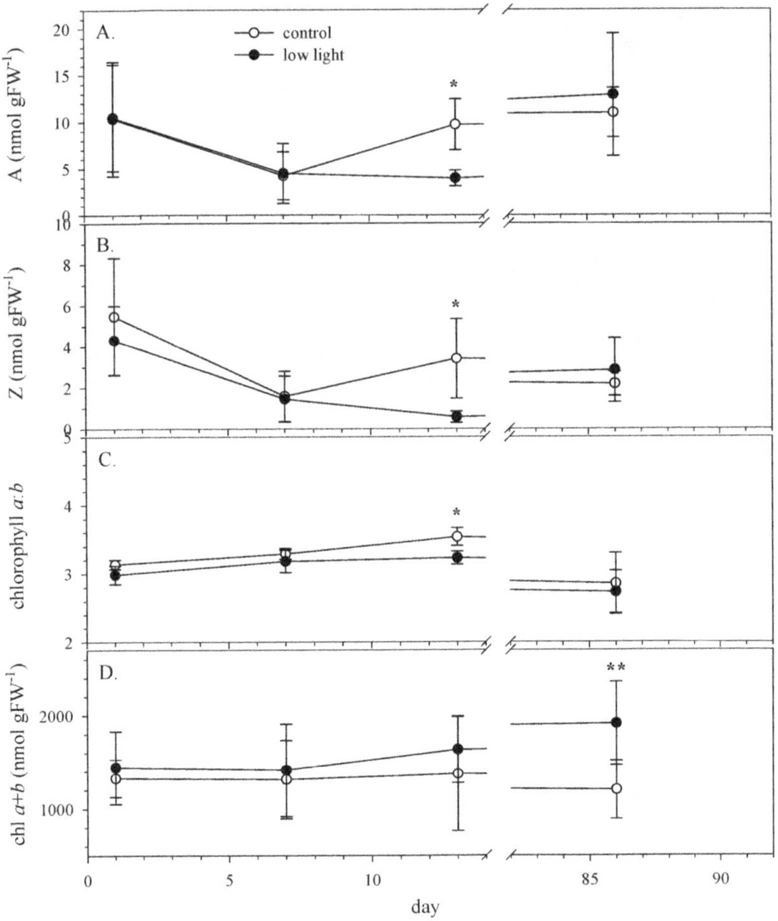

Figure 5.6. Pigment concentrations of *Zostera capricorni* leaves affected by low light stress in a laboratory shading experiment; (A) antheraxanthin, (B) zeaxanthin, (C) chlorophyll a:b, and (D) total chlorophyll concentration. Control and shadecloth treated plants were paired in tanks. All data are means ± SD (n = 5). $^*p < 0.05$, $^{**}p < 0.01$.

became significantly higher in control samples by day 13 (Figure 5.6C). There were no other significant differences in the concentrations of photosynthetic and photoprotective pigments measured in control and

Table 5.4. Paired t-tests to examine the short-term influence of low light stress on the photosynthetic and photoprotective pigment concentrations observed in *Zostera capricorni* leaves. See Appendix 1.1 for data values.

	INITIAL			DAY 7			DAY 13			DAY 86		
Variable	df	t	*p*	df	t	*p*	df	t	*p*	df	t	*p*
Tchls	4	0.654	0.549	4	0.827	0.455	4	0.686	0.531	4	5.525	0.005**
chl *a*:*b*	4	1.939	0.125	4	1.382	0.239	4	4.323	0.012*	4	1.606	0.184
lutein	4	1.413	0.230	4	1.853	0.137	4	0.529	0.625	4	4.384	0.012**
lutein:Tcars	4	0.924	0.408	4	1.048	0.354	4	1.474	0.215	4	0.433	0.687
neoxanthin	4	1.308	0.261	4	0.523	0.629	4	0.322	0.764	4	5.708	0.005**
neoxanthin:Tcars	4	0.160	0.881	4	0.308	0.774	4	0.290	0.786	4	0.880	0.429
Tcars	4	1.288	0.267	4	0.751	0.494	4	0.331	0.757	4	5.724	0.005**
Tcars:Tchls	4	0.495	0.647	4	0.999	0.374	4	2.014	0.114	4	1.305	0.262
T β-cars	4	0.707	0.519	4	0.532	0.623	4	0.162	0.879	4	6.718	0.003**
T β-cars:Tcars	4	1.497	0.209	4	0.524	0.628	4	1.736	0.158	4	1.232	0.285
α-carotene	4	1.357	0.245	4	0.443	0.681	4	1.170	0.307	4	0.021	0.985
α-car:β-cars	4	1.269	0.273	4	0.260	0.808	4	1.000	0.374	4	1.496	0.209
VAZ pool	4	0.949	0.396	4	0.140	0.896	4	0.166	0.877	4	2.462	0.070
VAZ:Tchls	4	0.120	0.910	4	2.425	0.072	4	1.343	0.251	4	4.476	0.011*
VAZ:Tcars	4	0.576	0.595	4	1.760	0.153	4	0.480	0.657	4	4.359	0.012*
V:Tcars	4	0.170	0.874	4	1.315	0.259	4	1.497	0.209	4	3.810	0.019*
A+Z:VAZ	4	0.834	0.451	4	0.374	0.728	4	2.260	0.087	4	0.624	0.567
Z:VAZ	4	1.191	0.300	4	0.010	0.993	4	1.762	0.153	4	0.162	0.879
EPS (V+0.5A)/VAZ	4	0.950	0.396	4	0.264	0.805	4	2.128	0.100	4	0.487	0.652
violaxanthin	4	1.014	0.368	4	0.121	0.909	4	0.547	0.613	4	3.075	0.037*
antheraxanthin	4	0.097	0.928	4	0.571	0.599	4	4.475	0.011*	4	0.671	0.539
zeaxanthin	4	0.927	0.406	4	0.146	0.891	4	3.178	0.034*	4	0.819	0.459
anthocyanins				3	0.553	0.619	3	0.006	0.996			

T = total, chl = chlorophyll, car = carotenoid, V = violaxanthin, A = antheraxanthin, Z = zeaxanthin, EPS = epoxidation state. Values significant at the $*p < 0.05$, $**p < 0.01$.

shaded *Z. capricorni* leaves over the first two weeks of the experiment but there was a trend toward increasing chlorophyll concentration in the low light stressed samples (Figure 5.6D) and a general decrease in the anthocyanin concentration and proportions of A+Z:VAZ and Z:VAZ across both treatments.

By day 86 of the experiment, however, several other significant pigment differences between paired control and shaded samples had become apparent; i.e. total chlorophyll, total carotenoid, lutein, total β-carotenoid, violaxanthin and neoxanthin concentrations had all increased significantly in the low light stressed samples (Table 5.4; Appendix 1.1). In contrast, VAZ:total chlorophyll, VAZ:total carotenoids and violaxanthin:total carotenoids were significantly lower for shaded samples than for control samples on day 86.

Shade treatment did not affect the spectral reflectance of *Z. capricorni* leaves sampled on the initial day of the experiment (Figure 5.7A-B). Significant differences observed at three single wavelengths can be explained as errors that occur as a consequence of performing a large number of consecutive t-tests (Zar 1984). Spectral reflectance differences were apparent between control and shaded samples by day 7 of the experiment (Figure 5.7C-D). Low light stressed *Z. capricorni* leaves reflected light more strongly at the green peak at 526 nm, 544 nm and between 553-570 nm. The reflectance of shaded leaf samples decreased (i.e. absorption increased) relative to control samples at 590 nm, 604 nm and particularly between 618-648 nm where the strongest differences occurred. These differences in spectral reflectance intensified by day 13 of the experiment so that shaded leaf samples displayed significantly higher leaf reflectance across the entire green peak (from 532-572 nm) and lower reflectance across a wider range of the orange shoulder (between

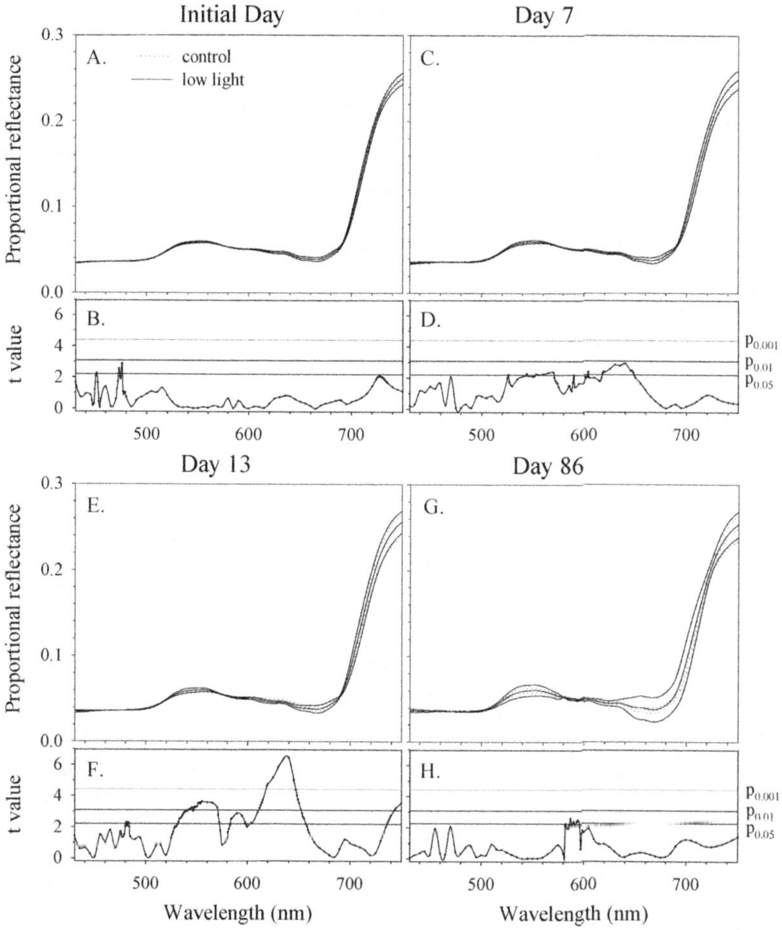

Figure 5.7. Mean ± SD spectral signatures of healthy and stressed *Zostera capricorni* leaves exposed to low light in a laboratory shading experiment and the wavelengths where significant reflectance differences between control and light stressed plants occurred in paired t-tests; (A-B) initial day, (C-D) day 7, (E-F) day 13 and (G-H) day 86 after shade treatment (n = 12, except day 86 n = 10). Critical p values are denoted by horizontal reference lines.

584-598 nm and 604-657 nm) (Figure 5.7E-F). In addition, the NIR reflectance of shaded leaves was significantly higher than that of control leaves between 736 nm and the limit of analysis in the NIR at 750 nm. The spectral reflectance of shaded *Z. capricorni* leaves was exceptionally variable by day 86 of the experiment and consequently, significant spectral reflectance differences between control and low light stressed samples only occurred over a short wavelength range (Figure 5.7G-H). Shaded samples displayed significantly lower reflectance than control samples at wavelengths 587, 589-590 and 592-595 nm.

5.4.2 Response of *Z. capricorni* to high light stress

The photosynthetic efficiency of *Z. capricorni* leaves stressed by exposure to high light decreased significantly below that of the control leaves from day 7 through to day 73 of the experiment (Table 5.5; Figure 5.8A).

Table 5.5. Results of paired t-tests used to examine the affect of high light exposure on the initial and maximal chlorophyll fluorescence, and on the photosynthetic efficiency of *Zostera capricorni* leaves. Leaves were analysed on the initial day of the experiment and on the subsequent days as specified after continual exposure to high light stress.

Day	F_o			F_m			F_v:F_m		
	df	t	*p*	df	t	*p*	df	t	*p*
initial	20	0.491	0.629	20	0.302	0.766	20	0.301	0.766
day 4	20	0.277	0.785	20	0.490	0.629	20	1.079	0.293
day 7	20	1.575	0.131	20	0.281	0.781	20	3.119	0.005**
day 10	23	0.078	0.939	23	1.882	0.073	23	5.206	<0.001***
day 13	23	0.085	0.933	23	1.670	0.108	23	4.244	<0.001***
day 17	23	2.492	0.020*	23	0.466	0.646	23	4.847	<0.001***
day 28	23	2.692	0.013*	23	0.161	0.874	23	4.048	<0.001***
day 41	23	0.550	0.588	23	1.174	0.253	23	2.843	0.009**
day 48	23	1.198	0.243	23	2.739	0.012*	23	6.675	<0.001***
day 73	23	0.447	0.659	23	3.921	<0.001***	23	5.383	<0.001***
day 87	23	1.156	0.260	23	1.511	0.144	23	0.470	0.643

Values significant at the *$p < 0.05$, ** $p < 0.01$, *** $p < 0.001$.

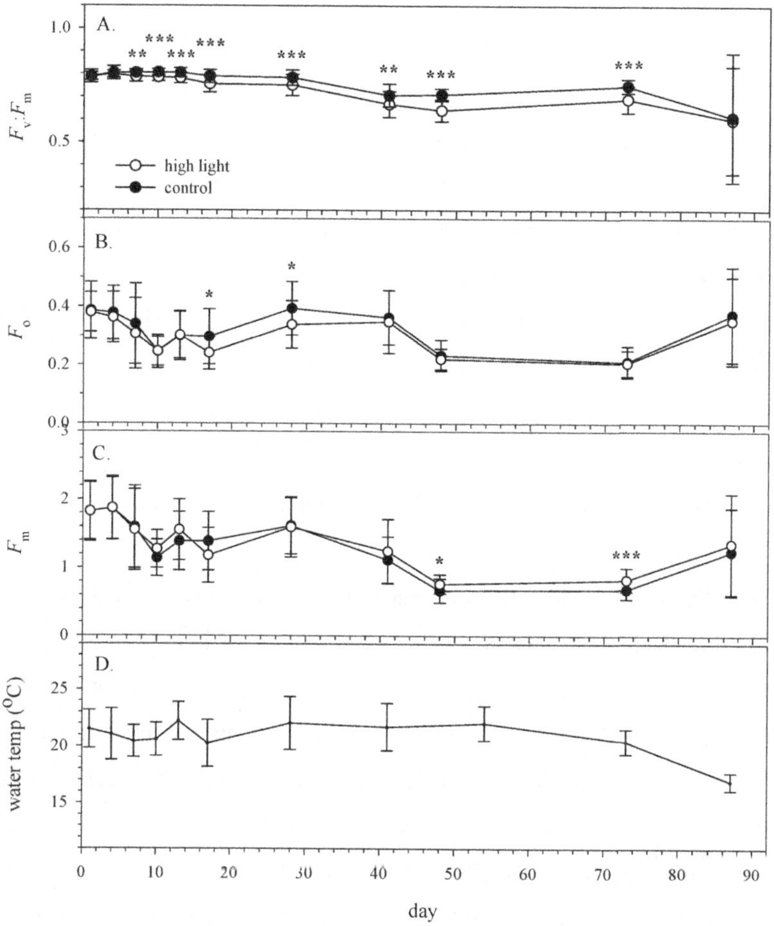

Figure 5.8. Fluorescence parameters of *Zostera capricorni* leaves affected by high light stress in a laboratory shading experiment; (A) photosynthetic efficiency (Fv:Fm), (B) initial fluorescence (Fo), (C) maximal fluorescence (Fm) and (D) tank water temperatures. Control (shadecloth covered) and high light treated plants were paired in tanks. (n = 21 for days 1-17, n = 24 for days 28-87). All data are means ± SD. $^*p < 0.05$, $^{**}p < 0.01$, $^{***}p < 0.001$.

The trend did not continue to the end of the experiment, since no significant difference was observed in F_v:F_m by day 87. The fluorescence parameters recorded for both treatments on day 87 were extremely variable and this could be associated with the drop in water temperature from an average of 20 °C or more to around 17 °C (Figure 5.8D) observed at this time. However, tank water temperatures were generally maintained at a comfortable level for this species and were unlikely to have imparted an additional stress on photosynthesis. The mean F_v:F_m values recorded for high light stressed *Z. capricorni* in this experiment (0.787 $\pm$ 0.023 on day 7 decreasing to 0.640 $\pm$ 0.048 on day 48) were higher than, or within the range of fluorescence values recorded for this species growing successfully in the field under midday summer conditions (see Section 5.3.1). The mean F_v:F_m value of the control samples ranged between 0.806 $\pm$ 0.014 and 0.706 $\pm$ 0.050, and were similar to the values recorded for healthy, non-photoinhibited *Z. capricorni* leaves in the field (see Section 5.3.2).

The values recorded for the minimal and maximal fluorescence appeared to be quite sensitive to tank water temperature, at least in the first two weeks of treatment (Figure 5.8B-C). The significant differences observed in the F_v:F_m of control and high light stressed samples can be related to changes in both F_o and F_m over the course of the experiment. High light stressed samples displayed significantly higher minimal fluorescence than control samples on days 17 and 28 when no difference was observed in their maximal fluorescence (Table 5.5). Conversely, on days 48 and 73 the high light stressed samples did not differ in their minimal fluorescence but displayed significantly lower maximal fluorescence.

The ability of *Z. capricorni* to grow new shoots and leaves did not appear to be diminished by the level of photosynthetic stress imparted by high light treatment in this experiment. Although biomass was not measured on completion of this experiment, no apparent difference could be observed in the shoot density or leaf biomass of the paired sides of each sample tank.

There were no significant differences in the concentrations of photosynthetic and photoprotective pigments measured in control and high light treated *Z. capricorni* leaves until day 13 of the experiment, when a significantly higher proportion of VAZ:total carotenoids was recorded in the light stressed samples (Table 5.6; Figure 5.9C). By day 17 the high light stressed leaves contained not only a higher ratio of VAZ:total carotenoids than control samples but also higher VAZ:total chlorophylls, total carotenoids:total chlorophylls and more zeaxanthin as a proportion of the VAZ pool (Figure 5.9A-D). The concentrations of chlorophylls (Figure 5.9E), β-carotenoids, neoxanthin and anthocyanins generally increased for both control and high light treated samples over the first three weeks of the experiment (Appendix 1.2). A drop in the ratio of VAZ:total chlorophyll observed for both treatments could be related to this trend since the VAZ pool remained at a relatively stable level over the first 17 days.

The spectral reflectance of samples assigned to control and high light treatments did not differ significantly on the initial day of the experiment, apart from at the wavelength 466 nm (Figure 5.10A-B). Once again, this single wavelength result can be disregarded as a probability error related to the large number of statistical analyses performed. High light stressed leaves displayed significantly stronger absorbance in wavelengths 481-482 nm and significantly higher reflectance of NIR light between 716-729 nm than did control leaves by day 7 of the experiment (Figure 5.10C-D). By day 13, significant differences in the absorption of blue wavelengths and in the reflectance of NIR had intensified (Figure 5.10E-F). The range of wavelengths at which the light stressed plants differed significantly from control plants widened by day 17 to include 437-438 nm, 442-443 nm, 446-466 nm and 471-472 nm in the blue region of carotenoid absorption, and far-red to NIR wavelengths from 692-733 nm indicative of

Table 5.6. Paired t-tests to examine the short-term influence of high light stress on the photosynthetic and photoprotective pigment concentrations observed in *Zostera capricorni* leaves. See Appendix 1.2 for data values.

	INITIAL			DAY 7			DAY 13			DAY 17		
Variable	df	t	*p*	df	t	*p*	df	t	*p*	df	t	*p*
Tchls	7	0.434	0.677	6	1.725	0.135	6	0.652	0.538	7	1.105	0.306
chl *a*:*b*	7	0.237	0.820	6	0.921	0.393	6	0.345	0.742	7	1.666	0.140
lutein	7	0.977	0.361	6	1.377	0.218	6	0.344	0.742	7	0.910	0.393
lutein:Tcars	7	0.858	0.419	6	1.574	0.167	6	1.006	0.353	7	1.516	0.173
neoxanthin	7	0.493	0.637	6	1.128	0.302	6	0.182	0.861	7	0.682	0.517
neoxanthin:Tcars	7	1.076	0.318	6	1.257	0.255	6	0.157	0.880	7	0.638	0.544
Tcars	7	0.956	0.371	6	0.795	0.457	6	0.144	0.890	7	0.589	0.575
Tcars:Tchls	7	0.925	0.386	6	1.244	0.260	6	1.527	0.178	7	2.390	0.048*
T β-cars	7	0.385	0.712	6	0.401	0.702	6	0.510	0.629	7	1.026	0.339
T β-cars:Tcars	7	0.559	0.593	6	1.023	0.346	6	1.371	0.220	7	1.340	0.222
α-carotene	7	0.456	0.663	6	0.922	0.392	6	0.526	0.618	7	1.527	0.171
α-cars:β-cars	7	0.723	0.493	6	0.960	0.374	6	0.573	0.587	7	2.278	0.057
VAZ pool	7	0.825	0.437	6	0.469	0.655	6	0.382	0.716	7	0.157	0.879
VAZ:Tchls	7	0.551	0.599	6	1.220	0.268	6	2.239	0.066	7	3.318	0.013*
VAZ:Tcars	7	0.606	0.564	6	0.290	0.782	6	2.832	0.030*	7	2.910	0.023*
V:Tcars	7	0.324	0.755	6	0.651	0.539	6	0.362	0.730	7	0.001	0.999
A+Z:VAZ	7	0.189	0.856	6	1.348	0.226	6	1.920	0.103	7	1.970	0.090
Z:VAZ	7	0.535	0.610	6	1.621	0.156	6	1.988	0.094	7	2.600	0.036*
EPS (V+0.5A)/VAZ	7	0.470	0.655	6	1.443	0.199	6	1.954	0.099	7	2.124	0.071
violaxanthin	7	0.934	0.381	6	0.758	0.477	6	0.304	0.772	7	0.330	0.751
antheraxanthin	7	0.347	0.739	6	0.361	0.731	6	2.038	0.088	7	2.067	0.078
zeaxanthin	7	0.511	0.625	6	0.650	0.150	6	2.287	0.062	7	2.866	0.024*
anthocyanins	7	0.208	0.841	7	0.257	0.805						

T = total, chl = chlorophyll, car = carotenoid, V = violaxanthin, A = antheraxanthin, Z = zeaxanthin, EPS = epoxidation state.
Values significant at the *$p < 0.05$.

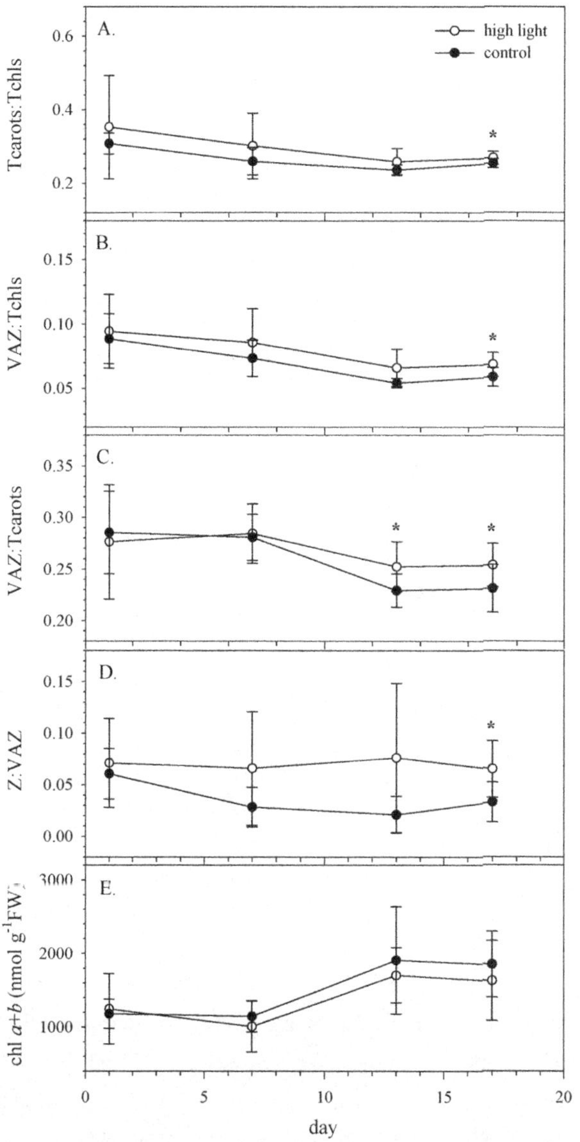

Figure 5.9. Pigment concentration of *Zostera capricorni* leaves affected by high light stress in a laboratory shading experiment; (A) total carotenoids:total chlorophylls, (B) violaxanthin + antheraxanthin + zeaxanthin (VAZ):total chlorophylls, (C) VAZ:total carotenoids, (D) zeaxanthin:VAZ and (E) total chlorophyll concentration. Control (shadecloth covered) and high light treated plants were paired in tanks (n = 8 for days 1 and 17, n = 7 for days 7 and 13). All data are means $\pm$ SD. $^*p < 0.05$.

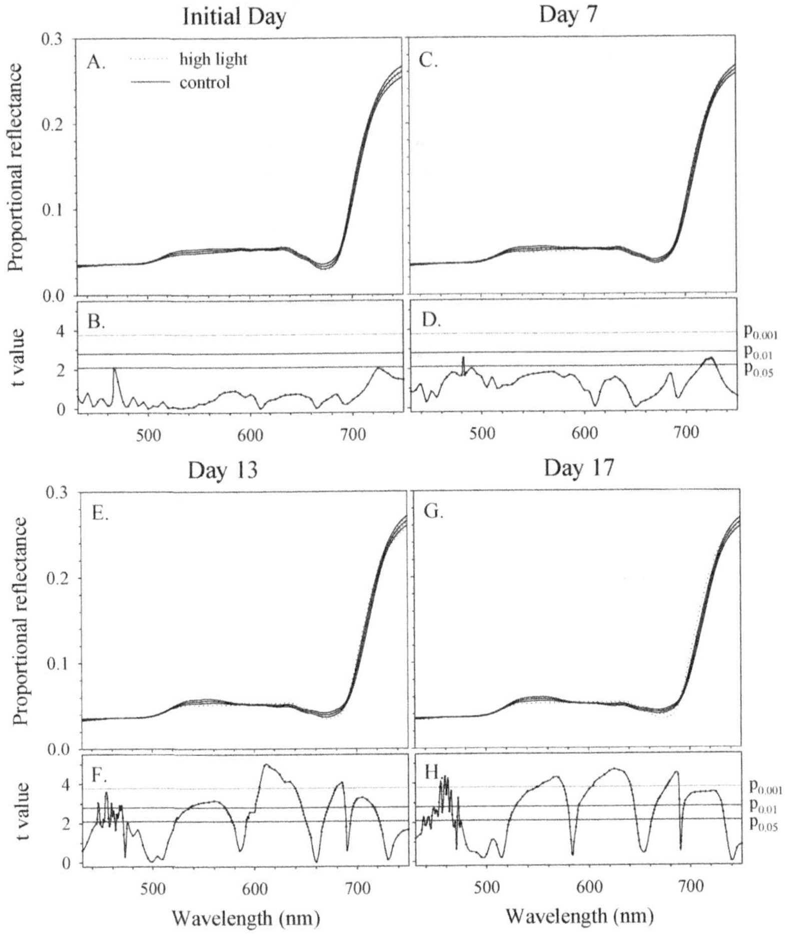

Figure 5.10. Mean $\pm$ SD spectral signatures of healthy and stressed *Zostera capricorni* leaves exposed to high light in a laboratory shading experiment and the wavelengths where significant reflectance differences between control and light stressed plants occurred in paired t-tests; (A-B) initial day, (C-D) day 7, (E-F) day 13 and (G-H) day 17 after high light treatment (n = 24). Critical p values are denoted by horizontal reference lines.

a shift in the red edge toward shorter wavelengths for light stressed samples (Figure 5.10G-H).

Significant differences in the spectral reflectance of control and treated samples were also apparent by day 13 of the experiment at the green peak, across the orange shoulder and in the red chlorophyll absorption trough (Figure 5.10E-F) and these differences likewise increased in range by day 17 (Figure 5.10G-H). High light stressed leaves reflected less strongly than control leaves in the green reflectance peak between 523-580 nm and displayed a corresponding shift in the green edge towards longer wavelengths. Spectral reflectance in the region between 588-646 nm was significantly higher for light stressed *Z. capricorni* than for control plants, while significantly stronger absorption of wavelengths 662-689 nm at the red chlorophyll absorption trough probably represents a narrowing of this trough for the stressed samples relative to that of control samples (Figure 5.10G-H).

5.5 Discussion

5.5.1 Summary of results

Laboratory experiments

Zostera capricorni grown at low light levels in laboratory experiments displayed the ability to photoacclimate so as to maintain high photosynthetic efficiency even at irradiance levels well below the compensation point. However, the only significant short term chromatic responses observed for shade grown plants were reduced zeaxanthin and antheraxanthin concentrations in the leaves and lower chlorophyll *a:b* relative to controls (Table 5.7). Over the longer term, photosynthetic pigment concentrations increased in light deprived *Z. capricorni* to maximise light harvesting.

The seagrass was also observed to photoacclimate to high light levels through an increase in the maximum photosynthetic rate of the leaves.

Table 5.7. Summary of the significant short term (within 2-3 weeks of treatment) physiological, biochemical and spectral reflectance responses recorded in the leaves of *Zostera capricorni* grown under conditions of low light and high light stress. (↑) increase, (↓) decrease, R reflectance, F_v:F_m photosynthetic efficiency, Tcarot total carotenoids, Tchl total chlorophylls, VAZ = V violaxanthin + A antheraxanthin + Z zeaxanthin.

RESPONSE	LOW LIGHT STRESS		HIGH LIGHT STRESS	
HEALTH & GROWTH	F_v:F_m	(no change)	F_v:F_m	(↓)
	biomass	(↓)	growth	(no change)
PIGMENT CONTENT	chl *a*:*b*	(↓)		
			Tcarot:Tchl	(↑)
	Z and A	(↓)	Z	(↑)
			VAZ:Tchl	(↑)
			VAZ:Tcarot	(↑)
			Z:VAZ	(↑)
SPECTRAL REFLECTANCE RESPONSE	green R	(↑)	green R	(↓)
			GE	(red shift)
	orange R	(↓)	orange R	(↑)
			red R	(↓)
			RE	(blue shift)
	NIR R	(↑)		

Excess levels of irradiance caused a decrease in photosynthetic efficiency of *Z. capricorni*, an increase in the concentration of xanthophyll cycle carotenoids relative to other leaf pigments and an increase in the proportion of the VAZ pool occurring in the de-epoxidised state (Table 5.7). Such changes are consistent with light-dependent down-regulation of the photosynthetic process under conditions of high light stress.

Significant spectral reflectance changes induced by both low light and high light stress were apparent within one week of treatment and had increased in significance after two weeks. The spectral shifts associated with low light stress were characteristically different from those brought about by high light stress (Table 5.7). Therefore, although the magnitude of spectral reflectance change in each case was small, the study has shown that there is some potential for the detection of physiological stress in seagrasses by remote sensing.

The less than ideal conditions that laboratory experiments were performed under implies that the above conclusions may be conservative in respect to the magnitude of chromatic, and therefore, spectral reflectance changes that would be recorded if experiments were carried out in a more controlled laboratory environment. In particular, the irradiance levels output by light sources used in experiments could not be standardised between tanks, or between the high and low light experiments, so the illumination of controls was highly variable. This probably led to greater variability in the photosynthetic, biochemical and reflectance parameters measured from the seagrass samples, so the possibility of achieving significant results would be much lower. Despite the variability, the use of paired samples did allow for the monitoring of relative changes in seagrass physiological condition, pigment content and spectral reflectance. Since seagrasses acclimate to widely varying environmental conditions in nature, experiments designed to measure relative change in response to stress are possibly more realistic because the same principles and conclusions will then be applicable to the remote sensing monitoring of stress under natural field conditions.

Patterns in combined laboratory and field data

There was a trend for the total concentration of major light harvesting pigments in *Z. capricorni* to increase, and for the total concentration of photoprotective pigments to decrease, in response to decreasing average irradiance levels over the range of light conditions experienced in the field and laboratory experiments carried out in this study (Figure 5.11). These

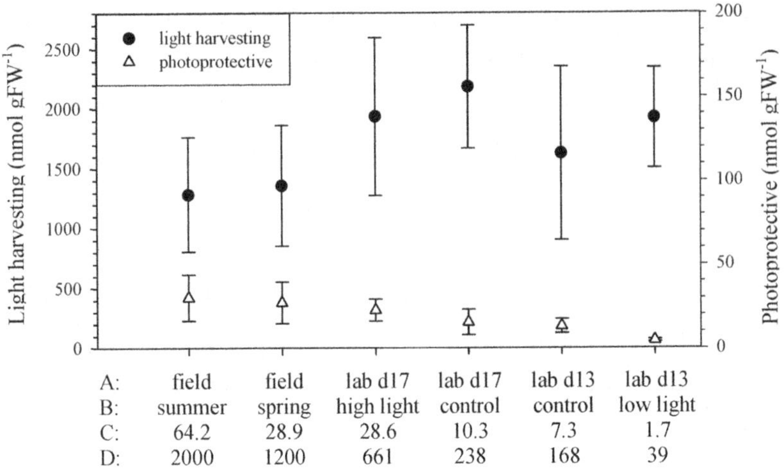

Figure 5.11. Comparison of the total concentrations of major light harvesting pigments (chlorophyll *a* + chlorophyll *b* + lutein + neoxanthin + violaxanthin) and photoprotective pigments (antheraxanthin + zeaxanthin) measured in *Zostera capricorni* leaves grown under a range of light environments in its natural habitat and in laboratory experiments. Data are means ± SD. A: collected from field or grown in lab for 13 or 17 days, B: treatment, C: total daily integrated PAR (mol photon $m^{-2}d^{-1}$), D: maximum PAR at midday (μmol photon $m^{-2}s^{-1}$). Note the difference in scale of the Y-axes.

patterns of chromatic acclimation were distinct given that allowances must be made for other differences in the growing environment of seagrass in the field and in different experiments. In the case of the photoprotective pigments, the total concentration in *Z. capricorni* leaves decreased linearly and significantly with a decreasing daily dose of PAR ($r^2 = 0.786$, $p = 0.019$).

Significant linear relationships were also observed between average daily irradiance and many of the photosynthetic pigment ratios that had been significantly affected by high light exposure or shading in laboratory experiments (Figure 5.12). Pigment ratios increased linearly with increasing daily integrated PAR for chlorophyll *a:b* (r^2 = 0.803, $p = 0.016$), VAZ:total carotenoids (r^2 = 0.968, $p < 0.001$), VAZ:total chlorophyll (r^2 = 0.871, p = 0.007) and A+Z:VAZ (r^2 = 0.674, $p < 0.045$), though the relationship was only significant for Z:VAZ at the 10% significance level (r^2 = 0.544, p = 0.094). The total carotenoid:total chlorophyll content of *Z. capricorni* leaves tended to increase with increases in total daily PAR (r^2 = 0.452) while the percent total chlorophyll content generally decreased (r^2 = 0.448), but neither of these relationships was significant.

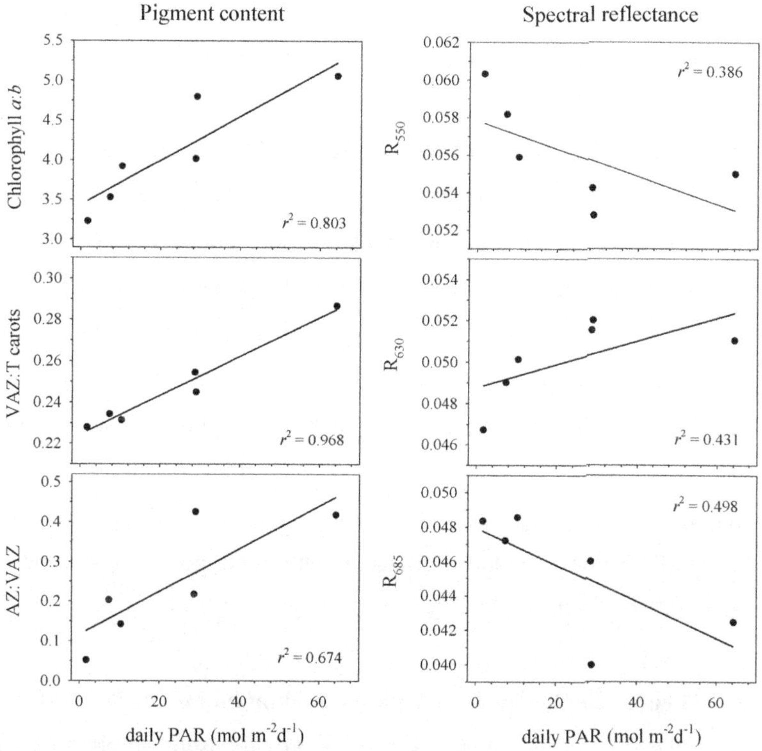

Figure 5.12. Influence of daily integrated PAR on the mean ratio of selected pigments and spectral reflectance at wavelengths 550, 630 and 685 nm for *Zostera capricorni* leaves grown under different irradiance levels in the laboratory and in the field.

The wavelength regions of the electromagnetic spectrum that were significantly affected by high light and low light treatments in laboratory experiments did not demonstrate significant linear relationships with daily growth PAR when all laboratory and field data points were combined, although trends were evident in the data (Figure 5.12). When the data for *Z. capricorni* measured in the field during summer were removed from the regression, however, the relationships for spectral reflectance at 550 nm and 630 nm became strongly significant. With no summer field data included, green peak reflectance at 550 nm decreased significantly with increasing average daily irradiance ($r^2 = 0.881$, $p = 0.018$) while orange shoulder reflectance at 630 nm increased with increasing daily PAR ($r^2 = 0.865$, $p = 0.022$). It appears that these results may better fit a nonlinear relationship and that reflectance at 550 and 630 nm reach an asymptote under high levels of daily PAR. While pigment concentrations change in a linear fashion with increasing PAR, the spectral reflectance response at wavelengths associated with the absorption of light by these pigments becomes saturated causing the relationship between reflectance and PAR to curve off and become nonlinear (Figure 5.12). Onset of saturation of the spectral signal can occur at quite low levels of pigment concentration in the centre wavelength of an absorption feature (Curran 1989). Removing the spring data 'outlier' from the regression of reflectance in the red chlorophyll absorption trough at 685 nm and daily integrated PAR improved the relationship but did not make it a significant one ($r^2 = 0.759$, $p = 0.054$).

5.5.2 P-I curves and evidence for photoacclimation in *Z. capricorni*

Photosynthesis-irradiance relationships can provide useful information on the light levels required to maintain healthy growth in a species. The P-I response of *Z. capricorni* was typical of plant species with the ability to photoacclimate in order to make the most efficient use of the ambient light climate. The photosynthetic parameters recorded in this laboratory-based study varied considerably with growing irradiance but are consistent with the results previously reported for field grown *Z. capricorni* (Table 5.8). For example, Pollard and Greenway (1993) found that *Z. capricorni* samples collected from turbid sites were 4-10 times more efficient at utilizing light in photosynthesis than those collected from higher light environments in the same estuary. The large difference in the α observed for laboratory grown samples in the current study from that observed in the previous studies of Pollard and Greenway (1993) and Flanigan and Critchley (1996), appears to indicate chronic photoinhibition in field grown *Z. capricorni*, particularly in shallow, tropical waters.

When *Z. capricorni* grows in a low light environment, light-harvesting will be maximised but the rate of carbon assimilation will be limited by reductions in enzyme capacity (reviewed in Björkman 1981; Robinson 1999). Sun grown leaves concentrate resources into transport chain components and Calvin cycle enzymes and therefore can maintain higher P_{max} (Robinson 1999). For a seagrass of temperate affinity, *Z. capricorni* can attain high rates of productivity (Duarte and Chiscano 1999). However, high photosynthetic capacity will be accompanied by increased respiration rates and therefore higher compensation points, so the growth of this seagrass can be dramatically affected by sudden light reductions. Touchette and Burkholder (2000b) noted that *Z. capricorni* reached compensation and saturating irradiance at much higher light levels than most other temperate seagrasses. For example, the I_c values recorded for *P. australis* (17-20 $\mu E\ m^{-2}\ s^{-1}$ over a range of temperatures; Masini and Manning 1997) are half those measured for low light acclimated *Z. capricorni* but the former species typically inhabits deeper water. In addition, the I_c of *Z. capricorni* from tropical regions will generally be higher than from temperate regions (Table 5.8), not only because ambient

Table 5.8. Photosynthetic-Irradiance parameters reported for *Zostera capricorni* including I_c (compensating irradiance), α (maximum quantum efficiency), P_{max} (maximum photosynthetic rate), I_{max} (minimum light for maximum photosynthesis) and the growing/measurement conditions reported for each study. Since respiration rates were not measured in the current study, I_c was estimated from the level of irradiance that caused significant or complete dieback of the seagrass after 3 months.

		current study	**Flanigan& Critchley (1996)**	**Pollard & Greenway (1993)**
I_c		50+ μmol photon $m^{-2}s^{-1}$	45 μmol photon $m^{-2}s^{-1}$	98 ± 6 μE $m^{-2}s^{-1}$
α	**low**	0.107	0.018	0.004
	mod	0.063		
	high	0.066		
P_{max}	**low**	30 μmol ē $m^{-2}s^{-1}$	4.2 μmol O_2 $m^{-2}s^{-1}$	0.21 ± 0.01 mL lacunal gas h^{-1} $shoot^{-1}$
	mod	36 μmol ē $m^{-2}s^{-1}$		
	high	44 μmol ē $m^{-2}s^{-1}$		
I_{max}	**low**	520 μmol photon $m^{-2}s^{-1}$	450 μmol photon $m^{-2}s^{-1}$	~ 900 μE $m^{-2}s^{-1}$
	mod	700 μmol photon $m^{-2}s^{-1}$		
	high	1130 μmol photon $m^{-2}s^{-1}$		
Conditions		measurements of chlorophyll *a* fluorescence from leaf segments; plants grown in temperate estuary but stabilised in laboratory for 1 month	measurements of gas release from leaf segments in artificial seawater using fresh leaves collected from field	measurements of gas release from whole plants growing *in situ* in shallow tropical bay

N.B. plants in the current study were grown in the laboratory for 1 month under low, moderate and high light regimes. The value of α varies according to the units used to measure photosynthesis, hence, an estimate of α in μmol O_2 m^{-2} s^{-1}/μE m^{-2} s^{-1} is provided for the current study to allow for comparison with Flanigan and Critchley (1996). Irradiance units μE $m^{-2}s^{-1}$= μmol photon $m^{-2}s^{-1}$.

light levels are lower in temperate regions (Touchette and Burkholder 2000b) but because water temperature is a major determinant of enzyme function and respiration rates in seagrasses (Bulthuis 1983). In contrast, *Z. capricorni* can survive at considerably lower light levels than that documented for the survival of *Z. marina*, the most common temperate seagrass of the northern hemisphere (Abal *et al.* 1994) and consequently, productivity rates for *Z. marina* have been measured at five times the rate of *Z. capricorni* (Duarte and Chiscano 1999).

Zostera capricorni is a hardy and adaptable seagrass and this is not surprising given that its geographic range extends from cool, turbid, temperate estuaries into warm, clear, tropical bays. However, it appears that *Z. capricorni* is more capable of acclimatizing to high light and other stresses associated with exposure (e.g. Pérez-Llorens *et al.* 1994) than to low light levels and this is reflected in its preference for shallow water or intertidal habitats. Pollard and Greenway (1993) saw no evidence of light stress or photoinhibition in tropical *Z. capricorni* exposed to irradiance levels of 1400 μmol photon m^{-2} s^{-1} although maximum photosynthetic rate did decline in temperate *Z. capricorni* from around 1100 μmol photon m^{-2} s^{-1} in the field (Flanigan and Critchley 1996) and from approximately 500 μmol photon m^{-2} s^{-1} in the laboratory in this study. The quantum efficiency of photosynthesis, α, was reduced for both the moderate and high light grown laboratory samples in comparison to low light grown *Z. capricorni* which suggested that plants were utilising photoprotection or had suffered photodamage (Osmond 1994; Robinson 1999). Photoinhibition could be observed in the lower photosynthetic efficiencies recorded for *Z. capricorni* at Port Hacking around midday in summer and in autumn sometime prior to daily irradiance levels reaching 900 μmol photon m^{-2} s^{-1}. Since the F_v:F_m measured from early morning samples at Gray's Point was as high as any recorded for *Z. capricorni* in this study, daily recovery of photosynthesis from midday depression indicated the dynamic down-regulation of photosynthesis rather than sustained photodamage (Ralph *et al.* 2002). Ralph (1996) and Ralph *et al.* (2002) observed similar diurnal patterns in F_v:F_m for *H. ovalis* and *Z. marina* and found that F_v:F_m changes in *Z. marina* were inversely related to the level of NPQ and to xanthophyll pigment changes (Ralph *et al.*

2002). Unlike *H. ovalis* and *Z. marina*, the F_v:F_m of *Z. capricorni* samples from Port Hacking became more variable and decreased slightly though not significantly as early as 8.30 am, before irradiance levels in the field had risen substantially. Maximum photoprotection was already observed in the samples well before midday.

Although photoacclimation has been recorded for several seagrasses, the ability is usually not as well developed as it is in some higher plants and algae (reviewed in Boardman 1977; Falkowski and Raven 1997). Not all seagrass species can photoacclimate to cope with light reductions, for example, alteration of photosynthetic parameters in response to ambient light conditions has not been seen in *Z. marina* (Dennison and Alberte 1982) or *H. wrightii* (Neely 2000). *Syringodium filiforme* endures sub-optimal conditions by apparent dormancy in the winter months and only achieves high photosynthetic rates when the temperature and irradiance levels are high (Major and Dunton 2000). Species with limited ability to acclimate to low light conditions such as these are less likely to survive the effects of persistent light stress in disturbed or polluted estuaries.

5.5.3 Response of *Z. capricorni* to low light stress

Effects of low light stress on health and growth

The condition of *Z. capricorni*, measured as the level of photosynthetic stress experienced by the plants, was not affected by shading in laboratory experiments. This result agrees with previous F_v:F_m measurements of light limited seagrass leaves (Dawson and Dennison 1996; Longstaff *et al.* 1999). However, low light levels in laboratory experiments led to severe or complete dieback of all *Z. capricorni* samples after three months of treatment. Clearly photosynthetic efficiency is not a good indicator of seagrass 'health' when dieback is caused by light deprivation. While the photosynthetic *efficiency* of the plant was not reduced by shading, the actual photosynthetic *rate* was consistently below the compensation point where carbon would be fixed in sufficient quantity to meet respiration needs. The seagrass ultimately used up all available stored carbon resources from the leaves, roots and rhizomes and wasted away. P-I curves have shown that *Z. capricorni* is able to photoacclimate to a certain extent to maximise light harvesting and quantum yield, and minimise the irradiance required for maximal photosynthesis when grown in a low light regime. However, despite having optimal photosynthetic efficiency, the seagrass can not survive if the quantity of irradiance received is consistently below compensation irradiance. The compensation point is a function of respiration rate which is primarily controlled by water temperature (Bulthuis 1987; Perez and Romero 1992; Masini *et al.* 1995; Masini and Manning 1997). It is not apparent from this or other studies whether the I_c of *Z. capricorni* has any ability to acclimate to changes in light climate although I_c acclimation to *in situ* light reduction has been reported for other seagrass species (Dennison *et al.* 1993; Alcoverro *et al.* 2001). In fact, the relatively high I_c of *Z. capricorni* compared to that of other temperate seagrasses may be genetically determined for this species (Touchette and Burkholder 2000b) and might severely restrict the depth to which this plant can grow, particularly if the water quality is poor.

The F_v:F_m of control and shade treated *Z. capricorni* samples was maintained at a generally high level (~ 0.750-0.800) throughout the 3 months of the experiment. In contrast, while the initial and maximal fluorescence values recorded for both treatments were relatively steady for the first two weeks of the experiment, by day 86 both F_o and F_m values had fallen significantly. Longstaff *et al.* (1999) noted that a similar decline in the F_o and F_m of light deprived *H. ovalis* was proportional so that the photosynthetic efficiency (F_v:F_m) did not change. These authors linked the fall in fluorescence values to the breakdown of chlorophyll in light stressed seagrass leaves but in the current experiment, irradiance levels sustained healthy growth of control *Z. capricorni* and there was no significant loss of chlorophyll from the leaves of either control or shade treated samples. By day 86, however, the chlorophyll *a*:*b* ratio of both control and treated samples had dropped substantially consistent with shade acclimation and this may reflect an increase in LHCII complexes relative to PSII reaction centres given that chlorophyll concentration did not fall (Schiefthaler 1999).

Low light and slower growth rates should ultimately result in less below-ground biomass compared to above-ground biomass since carbon is better invested in light harvesting organs (Abal *et al.* 1994). This was not the case for *Z. capricorni* in the laboratory after three months of shade treatment; low-light stressed *Z. capricorni* had an above-ground:below-ground biomass ratio of approximately 1:3 compared to 1:1.5 for control grown seagrass. This response appears to represent short term shedding of leaves by low light stressed seagrasses which will reduce self-shading (van Lent *et al.* 1995; Longstaff *et al.* 1999) without any further acclimatory response to the lower light regime. Higher root:shoot ratios are usually characteristic of rapidly growing seagrasses that need to exploit a wider expanse of sediment to maintain the supply of nutrients (Abal *et al.* 1994). In this case, the lower above-ground:below-ground biomass ratio is most likely due to the almost complete loss of leaf material while rhizomes probably persist a little longer during dieback events.

It is interesting to note that there appeared to be no major translocation of nutrients and sugars through the rhizomes from control grown shoots to support the growth of adjacent shaded shoots. Longstaff and Dennison (1999) severed connecting rhizomes in field experiments so as to avoid translocation between control and shaded plots but no division was made between paired treatments in this experiment, despite the fact that seagrass turfs generally overlapped both ends of the tanks. In fact, the line of dieback between control and shaded treatments was quite distinct in each replicate tank. It is possible that translocation occurred instead from the low light stressed shoots into the control grown shoots since it would be more advantageous for the plant to invest resources in leaves situated in the light rather than attempt to acclimate to very low light levels.

The mean percent nitrogen content of control and shaded leaves was 2.13 $\pm$ 0.21% and 2.30 $\pm$ 0.41% of dry weight respectively after 13 days of the experiment. These levels are similar to those measured for *Z. capricorni* in Moreton Bay by Abal *et al.* (1994) and Grice *et al.* (1996) under low and moderate light conditions and are both above the level considered by Duarte (1990) to indicate nitrogen limitation in *Z. marina* (less than 2% of DW). A nitrogen deficiency will limit the ability of the seagrass to produce chlorophyll and other photosynthetic components (Touchette and Burkholder, 2000a; Alcoverro *et al.* 2001) and this can reduce photosynthetic efficiency. In this experiment, *Z. capricorni* were collected from a highly eutrophic estuary with sediments intact, so nutrients were unlikely to be limiting factor for the samples. In contrast, the percent carbon content of leaves in both treatments after 13 days of the experiment were lower overall than those recorded in outdoor aquaria for this species by Grice *et al.* (1996), and somewhat lower than recorded by Abal *et al.* (1994). This could suggest that the *Z. capricorni* leaves in the current laboratory experiments may be CO_2 limited and as a result, photosynthetic rates would also be limited. On the other hand, tissue carbon levels recorded by Moore and Wetzel (2000) for *Z. marina* were similar to those of the current experiments and these authors considered these levels to indicate that no carbon limitation was occurring. The carbon content of shaded leaves was significantly less than that of light grown *Z. capricorni* after 13 days of treatment. While Grice *et al.* (1996) also found that the carbon content of light deprived *Z. capricorni* leaves was distinctly lower than for leaves grown in moderate to high light, Abal *et al.* (1994) found that these plants had significantly higher carbon contents when grown in 30% or less ambient surface irradiance compared to 50-100% ambient light. Due to relative leaf carbon levels, the C:N ratios (16.83 $\pm$ 1.96 for control and 15.04 $\pm$ 2.47 for low light samples) measured in this experiment are also much lower than the C:N of 20-29 recorded by Grice *et al.* (1996) and somewhat lower than Abal *et al.* (C:N 17–19; 1994) under a range of light conditions in outdoor aquaria. Higher light intensities lead to higher rates of photosynthesis and CO_2 fixation, so if the respiration rate does not change at higher irradiance levels, then leaves are likely to have more carbohydrate available for storage and this appears to be the case for *Z. capricorni.* Translocation of carbohydrates out of shaded shoots may also contribute to the lower carbon levels observed in low light stressed leaves.

Effects of low light stress on pigment content

The total chlorophyll concentration of laboratory grown *Z. capricorni* leaves was highly variable and, although there was a trend in low light stressed samples toward an increase in total chlorophyll content over the first two weeks of the experiment, the difference did not become significant until some weeks later. By day 86, the chlorophyll *a* + *b* of shaded samples had increased by 33% and was significantly different to that of controls, however, pigment content was not measured in the intervening period between day 13 and day 86. This result confirms that of Abal *et al.* (1994) who used shading screens of different density to reduce the irradiance to *Z. capricorni* plants grown in outdoor tanks. After two months, Abal *et al.* (1994) observed significant increases in the total chlorophyll concentration of *Z. capricorni* leaves grown at 5, 15 and 30% of ambient irradiance with a maximum increase of 83% observed in samples grown at 20% of the ambient irradiance.

Previous studies have shown that the chlorophyll concentration of many seagrass species reacts rapidly and markedly to changes in the irradiance intensity. For example, increases in total chlorophyll content associated with decreased light availability have been recorded for a number of temperate and tropical seagrass species (Wiginton and McMillan 1979; Williams and Dennison 1990; Abal *et al.* 1994; Masini and Manning 1997; Longstaff *et al.* 1999). On the other hand, the chlorophyll content of *H. wrightii* and *Z. marina* did not change significantly (van Lent *et al.* 1995; Neely 2000), or increased only occasionally (Dennison and Alberte 1982; 1985), when shaded with screens *in situ*. There was no observable relationship between chlorophyll content and depth for *P. sinuosa* (Masini *et al.* 1995), *H. pinifolia* (Longstaff and Dennison 1999) or three US seagrass species (Wiginton and McMillan 1979). The total chlorophyll content of deep growing *H. ovalis* was very responsive to light reductions caused by dense cloud cover but shading with screens only increased chlorophyll concentrations in shallow grown *H. ovalis* (Longstaff *et al.* 1999). At the same time, the total chlorophyll content of *H. ovalis* grown in total darkness fell by 50% by the process of etiolation (Longstaff *et al.* 1999). It is likely that the chlorophyll concentration of *H. ovalis* (and possibly other seagrasses) increases with decreasing irradiance up to a threshold irradiance below which chlorophyll synthesis is restricted. Williams and Dennison (1990) noted that chlorophyll synthesis was light limited for deep growing *H. decipiens* since chlorophyll concentrations increased linearly with the daily period of saturating irradiance. Alternately, the synthesis of chlorophyll and other pigments may be preferentially stimulated by certain wavelengths of light. If light quality is as important as intensity, then shade screens may not simulate the affects of other types of light deprivation. The chromatic response of a seagrass to shading by suspended sediment in the water, for example, may differ from that induced by algal blooms, seasonal changes or differing water depths. In addition, Alcoverro *et al.* (2001) noted that plants growing in oligotrophic waters presented 30% lower chlorophyll and nitrogen concentrations than in eutrophic waters where nutrients were not limiting. Hence, seagrasses in oligotrophic bays may not have the capacity to express pigment changes they may demonstrate in response to irradiance changes in eutrophic estuaries.

Chlorophyll *a*:*b* increased from initial levels for both control and shade treated *Z. capricorni* but by day 13 control samples had significantly higher chlorophyll *a*:*b* than the plants grown in low light. An increase in proportion of chlorophyll *b* to chlorophyll *a* is typical of shade adaptation in plants because this will enhance absorption in the longer blue wavebands to some degree and broaden the range of chlorophyll absorption of red wavelengths. Abal *et al.* (1994) observed significant differences in the chlorophyll *a*:*b* of *Z. capricorni* grown at different irradiance levels but found that ratios did not decrease consistently with decreasing light levels. Furthermore, chlorophyll *a*:*b* did not change with depth for four US seagrass species (Wiginton and McMillan 1979) despite the obvious advantage that increasing absorption of shorter wavelengths would have for an aquatic plant where longer wavelengths are increasingly attenuated with depth of water.

The general increase in the total chlorophyll content of low light grown *Z. capricorni* was accompanied by a corresponding increase in total carotenoid content, due mainly to increases in the concentrations of the light harvesting carotenoids neoxanthin, lutein and violaxanthin and the photoprotective β-carotenoids. This general increase in the photosynthetic pigment content of light deprived seagrass leaves would act to maximise light harvesting across as broad a range of visible wavelengths as possible since large amounts of any pigment will act to broaden the wavelengths at which absorption occurs (Curran *et al.* 1991).

After seven days of treatment the zeaxanthin and antheraxanthin concentration of both shaded and control *Z. capricorni* samples dropped to less than half that of the initial day. The Z and A concentrations of shade treated samples remained low while concentrations in control samples rose significantly higher than that of the low light stressed plants by day 13 of the experiment. From these results it appears that there is some requirement for photoprotection even at the relatively low irradiance intensities experienced by control grown *Z. capricorni* in the laboratory (168 ± 49 μmol photon $m^{-2}s^{-1}$).

5.5.4 Response of *Z. capricorni* to high light stress

Effects of high light stress on health and growth

The photosynthetic efficiency of *Z. capricorni* grown at high irradiance levels was consistently lower than that of control grown plants from day 7 through to day 73 of the laboratory experiment. The difference in the F_v:F_m of high light and control samples was highly significant on each measuring date during this period, and was of a similar magnitude to that associated with midday depression measured in *Z. capricorni* at Port Hacking.

The irradiance intensity at the control tanks (238 ± 127 μmol photon $m^{-2}s^{-1}$) did not approach the level required to saturate photosynthesis in *Z. capricorni* acclimated to a moderate light regime but high light treatments were exposed to a PAR (661 ± 190 μmol photon $m^{-2}s^{-1}$) well above that required for the onset of saturation, and close to the I_{max} for high light grown plants (700 μmol photon $m^{-2}s^{-1}$) (Figure 5.3). The excess light energy harvested above the saturating irradiance must therefore be diverted by the processes of NPQ or photodamage will occur, and in either case, the maximum quantum yield (F_v:F_m) will decline. Photoinhibitory responses such as this have previously been recorded for sun and shade-grown seagrasses exposed to excess light (e.g. Ralph and Burchett 1995; Dawson and Dennison 1996; Ralph *et al.* 1998; Beer and Björk 2000; Major and Dunton 2002) including *Z. capricorni* (Flanigan and Critchley 1996). For some species, photosynthetic efficiency was particularly sensitive to high intensities of PAR, e.g. *C. serrulata* and *S. isoetifolium*, however the F_v:F_m of *Z. capricorni* did not decline at high irradiance intensities if the plants were screened from the UV component of the light (Dawson and Dennison 1996).

A decline in the photosynthetic efficiency of *Z. marina* in laboratory experiments and in the field was considered to be associated with the dynamic down-regulation of photosynthesis because F_v:F_m was inversely related to the level of NPQ and to xanthophyll pigment changes (Ralph *et al.* 2002). Since F_m declined but F_o did not change, Ralph *et al.* (2002) suggested that photodamage did not occur in *Z. marina* during high light exposure. Photoprotective down-regulation is characterised in sun adapted plants by a relatively constant F_o and fluctuating F_m (Demmig and Björkman 1987; Osmond 1994). A depression in F_v:F_m accompanied by a decline in F_o may also suggest that NPQ, and in particular zeaxanthin synthesis, has circumvented photodamage (Kraus and Weis 1991; Müller *et al.* 2001). However, if photoinhibition is accompanied by a rise in F_o (and a decrease in F_m), then significant photodamage to the PSII reaction centres has occurred (Demmig and Bjorkman 1987; and see review of Osmond 1994). UV damage to PSII reaction centres has been observed to cause an increase in the F_o of *Z. capricorni*, *H. ovalis*, *P. australis* and other seagrasses (Larkum and Wood 1993; Dawson and Dennison 1996).

In the current experiment, the F_o and F_m of all samples fluctuated considerably and in synchrony for at least five weeks in response to small changes in the water temperature of the tanks. However, since high light and control samples were paired in tanks, smaller but nonetheless significant differences between the F_o and F_m of high light stressed and control *Z. capricorni* could be detected. The F_o of high light treated *Z. capricorni* leaves was always slightly below, or equivalent to, that of control samples but it was significantly lower on days 17 and 28 of the experiment when F_m levels did not differ between treatments. On this basis, the significantly greater level of photoinhibition observed in high light treated *Z. capricorni* could be attributed to a higher level of photoprotection rather than photodamage. In addition, the F_o values recorded for both control and high light samples were always lower than those recorded on the initial day of the experiment, so *Z. capricorni* probably did not suffer any photodamage at the irradiance levels that they were exposed to in this experiment.

In general, the dark adapted F_v:F_m of both the controls and high light treated *Z. capricorni* began to decline after two weeks of the experiment and continued to fall gradually over the remaining time period. While photodamage may lead to sustained depression of F_v:F_m which does not recover in the evening of a diurnal cycle (Ralph *et al.* 2002), photodamage was not considered to have occurred in the current experiment. Presumably the seagrass plants were subjected to some additional stress in the laboratory that had not affected the plants in the low light stress experiment. Combined environmental stresses are known to have an additive effect on photoinhibition in seagrasses (Ralph 1999) and many terrestrial plants (e.g. Björkman 1981). Elevated water temperatures may have enhanced the photosynthetic demand for inorganic carbon and other nutrients, which could then have been a limiting factor for photosynthetic rate and efficiency since the concentration of dissolved CO_2 in the tank water may also have been affected. Laboratory water temperatures were somewhat higher (~20-24 °C) than observed in the low light experiment (~18-21 °C) and would certainly be higher than October water temperatures in Lake Illawarra (~15-18 °C) from where the seagrass was collected. Notably, the plants used in this experiment were only acclimated to laboratory conditions for one day prior to commencing treatment.

Despite a consistent decrease in the photosynthetic efficiency of all *Z. capricorni* samples over the three months of the experiment and in spite of the significantly higher stress levels experienced by high light treated plants, photosynthetic 'health' had no observable impact on the density or biomass of seagrass remaining in the tanks at the end of the experiment. High meadow biomass is typical of seagrasses growing in high irradiance regimes because photosynthetic rate and productivity are maintained at an optimal level by dynamic down-regulation of photosynthesis, so long as NPQ is adequate to prevent significant photodamage (Alcoverro *et al.* 2001).

Effects of high light stress on pigment content

In previous studies where sustained F_v:F_m depression was observed in seagrasses exposed to enhanced PAR (Major and Dunton 2002) or UV radiation (Dawson and Dennison 1996), photodamage did occur and it resulted in significant chlorophyll bleaching. However, there was no significant difference in the chlorophyll concentrations of high light stressed and control *Z. capricorni* leaves during this experiment. In fact, the chlorophyll content of both treatments increased after two weeks of treatment by about 40% for the high light samples and around 60% for controls. This enhanced chlorophyll content plus the observed activity of the xanthophyll cycle provides further evidence that photoprotection was well developed in *Z. capricorni* and that photodamage did not occur during laboratory experiments.

Small differences in VAZ pigment concentration combined with small differences in the chlorophyll content of *Z. capricorni* leaves would almost entirely account for the significant difference observed in the total carotenoid:total chlorophyll content of high light and control grown samples. Demmig-Adams and Adams (1992b) observed that the sun leaves from a terrestrial plant had around 1.4 times the total carotenoid:total chlorophyll content of shade leaves from the same plant. High light grown *Z. capricorni* leaves displayed slightly higher total carotenoid:total chlorophyll throughout the course of the experiment but the difference was only significant on day 17 when the ratio for high light stressed samples was just 1.06 times greater than that of controls. The α:β-carotene ratio of *Z. capricorni* samples grown under high irradiance levels did not decrease as suggested by Thayer and Björkman (1990) and Demmig-Adams and Adams (1992b) for terrestrial plants. High light levels in leaves should, in theory, specifically promote β-carotenoid synthesis, resulting in accumulation of both xanthophyll cycle carotenoids and β-carotene (Thayer and Björkman 1990; Demmig-Adams and Adams 1992b). In fact, high light grown *Z. capricorni* had proportionally more α:β-carotene than control samples and the difference was almost significant by day 17 ($p = 0.057$).

Since *Z. capricorni* were not acclimated to laboratory conditions prior to commencing this experiment, the initial VAZ pool concentrations in the leaves of laboratory samples (high light 117 $\pm$ 54 nmol g^{-1}FW; control

102 ± 10 nmol g^{-1}FW) were generally within the range of field samples collected at midday in spring from Lake Illawarra (i.e. 90 ± 53 nmol g^{-1}FW). There were no significant differences between high light stressed and control *Z. capricorni* in the magnitude of the VAZ pool, and VAZ pool concentration did not change significantly for either treatment during the 2 ½ week experiment except for a temporary decline that affected both treatments equally on day 7. However, control grown *Z. capricorni* did contain a slightly lower concentration of VAZ than high light stressed seagrass, which, when combined with perceptibly (but not significantly) higher levels of total chlorophyll and total carotenoid, was sufficient to induce a significant difference in VAZ:total chlorophyll (day 17) and VAZ:total carotenoid (days 13 and 17) for control and treated samples. High light grown *Z. capricorni* however, contained only 16.5% more VAZ:chlorophyll and 10% more VAZ:carotenoid than the control samples. In contrast, terrestrial species can experience up to 400% increases in the VAZ pool relative to total chlorophyll content when plants are grown in the sun (Thayer and Björkman 1990) and increases in the VAZ pool of sun compared to shade leaves from the same plant can be around 250% (Demmig-Adams and Adams 1992b). These changes in total VAZ pool size can occur very rapidly, for example, Ralph *et al.* (2002) observed increases in the VAZ pool of the seagrass *Z. marina* with increasing irradiance during the course of a single day. Hence, it is unlikely that larger changes in the size of the VAZ pool of *Z. capricorni* would have occurred if this experiment had been run for a longer period of time.

The VAZ pool recorded by Ralph *et al.* (2002) for *Z. marina* was 120-150 mmol VAZ per mol chlorophyll *a+b* which is within the range of terrestrial sun-acclimated plants (100-200 mmol mol^{-1} Tchl; Thayer and Björkman 1990; Adams and Demmig-Adams 1992). Laboratory grown *Z. capricorni* contained only 68 ± 9 mmol mol^{-1} Tchl when acclimated to high light stress and 63 $\pm$ 8 mmol mol^{-1} Tchl when light deprived. The VAZ:total chlorophyll of field grown *Z. capricorni* were similar to laboratory grown plants in spring (64 $\pm$ 25 mmol mol^{-1} Tchl) but reached a maximum mean value of 101 $\pm$ 64 mmol mol^{-1} Tchl during summer at Lake Illawarra.

The level of de-epoxidation of the VAZ pool measured in these experiments was somewhat dependent on the dark adaptation of samples prior to pigment analysis and may be underestimated. Despite this, control grown *Z. capricorni* samples began to display lower A+Z:VAZ and Z:VAZ than high light stressed samples from day 7 of the experiment and, although differences in these ratios increased over time, the only significant difference observed was for Z:VAZ on day 17. Dawson and Dennison (1996) observed a similar fall in Z:VAZ for *Z. capricorni* shaded to 50% ambient outdoor PAR but in that instance, the difference between full sunlight and shade treatments was not significant. Individually, the antheraxanthin and zeaxanthin concentrations in the current experiment fluctuated from day to day for both treatments. The main difference was that the level of Z was consistently lower in control samples than high light treated samples from day 7 and significantly lower by day 17. The concentration of A was also lower in the control grown *Z. capricorni* from day 13 to day 17 but the difference was not significant. Consequently, A+Z:VAZ and Z:VAZ for high light treated samples remained relatively stable over the course of the experiment (at around 20% and 7% respectively), displaying values that were approximately ½ - ⅓ of the ratios observed in field grown *Z. capricorni* collected at midday from Lake Illawarra during October 2000 (42% and 24% respectively). The mean A+Z:VAZ of control grown samples dropped to around 10% by day 13 with only 2% of the VAZ converted to Z. Presumably the requirement for NPQ quenching of excess light was greater in the field because, even though the daily integrated PAR received by high light treatments was much the same, the maximum irradiance at midday in the field was almost twice the level experienced indoors in high light treated tanks. In addition, while laboratory plants are grown under constant stable light conditions, field grown plants experience considerable fluctuations in irradiance and this would increase the need for photoprotection via A+Z.

Significantly higher Z concentrations, and therefore Z:VAZ, were associated with significantly lower photosynthetic efficiency in high light treated *Z. capricorni* and this is typical of terrestrial plants exposed to excess irradiance (Thayer and Björkman 1990; Adams and Demmig-Adams 1992; Demmig-Adams and Adams 1992b). However, *Z. capricorni* leaves exposed to high light stress displayed relatively smaller depressions in F_v:F_m compared to other seagrass species when subjected to excess light and related stresses (see review of Touchette and Burkholder 2000b) and relatively lower levels of de-epoxidation compared to other seagrass and terrestrial plant species. Despite significant photoinhibition, the minimum EPS observed for high light treated laboratory *Z. capricorni* samples was 0.865 $\pm$ 0.102 on day 13, which suggests a very high rate of photosynthetic electron transport. Even the *Z. capricorni* sampled during extremely high midday PAR in summer (mean EPS 0.664 $\pm$ 0.100) displayed EPS values well above those recorded at noon for sun grown terrestrial plant species (EPS range 0.145-0.605; Adams and Demmig-Adams 1992) and *Z. marina* (around 0.550; Ralph *et al.* 2002). Unlike the laboratory samples, A and Z concentrations would not have been underestimated in field samples because these samples were not dark adapted prior to freezing in liquid N. If, as Königer *et al.* (1995) suggest, the maximum content of Z in the plant indicates the maximum amount of photoprotection afforded by qE, then *Z. capricorni* may not depend solely on the xanthophyll cycle. Dawson and Dennison (1996) did not observe any increase in the zeaxanthin content of *Z. capricorni* when exposed to high light. Ralph *et al.* (2002) noted that *Z. marina* utilised A more than Z for photoprotection (zeaxanthin-independent quenching) and considered that EPS did not represent the degree of de-epoxidation particularly well for this species. While *Z. capricorni* leaves consistently carried around twice as much A as Z, a feature also observed by Flanigan and Critchley (1996), this seagrass does not sustain the very high A concentrations measured in *Z. marina* (150 mmol mol^{-1} Tchl compared with less than 50 mmol mol^{-1} Tchl for terrestrial plants and less than 30 mmol mol^{-1} Tchl for *Z. capricorni* in this study). Both *Zostera* spp. appear to be very capable of dealing with the effects of high light exposure as well as rapidly fluctuating irradiance, high temperatures and desiccation when they inhabit intertidal sites. Ralph (1999) found that NPQ in the seagrass *H. ovalis* was elevated and not modulated like that of *Z. marina* (Ralph *et al.* 2002) or *Z. capricorni.* Plants exposed to high PAR and frequent sunflecks generally maintain elevated levels of A and Z so energy dissipation pathways are always engaged and ready to protect (Demmig-Adams *et al.* 1999) but this was not the case for *Z. capricorni* growing in full sun in shallow water where wave-focussing is almost continuous.

It is possible that the photoacclimation of *Z. capricorni* to varying light conditions is so effective and rapid that this very productive species rarely experiences much in the way of *excess* light above that which it can utilise in photochemistry. A plant that has an inherently high photosynthetic capacity quenches large amounts of light energy through photosynthetic electron transport and may not need to dissipate a great deal of excess energy via NPQ. This would be especially true of plants growing under constant conditions of high light. The VAZ pool size and degree of de-epoxidation will therefore indicate potential accumulation of excess light that needs dissipating rather than the light intensity in itself (Thayer and Björkman 1990; Demmig-Adams and Adams 1992b). EPS depends not only on the VAZ pool size which determines the potential for NPQ but also on the photosynthetic rate which controls qP. Demmig-Adams and Adams (1992b) observed that the product of the efficiency of photosynthetic energy conversion and VAZ pool size was closely related to the epoxidation state for a range of terrestrial plants. Alternately, *Z. capricorni* may depend more on other forms of photoprotection such as the accretion of secondary metabolites, morphological or anatomical modifications. Notably, the levels of self-shading within the canopies of seagrass meadows are very high so surface irradiance may not represent the light levels that most of the leaf area is exposed to, particularly if the seagrass is submerged to any depth. Incident PAR was reduced by approximately 85% by moving 100 mm into a relatively dense *P. sinuosa* canopy (Masini *et al.* 1995). Even when seagrass leaves are exposed at the water surface, they lie across each other in dense layers so that only the uppermost leaves would be stressed by high light exposure at any point in time.

The amount of anthocyanin pigment produced by a plant is affected by a number of environmental factors including nutritional status, water status and other forms of stress, wounding, disease, age and seasonal effects, temperature and light (reviewed in Mancinelli 1985; Chalker-Scott 1999). For the seagrass *Z. marina,* phenolic compounds have also been reported to provide defence against predation and fouling by epibionts (Todd *et al.* 1993). Light is particularly important because little or no anthocyanin is formed in the dark and exposure to light will always enhance production, regardless of the functional significance of the anthocyanin colouration (Mancinelli 1985). Anthocyanin production is wavelength dependent but the spectral sensitivity is different for different species; in some plants anthocyanin synthesis is induced by UV-blue, red and far-red light, other plants respond primarily to red wavelengths with some response in UV-blue but none in the far-red (anthocyanin production in these systems is related to photosynthetic activity) while anthocyanin production in yet other plants is only induced by UV-blue light. The anthocyanin activity so far reported for *Z. capricorni* appears to fall into the third group. Abal *et al.* (1994) and Dawson and Dennison (1996) attributed the very sensitive UV absorption response of *Z. capricorni* leaves to the accumulation of anthocyanins and other flavonoid-like UV-blocking pigments in the leaves. This photoprotective activity seemed to be restricted to UV light, however, because reducing PAR levels with shade screens did not induce significant changes in the anthocyanin content of *Z. capricorni* leaves in the current experiment or in outdoor aquaria (Dawson and Dennison 1996). Whilst the anthocyanin content of both shaded and control leaf samples dropped consistently over the first two weeks of the low light stress experiment (although not significantly since the variability in samples was very high), in the high light stress experiment, control and treated *Z. capricorni* samples both experienced a significant increase in anthocyanin content of around 20% between day 1 and day 13 of the experiment ($t = 3.224$, $df = 15$, $p = 0.006$). This increase in anthocyanin content is difficult to explain since the artificial laboratory light source is deficient in the shorter wavelength blue light (and probably in UV) compared with natural sunlight (Figure 3.2). Presumably there was sufficient light of suitable wavelength to allow anthocyanin synthesis in both the control and high light treatments, but the trigger that caused increased anthocyanin production is unknown.

5.5.5 Evidence for chromatic acclimation in *Z. capricorni*

Marine algae are renowned for their ability to chromatically adapt to the light quality at different water depths by varying either total pigment content or the relative proportions of their photosynthetic pigments. For example, in the Rhodophyta, Cryptophyta and Cyanophyta, biliprotein synthesis is enhanced in response to the colour of the light field (Jeffrey 1981; Kirk 1994). Chromatic adaptation is biochemically efficient for small cells and consequently the unicellular phytoplankton that move continuously through waters of varying spectral quality and intensity contain the greatest diversity of pigments (Raven 1995). Optically thick macroalgae have the same light harvesting capacity (approaching 100% at all visible wavelengths) no matter what combination of pigments they contain due to the package effect that occurs in multicellular plants (Kirk 1994; Raven 1995) including higher plants. A mature ivy leaf with high chlorophyll concentration (60 $\mu g\ cm^{-2}$) absorbs not only ~100% incident red wavelengths but ~70% of green wavelengths at 550nm (Kirk and Goodchild 1972).

Seagrasses are angiosperms so they do not have the flexible biochemistry of the algae and are dependant on chlorophyll *a* as the major light-harvesting pigment. The carotenoid composition of higher plant leaves is highly conserved but the relative levels of these carotenoids in leaf tissues can be affected by environmental and developmental factors such as greening, senescence, and the ambient light climate (Young 1993b; Thayer and Björkman 1990; Demmig-Adams and Adams 1992b). Kirk (1994) suggested that plants that did not contain biliproteins could only respond to changes in light quality or intensity with relatively small changes in pigment stoichiometry, if they occurred at all. In terms of increasing light harvesting under conditions of light deprivation, the most effective chromatic strategy for a seagrass would therefore be to increase the concentrations of all light harvesting chlorophylls and carotenoids, and this strategy was observed for *Z. capricorni* leaves grown at decreasing daily irradiance levels (Figure 5.11). A change in irradiance regime from extreme summer field exposure to light deprivation sufficient to cause the death of many plants resulted in an increase of 80% in the total light harvesting pigment content per gram fresh weight of *Z. capricorni* leaf. This is fairly similar to pigment increases experienced in terrestrial plants, for example, shade grown Engelmann spruce needles contained about 70% more chlorophyll than sun grown needles (Moran *et al.* 2000).

Changes in the pigment stoichiometry of *Z. capricorni* also occurred in response to changes in the light environment, but they were relatively small in terms of the concentrations of pigments involved (Figure 5.13). Growth at lower irradiance intensities caused a consistent reduction in the chlorophyll *a:b* ratio of *Z. capricorni* (Figure 5.12). Chlorophyll *a:b* decreased from around 5.0 for field grown plants to 3.5-4.0 for control samples after two weeks in the laboratory, to less than 3.0 for low light stressed plants after three months of treatment. This increase in the relative proportion of chlorophyll *b* would further increase the efficiency of light harvesting, particularly once chlorophyll *a* absorption had become saturated. All other notable changes in the pigment stoichiometry of *Z. capricorni* as a result of fluctuating irradiance levels were linked to photoprotection by the xanthophyll cycle pigments (Figure 5.12). These adaptations appeared to occur on a much shorter time scale than changes in light harvesting pigments. VAZ pools were generally higher in high light treated *Z. capricorni* compared to controls, in controls compared to low light stressed samples, and outdoors in summer compared to spring, because of the higher requirement for photoprotection in plants exposed to higher light intensities. The de-epoxidation of V to A and Z in the process of NPQ involved only a very small fraction of the total pigment content (Figure 5.13) but it appeared to be very responsive to differences in light regime and may offer a sensitive and rapid indicator of physiological stress (Figure 5.12).

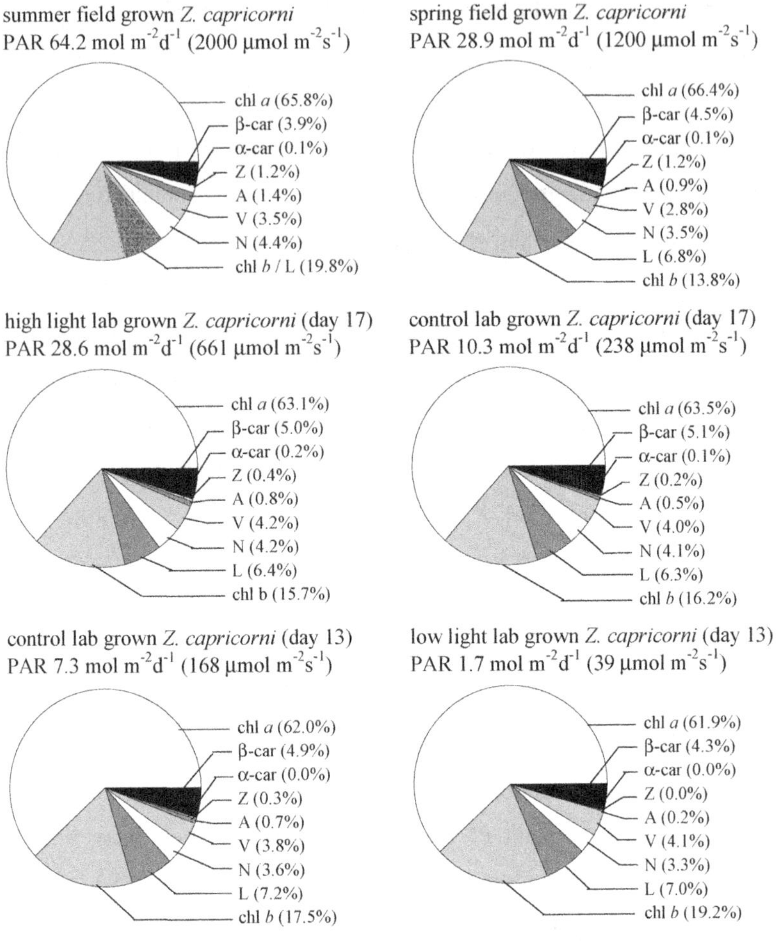

Figure 5.13. Comparison of the mean pigment (chlorophyll and carotenoid) composition of *Zostera capricorni* grown under a range of light environments in its natural habitat and in laboratory experiments. PAR is given as integrated daily PAR with maximum midday PAR in brackets. Pigments are; chl *a*, chlorophyll *a*; ß-car, ß-carotene; α-car, α-carotene; Z, zeaxanthin; A, antheraxanthin; V, violaxanthin; N, neoxanthin; L, lutein and chl *b*, chlorophyll *b*. Since chlorophyll *b* and lutein were not reliably separated by HPLC for the summer field samples, their combined value has been given and their approximate proportions are shown dotted.

Research has shown that temperate seagrasses have great potential for photosynthetic acclimation because they tend to grow in a highly variable and dynamic environment (e.g. Dennison and Alberte 1985; Abal *et al.* 1994; Goodman *et al.* 1995; Ralph *et al.* 1998; Alcoverro *et al.* 2001) although many studies display little evidence of chromatic adaptation to low light (e.g. Masini *et al.* 1995; van Lent *et al.* 1995; Dawson and Dennison 1996; Longstaff and Dennison 1999; Neely 2000). Changes in chlorophyll content, chlorophyll *a:b* ratios (e.g. Dennison and Alberte 1982; 1985; Jimenez *et al.* 1987; Abal *et al.* 1994; Agawin *et al.* 2001) and the xanthophyll cycle pigments (Dawson and Dennison 1996; Flanigan and Critchley 1996; Ralph *et al.* 2002) do occur but they do not seem to be of the same magnitude as the analogous chromatic adaptations of terrestrial plants to sun and shade environments (e.g. Thayer and Björkman 1990; Adams and Demmig-Adams 1992; Demmig-Adams and Adams 1992b; Björkman and Demmig-Adams 1995) and do not compare in magnitude or diversity to the pigment changes seen in marine algae in response to changing light intensity and quality (e.g. Kirk 1994; Raven 1995). It is possible that the light environment in which deeper growing (wholly submerged) species of seagrasses exist is more stable and moderated than that to which a terrestrial plant is exposed daily so the need for chromatic adaptation and photoprotection is less. This would not be the case for many of the *Zostera* spp. including *Z. capricorni* which often grow in intertidal habitats and are therefore exposed to daily extremes of irradiance of at least the same magnitude as a terrestrial plant.

5.5.6 Sun-adapted or shade-adapted?

Seagrasses are generally considered to be shade-adapted plants, that is, they are thought to be genetically predisposed for survival in low light environments. There are many species that live completely submerged, some to depths of up to 40-50 m (e.g. *P. oceanica*, Duarte 1991;

Halophila spp., Kirkman 1997), and for these species selective pressures would emphasise light harvesting and efficient use of absorbed energy. Indeed, only a few terrestrial shade species share the characteristic of epidermal chloroplasts with seagrasses (McComb *et al.* 1981). While all sun plants have some ability to acclimate to shade, obligate shade plants have an intrinsically low potential for photosynthetic light acclimation to high light conditions and are therefore highly susceptible to photoinhibition and photodamage (Björkman 1981). Many of the physiological and morphological features that characterise sun and shade adapted plants were not examined in this study, but the parameters that were investigated suggested that *Z. capricorni* was not an obligate shade plant.

Shade plants have lower chlorophyll *a*:*b* ratios, for example, Thayer and Björkman (1990) recorded an average chlorophyll *a*:*b* value of 2.46 $\pm$ 0.21 for nine terrestrial species grown in shade and 3.06 $\pm$ 0.29 for 10 sun grown species. The mean chlorophyll *a*:*b* measured for *Z. capricorni* across this set of experiments ranged from 2.7 for plants grown three months in extreme shade to 5.1 for field grown plants in full sun. Hence, the lowest values of chlorophyll *a*:*b* recorded for *Z. capricorni* were equivalent to the upper range seen in typical shade grown plants, while field sampled ratios greatly exceeded those typical of terrestrial sun adapted plants.

The difference between sun and shade leaves is mainly in the capacity for photosynthetic electron transport and for photoprotective responses (Demmig-Adams and Adams 1992a). *Zostera capricorni* is very productive for a temperate seagrass (Duarte and Chiscano 1999) and seagrasses are, in general, exceptionally productive plants (Zieman and Wetzel 1980). This seagrass has a relatively high minimum light requirement for survival and can photosynthesise at a high maximum rate (Touchette and Burkholder 2000b), displaying little evidence of light stress or photoinhibition during high irradiance conditions (Pollard and Greenway 1993; Flanigan and Critchley 1996) and only a small dependence on xanthophyll mediated photoprotection. These features are characteristic of sun adapted plants (Schiefthaler 1999; Björkman 1981). The high EPS and relatively low Z content recorded for *Z. capricorni* in this study is typical of plants with very high photosynthetic rates (Adams and Demmig-Adams 1992).

Zostera capricorni can therefore be considered a sun-adapted plant species and its maximum depth distribution in south eastern NSW (< 3 m) is an indication of its high requirement for light. Hootsmans *et al.* (1995) considered photosynthesis in the intertidal eelgrass, *Z. noltii* to resemble that of sun-adapted terrestrial plants given the relatively high light compensation and saturation points of this species. *Z. noltii* has to survive under both high and low light levels depending on the tides, and since it is not possible for a plant to acclimate to both, sun acclimation is apparently the most effective strategy. Similarly, the tropical seagrass *T. testudinum* shows the photosynthetic character of a sun-adapted plant (Enríquez *et al.* 2003).

Self-shading within a seagrass canopy can dramatically reduce the irradiance available to the meadow but wave-focussing can offset the effects of depth and canopy shading to some extent by supplying high intensity light to the leaves for brief, repeated intervals. This situation might be likened to sunflecks in the understorey of a forest canopy. Photoadaptation in some understorey species maximises their utilization of the short high light periods during sunflecks and consequently they maintain high photosynthetic rates and did not suffer photodamage when exposed to full sun (Pearcy 1994; Schiefthaler *et al.* 1999). Effective utilization of sunflecks requires a plant to have a high P_{max} plus rapid induction of the photosynthetic carbon reduction cycle so that the maximum carbon assimilation is achieved during the limited period of light (Robinson 1999). Since induction typically takes 10-60 minutes for terrestrial plants, the 7 minute induction of *Z. capricorni* would be an advantage in terms of sunfleck use. Inorganic carbon is always available for immediate use in the lacunae of seagrasses so they are not limited by light-dependent opening of stomata, a rate-limiting step for terrestrial plants (reviewed in Robinson and Osmond 1994). The 'flashing' light supplied by wave-focussing is a more effective light source than sun flecks because the induction gained during one light fleck enhances utilisation of the next (Pearcy 1994). In addition, carbon gain exceeds that expected because carbon fixation continues in the low light period following each sunfleck (post-illumination CO_2 fixation).

5.5.7 Effects of light stress on the spectral reflectance of *Z. capricorni*

The chromatic changes that occurred in the leaves of *Z. capricorni* during laboratory experiments were intrinsically the same whether they were induced by a difference between high light and moderate (control) light treatment, or moderate light and low light treatment. Reductions in light intensity led to increases in the concentrations of light harvesting pigments (chlorophylls *a*, *b*, lutein, violaxanthin, neoxanthin) plus β-carotene, and decreases in chlorophyll *a*:*b*, total carotenoids:total chlorophyll, VAZ: total carotenoids, VAZ:total chlorophyll and in the proportion of VAZ occurring in the de-epoxidised state. The reverse patterns occurred when irradiance intensity was increased. However, the most important and significant short term pigment change characterizing low light stress was the decrease in chlorophyll *a*:*b* while xanthophyll cycle pigment responses provided a rapid and significant indicator of high light stress in *Z. capricorni*.

Low light stress

The most distinctive change in the spectral signature of *Z. capricorni* after exposure to low light was increased absorption of the shorter red wavelengths, initially between 618-648 nm and widening to 604-657 nm by day 13 of the experiment. This clearly represented a broadening of the red absorption trough due to increasing chlorophyll *b* concentration. The fact that this spectral reflectance difference became statistically significant before the pigments had significantly adjusted may be due to the lower sample sizes used for pigment analysis ($n = 5$) compared to spectral reflectance measurements ($n = 12$), or it may indicate that the spectral response is extremely sensitive to this type of stoichiometric change. Chlorophyll *a* concentration also increased in shaded seagrass leaves (though not significantly in the first two weeks) but there was no corresponding increase in absorption of longer red wavelengths (660-690 nm) or any movement of the red edge. Chlorophylls are highly efficient absorbers of red and blue light so the main effect of increased chlorophyll is the enhanced absorption of green and far-red wavelengths where chlorophyll has a low absorption coefficient (Björkman 1981). Presumably the chlorophyll *a* absorption features were already saturated at the pigment concentrations found in *Z. capricorni* acclimated to a moderate light climate at the start of the experiment. A further significant decrease in the spectral reflectance of low light grown *Z. capricorni* occurred in the wavelengths 584-598 nm after 13 days of shade treatment. This change was probably also associated with the general increase in chlorophyll *a* and especially chlorophyll *b* enhancing absorption across all visible wavelengths.

The spectral reflectance of low light stressed *Z. capricorni* increased significantly in the green peak wavelengths. The spectral reflectance changes were again apparent after only seven days (in the region

553-570 nm) but had increased into the shorter green wavelengths (532-572 nm) after two weeks of low light treatment. This change is difficult to explain in terms of leaf pigment content because theoretically an increase in chlorophyll concentration should decrease reflectance of the green wavelengths, particularly when associated with somewhat lower total carotenoid:total chlorophyll. Even though the quantity of light harvesting carotenoids did increase, the increase in chlorophyll was proportionally greater and would have masked the 'yellow' reflectance of lutein and violaxanthin. Elevated green peak reflectance was accompanied by increased NIR reflectance after two weeks of shading so it seems more likely that these significant spectral reflectance differences were associated with structural or anatomical differences in the seagrass' leaf, cell or chloroplast structure. Shade grown plants tend to have smaller cells but larger chloroplasts and thicker grana stacks (Nishio *et al.* 1994; Robinson 1999; Schiefthaler 1999). Compositional and structural changes in the pigment-protein complexes and chloroplast membranes (Björkman and Demmig-Adams 1995; Nyitrai 1997) may also influence reflectance from a leaf grown in sun or shade. Chloroplast streaming in response to light intensity has been observed in the leaves of the seagrass *H. stipulacea* and probably occurs in other seagrasses. Drew (1979) observed that the chloroplasts of *H. stipulacea* assembled in a cluster around the nucleus in response to high light. In low light, the chloroplasts maximise absorption by spreading themselves out against the cell walls facing the light source (Jeffrey 1981). In addition, Abal *et al.* (1994) and Grice *et al.* (1996) recorded marked differences in the thickness of *Z. capricorni* leaves and in the size and number of the lacunae of leaves grown under high and low light conditions. Light scattering within the leaves and consequently, surface reflectance, must be considerably altered by such large changes in internal structure. Unfortunately the effects of morphological change are difficult to interpret and a great deal more research would be required to determine the source of spectral reflectance differences that might be related to differences in scattering.

After three months of the experiment, many of the shade treated samples had died back or deteriorated to such an extent that remaining tanks contained a mixture of leaves in various states of senescence or decay. Hence, data were too variable to show any consistent spectral reflectance differences between control and shaded *Z. capricorni* even though significant pigment differences were recorded on day 86.

High light stress

Chlorophyll dominated the spectral reflectance response of *Z. capricorni* leaves in the high light stress experiment as it did in the low light experiment, in spite of the fact that there were no significant differences recorded in the concentrations of chlorophyll *a* or *b* between control and high light treated samples. The first significant spectral reflectance difference observed was an increase in the NIR reflectance of high light exposed *Z. capricorni*, probably linked to the onset of a shift in the red edge. This was caused by a broadening of the chlorophyll *a* absorption trough for control samples that was already apparent in the spectral signature by day 7. Total chlorophyll increases in control samples were greater than the increases seen in high light treated samples, even though the difference between them was not significant. The chlorophyll absorption trough continued to widen in the spectra of controls as chlorophyll concentrations, especially chlorophyll *a*, increased through to day 17 and this led to an obvious shift in REP toward longer wavelengths. In addition, higher chlorophyll content significantly enhanced the absorption of far-green and red wavelengths (588-646 nm), the region where the strongest significant differences in the spectral reflectance of control and high light stressed samples were observed. As observed for low light stressed *Z. capricorni*, the increase in total chlorophyll concentration of control samples was also associated with significantly higher reflectance of the green peak (523-580 nm) wavelengths that are weakly absorbed by photosynthetic pigments. Presumably, this was again due to changes in internal leaf and cell structure.

While adjustments in the size of the VAZ pool and in the proportion of VAZ occurring as Z were the most significant pigment indicators of high light stress in *Z. capricorni,* these changes had very little impact on the spectral reflectance of seagrass leaves. Pre-darkened leaves of terrestrial plants show increased absorption of wavelengths around 500-560 nm upon exposure to strong irradiance (Björkman and Demmig-Adams 1995). The absorbance increase is made up of two elements; the maximum change occurs around 510 nm and is caused by de-epoxidation of V to Z because V, A and Z differ in their absorption spectra. The second change that peaks around 540 nm is thought to reflect conformational changes in the thylakoid membrane related to non-radiative dissipation (Björkman and Demmig-Adams 1995). Reflectance was significantly lower at 531 nm for high light stressed samples and this may be due to greater conversion of the xanthophyll cycle pigments compared to control samples (Gamon *et al.* 1992) but since there was no significant increase in absorption below 523 nm, it is more likely to be a consequence of the enhanced green peak of control samples instead. A green shift to shorter wavelengths occurs when the shoulders of the green peak move out due to increased reflectance in this region but this kind of spectral change is more typically associated with increased carotenoid:chlorophyll as a result of chlorosis in stressed plant leaves (Gamon *et al.* 1990; 1992). In the current experiment, significantly higher total carotenoid:total chlorophyll was a feature of high light stressed *Z. capricorni* but these samples actually displayed lower green peak reflectance. Furthermore, there was no sign of chlorophyll degradation that would indicate chlorosis in any of the *Z. capricorni* samples.

Non-significant modifications to the concentrations of carotenoids did make a significant difference to the spectral reflectance of *Z. capricorni* in the blue wavelengths. High light stressed samples contained slightly less of the light harvesting carotenoids (as well as chlorophyll) but slightly more of the VAZ pigments and this may have been sufficient to have increased spectral reflectance in several blue wavelength regions, particularly where carotenoids absorb maximally between 446-466 nm. However, these spectral reflectance differences were of extremely small magnitude (around 0.1-0.5% difference in reflectance).

Anthocyanins absorb green wavelengths and reflect red wavelengths. This distinctive pattern is clearly evident in field grown *Z. capricorni* (Figure 4.2) and even more so in the reflectance of *Z. capricorni* on the initial day of the high light stress experiment where red reflectance at 632-635 nm actually exceeded green reflectance (around 550 nm). The minor peaks and troughs that can be seen across the orange shoulder equate to the wavelengths of minimum and maximum anthocyanin absorbance recorded *in vitro* for this species, if the typical *in vivo* spectral shift toward longer wavelengths is taken into account. Despite the contribution that anthocyanins apparently make to the general shape of the spectral signature, the relative concentration of these accessory pigments did not appear to have any influence on the reflectance of differently treated *Z. capricorni* samples. There was no difference in the anthocyanin content of high light treated and control leaves though variability was high for both (predominantly because of low sample size, $n = 4$) but anthocyanin content increased from around 6 absorbance units at day 1 to 10 absorbance units at day 13 for both treated and control *Z. capricorni*. Concurrent with the anthocyanin increase was a counter-intuitive rise in green peak reflectance and fall in red reflectance for control (i.e. shadecloth covered) samples while high light treated *Z. capricorni* remained stable at close to initial day reflectance in these wavelengths. The same reflectance change was observed in shade treated *Z. capricorni* samples during the low light stress experiment, yet the anthocyanin content of both treatments in this experiment dropped from around 23 to 18 absorbance units by day 13. Furthermore, the green peak was far more prominent and red reflectance was lower for *Z. capricorni* leaves throughout the whole of the low light stress experiment compared to the high light experiment and field samples, even though anthocyanin contents were two to four times higher in those samples. Neill and Gould (1999) and Sims and Gamon (2002) previously noted that anthocyanins had negligible impact on red reflectance which was better correlated with chlorophyll concentration, although their presence did reduce green reflectance to some extent. Spectral reflectance therefore appears to be controlled mostly by chlorophyll content and the structural characteristics of the leaf while, in comparison, accessory pigments have little influence on spectral reflectance differences between sun and shade adapted seagrass leaves. Short term interconversion of the VAZ cycle pigments (within minutes or hours) might be detected in the spectra of plant leaves as suggested by Bilger *et al.* (1989) and Gamon *et al.* (1990; 1992), e.g. on exposure to sudden strong irradiance or during a diurnal cycle. However, when comparing plants grown under different light conditions for days, weeks or months the effect of changing chlorophyll content on the spectral response would swamp the input from other pigments (see Peñuelas *et al.* 1994).

Dawson and Dennison (1996) make the observation that species exhibiting high morphological plasticity such as *Z. capricorni* are more tolerant of high light conditions. Thicker leaves and larger epidermal cells allow more sensitive organelles to move into deeper, more protected layers. The greater volume of lacunal air space that occurs in the thicker leaves of high light grown seagrass should result in significant reflectance differences from shade grown plants when combined with chloroplast movements and lower chlorophyll concentrations.

5.5.8 Application to remote sensing

There were characteristic chlorophyll *a*:*b* changes that signaled low light stress and characteristic VAZ pigment changes that indicated high light stress in *Z. capricorni* but the pigment changes that were significant were not always responsible for the spectral reflectance differences observed between light grown and shaded leaves. In each case, the spectral reflectance of *Z. capricorni* was very sensitive to differences in the chlorophyll concentration of the leaves. Seagrass leaves grown in a higher light climate could always be discriminated from those grown under lower light conditions on the basis of reflectance changes linked to lower chlorophyll concentrations within 1-2 weeks of the change in irradiance regime. Morphological changes in the leaves associated with photoacclimation also appeared to produce consistent differences in reflectance. There is clearly some potential for the spectral reflectance monitoring of light stress in seagrasses. Agawin *et al.* (2001) found that the total chlorophyll contents of three tropical seagrass species varied in a similar pattern with seasonal changes and were not subject to species specific modulation. Developments in the spectral reflectance monitoring of stress may therefore be generally applicable to any seagrass species. However, even the most significant spectral responses induced by light stress were of relatively small magnitude (e.g. compared to the differences observed between species; Figure 4.2) and it is yet to be seen whether they can be discriminated by operational remote sensing instruments.

Detection of small spectral reflectance changes symptomatic of plant stress will depend on a wide range of factors including the resolution and SNR of the sensor, the depth and composition of the water column and atmosphere, and the homogeneity of the seagrass meadow (Hochberg and Atkinson 2000; Lubin *et al.* 2001). An airborne sensor of high spectral and radiometric resolution would be best suited to this task. Such a sensor would require either programmable bands or the appropriate set of band centres, band widths and interband centre distances to collect image data in the specific set of narrow wavebands required to detect stress-related spectral reflectance changes, i.e. a sensor of similar specifications to that recommended for the discrimination and mapping of seagrass species (see Section 4.5.5). An appropriate band set for remote sensing light stress in seagrasses could include narrow bands centred around;

- the green peak at 550 nm (bandwidth 5-15 nm)
- orange shoulder reflectance at 635 nm (bandwidth 5-10 nm)
- a spectrally invariant band located in the blue wavelengths that are minimally affected by attenuation in a water column, e.g. at 495 nm (bandwidth 5-15 nm)
- the chlorophyll absorption region at 682 nm which may also act as a spectrally invariant band (bandwidth 5 nm)
- and possibly the green edge at 530 nm (bandwidth 5 nm).

The results of laboratory experiments may not accurately represent the pigment concentrations, stoichiometry, photosynthetic rates or level of photoinhibition that occurs in seagrass plants growing under natural conditions in the field. If these measurements are to be ecologically meaningful, they should be carried out on naturally occurring material in the field or under conditions approximating these (Major and Dunton 2000). Environmental factors such as temperature, CO_2 concentration, nutrient availability, etc, may have a great effect on photosynthesis and a natural, variable light regime is more effective than an artificial one in inducing chromatic changes in plant pigments (Thayer and Björkman 1990). However, this study did not intend to report the absolute values of photosynthetic parameters or the intensities at which light stress will occur in the field but instead aimed to identify spectral reflectance responses related to physiological and biochemical changes that occur when seagrasses are stressed by changes in irradiance. For comparative purposes, the paired replicates in these laboratory experiments (in which all factors other than irradiance level are identical for control and treated samples) were sufficient to determine the general response to 'high' 'low' and 'comfortable' light climate for *Z. capricorni*. The spectral reflectance changes identified in this chapter were applied in the development of spectral indices that may be useful for the detection of light stress in seagrasses *in situ* using remotely-sensed data (see Chapter 6).

Chapter 6
Spectral Indices for the Prediction of Light Stress

S.K. Fyfe

6.1 Introduction

6.1.1 Vegetation indices

The advantages of collecting accurate and timely information on vegetation condition using remote sensing have been recognised since the 1970's when the first of the Landsat environmental satellites was launched (Jensen 1996). Vegetation indices (VI's) use simple algorithms to reduce information contained in multiple bands of spectral reflectance data down to a single number per pixel that predicts or assesses a required plant canopy characteristic. The simplest and best known VI's are the broad band Simple Vegetation Index (SVI = NIR/red) and Normalised Difference Vegetation Index (NDVI = NIR – red/NIR + red) (e.g. Rouse *et al.* 1973; Tucker 1979; Goward *et al.* 1991). These linear combinations of the red and NIR image wavebands are commonly used in remote sensing to predict vegetation biomass and LAI, but they have also been used to estimate yield, canopy photosynthetic capacity and productivity (Field *et al.* 1995). In particular, the NDVI has been applied weekly to NOAA AVHRR meteorological satellite data to monitor the condition of global vegetation since the early 1990's (Brown *et al.* 1993 cited in Jensen 1996). However, since the NDVI principally assesses the amount of 'green' plant biomass, it does not necessarily indicate vegetation condition or stress *per se* (Gamon *et al.* 1995a). Many types of healthy vegetation are not necessarily green or high in biomass (e.g. arid zone plants) and therefore variations on the basic VI theme including orthogonal and soil-adjusted VI's have been developed to account for the confounding effects of canopy closure and background reflectance on the signal (e.g. Kauth and Thomas 1976; Richardson and Wiegand 1977; Heute 1988; Spanner *et al.* 1990; Baret and Guyot 1991; Malthus *et al.* 1993).

6.1.2 Narrow band vegetation indices

If VI's are constructed from narrow rather than broad wavebands it may also partly 'remove' (or avoid) the spectral disturbances caused by external factors such as soil background, illumination or canopy geometry (Peñuelas and Fillela 1998). More importantly, the availability of individual wavelengths or very narrow wavebands allows the user to target very specific absorption features for use in VI's that would be otherwise unavailable in broad band spectral data. Researchers have therefore moved towards the application of narrow band indices to determine the absolute and relative concentrations of chlorophyll *a*, *b*, carotenoids and other accessory pigments in plants. A vast array of hyperspectral reflectance indices have been developed and tested to predict the biophysical characteristics (e.g. Malthus *et al.* 1993; Thenkabail *et al.* 2000) and physiological status (reviewed in Peñuelas and Fillela 1998) of agricultural crops. This drive to produce robust and generally applicable hyperspectral VI's has not only come from their potential application in high spectral resolution airborne or satellite remote sensing but from their current use in ground based spectrometry at the leaf and canopy scale. At the ground level, spectroradiometers supply a rapid, non-invasive and cost effective alternative to biochemical methods of obtaining information on plant photosynthesis, pigment content, nutrition and stress.

There are strong relationships between agricultural crop biomass, LAI, canopy structure, pigment content and specific narrow bands in the far red (650-700 nm), green (500-550 nm) and NIR (Thenkabail *et al.* 2000).

Narrow band SVI's and NDVI's have been widely applied to predict green biomass and chlorophyll content, often considered as surrogates for plant vigour and photosynthetic capacity (e.g. Peñuelas *et al.* 1993; 1995b; Carter 1994; Carter and Miller 1994; Carter *et al.* 1998; Luther and Carroll 1999; Thenkabail *et al.* 2000). Researchers invariably find the strongest relationships are achieved when one or more NIR wavelengths are included in the VI to act as a reference wavelength (e.g. Kupeic 1994; Curran *et al.* 2001) to the visible pigment absorbance or reflectance features of interest (e.g. Buschmann and Nagel 1993; Maccioni *et al.* 2001). The reference band is ideally positioned in the NIR to avoid absorption by photosynthetic pigments and therefore normalise the index band for illumination or scatter effects, including the effects of variable leaf cell structure on reflectance (Datt 1998b; Moran *et al.* 2000). Since reflectance at the centre of a wavelength feature will reach an asymptote when saturated, and this occurs at quite low chlorophyll concentrations, most narrow band VI's use an off centre wavelength for the pigment of interest. For example, NDVI's built using index wavelengths around the RE (695-705nm) are more commonly used to estimate chlorophyll concentration than those using the region of maximal chlorophyll *a* absorbance around 660 nm (Curran 1989; e.g. Carter 1998; Gitelson and Merzlyak 1996; Blackburn and Steele 1999). Datt's (1998a) best index for prediction of eucalypt leaf chlorophyll content combined red edge reflectance with NIR (R_{850}/R_{710}) but when applied to CASI remote sensing imagery the index was strongly influenced by soil background (Coops *et al.* 2003). To improve the generality of SVI's and NDVI's in estimating chlorophyll content for different species from widely varying functional groups, Sims and Gamon (2002) modified the ratios by adding a constant to the reflectance values that would compensate for differences in leaf surface reflectance. The modified ratio, $mND_{705} = (R_{750} - R_{705})/(R_{750} + R_{705} - 2R_{445})$, was highly correlated with chlorophyll concentration measured from different aged leaves of 53 plant species. A further alternative is the use of the green peak which is possibly more responsive to chlorophyll content than the RE. The Green NDVI $((R_{801} - R_{550})/(R_{801} + R_{550}))$ and combination indices such as the Modified Chlorophyll Absorption in Reflectance Index (MCARI= $[(R_{700} - R_{670}) - 0.2(R_{700} - R_{550}) \times (R_{700}/R_{670})]$) have been shown to respond to both leaf chlorophyll content and background reflectance in the remote sensing of plant canopies (Daughtry *et al.* 2000). A possible concern with the use of green VI's to predict chlorophyll concentration is that green reflectance is controlled by a range of factors other than chlorophyll including accessory pigments and leaf structure. If plant stress results in chlorosis, then such indices will probably perform well but stressed plants are not always chlorotic (e.g. Malthus and Madeira 1993; Stone *et al.* 2003).

Much research has been carried out on methods to develop 'optimal' narrow band indices for predicting the individual concentrations of chlorophylls *a*, *b* and carotenoids in plant leaves. For example, correlograms have been applied to choose the best and worst correlated reflectance or derivative wavelengths for a particular pigment or measure of plant stress and these are subsequently utilised in SVI or NDVI algorithms (e.g. Carter 1993; Blackburn 1998; Datt 1998b; Luther and Carroll 1999). Chappelle *et al.* (1992) used ratio analysis to select the optimal visible wavelengths for prediction of chlorophyll *a*, *b* and carotenoid concentration and applied these in SVI type algorithms with a treatment insensitive NIR wavelength. Small differences in absorbance will be amplified when the reflectance spectrum of a stressed plant leaf is divided by a reference spectrum collected from healthy plant leaves and multiplied by 100 to give a percent value. The 'sensitivity analysis' of Carter (1993; 1994) also used by Moran *et al.* (2000) applies the same principle but employs the difference spectrum (healthy - stressed leaf spectrum) which is then normalised as a ratio of the reference spectrum. In each case, however, strong intercorrelation between chlorophyll *a*, *b* and the carotenoids impedes the selection of unique wavelengths that might be used to independently predict the concentration of each of the photosynthetic pigments (Curran 1989; Datt 1998a; 1998b; Blackburn 1998; Sims and Gamon 2002). While many authors use wavelengths from the regions 675-680, 635-650 and 450-500 nm to estimate the concentrations of chlorophyll *a*, *b*, and the carotenoids respectively, the best correlations with each of the pigments usually occurs in the chlorophyll *a* region.

Stepwise regression has been applied to select the best multiple combination of independent wavelengths from the available reflectance or derivative reflectance bands that will optimise the strength of the correlation with a biochemical parameter of interest (e.g. Kupeic 1994; Thenkabail *et al.* 2000; Curran *et al.* 2001). Curran and co-workers have achieved some of the best reported results for chlorophyll determination using variations on this methodology to reduce additive and multiplicative scattering scatter effects. Their results always combine 2-3 blue, green-red shoulder and/or red absorption wavelengths with an NIR wavelength (e.g Curran *et al.* 2001). However, a critical problem with this method is 'overfitting' (i.e. using more spectral channels than experimental samples to obtain a highly maximum r^2 value), plus there are problems associated with the selection of noncausal wavelengths due to pigment intercorrelation and with collinearity among the wavelengths (Curran 1989; Curran *et al.* 2001). Thenkabail *et al.* (2000) tested this method and found that combinations of 2-4 narrow bands in OMNBR (optimum multiple narrow band reflectance) models were sufficient to explain most (64-92%) of the variability in crop biophysical parameters. Consequently, they investigated all possible two-band combinations of the 490 available channels (of various band widths as well as positions) in narrow band NDVI's. The best of these was further tested to see if soil adjustment (SAVI) or nonlinear fitting could improve their predictive accuracy. The four-band OMNBR models performed only marginally better than two-band NDVI type models in predicting crop biophysical variables and Thenkabail *et al.* (2000) noted that OMNBR models do not offer the simplicity and flexibility of use with different types of spectral reflectance data that the NDVI indices provide. Other multivariate approaches include the derivation of principal components, discriminant analysis based on a combination of narrow band spectral index inputs (e.g. Peñuelas *et al.* 1994) and canonical correlation (Korobov and Railyan 1993). Neural network models are also under development for the assessment of plant stress and have shown promise in the prediction of photosynthetic efficiency in seagrass leaves (e.g. Ressom *et al.* 2003; Sriranganam *et al.* 2003). Such models have the advantage that they do not assume a linear relationship between spectral response and the parameter of interest.

It is often difficult to distinguish the spectral contribution of chlorophyll content and/or plant stress from that of canopy biomass and LAI using SVI and NDVI style ratios. Derivative analysis was initially proposed as a method to normalise for the effects of scattering from remote sensing data but it was also found to be useful for eliminating canopy architecture, soil background, view angle and atmospheric effects from spectral reflectance data to provide more sensitive measures of plant stress than the classic VI's (Demetriades-Shah *et al.* 1990; Blackburn and Pitman 1999). For example, the red edge is represented by the maximum first derivative wavelength in the spectrum of vegetation. The red edge has been considered the best measure of chlorophyll concentration for plant leaves and canopies (Field *et al.* 1995) and therefore it has proven a useful indicator of many forms of plant stress in both ground spectroradiometer and remote sensing studies (e.g. Horler *et al.* 1983; Boyer *et al.* 1988; Rock *et al.* 1988; Boochs *et al.*

1990; Vogelmann *et al.* 1993; Peñuelas *et al.* 1995b; Curran *et al.* 1998; Blackburn and Steele 1999). However, REP is not an infallible measure since saturation can be reached at high chlorophyll concentrations plus the relationship at lower chlorophyll concentrations will only hold if canopy cover is high because background reflectance and LAI can confound the magnitude of far red-NIR reflectance (Curran *et al.* 1990; Blackburn and Pitman 1999). Canopy models may help overcome many of these problems in the operational use of REP (Sampson *et al.* 2003).

6.1.3 Vegetation indices utilizing only visible wavelengths

Vegetation indices that utilise NIR reflectance have been applied to estimate the biomass of intertidal macroalgae (e.g. Guillaumont *et al.* 1993) and wetland vegetation (e.g. Smith *et al.* 1998), and a variety of biophysical and physiological parameters from emergent aquatics (e.g. Peñuelas *et al.* 1993). However, they can only be successfully utilised on aquatic vegetation when the canopy lies at, or emerges above, the water surface. Near infrared radiation is rapidly attenuated by water so any methodology for the remote sensing of submerged vegetation must instead concentrate on the water penetrating wavelength regions in the visible spectrum (see Figure 4.24; Kirk 1994; Mobley 1994). The best known examples of this are the 'ocean colour' indices that are regularly applied to satellite data from sensors such as the CZCS (Coastal Zone Colour Scanner) to produce oceanic maps of chlorophyll concentration. Chlorophyll in a water body increases the ratio of green to blue upwelling irradiance and this important relationship has enabled the development of algorithms to quantify the phytoplankton concentrations of aquatic systems from satellite and airborne data (Kirk 1994).

Changes in the visible wavelengths can be clearly observed in the spectra of stressed vegetation (e.g. Gates *et al.* 1965; Gausman 1984; Carter 1993).

Chlorophyll degradation or modifications to the ratio of chlorophylls to carotenoids are typical symptoms of stress that have been correlated with reflectance at, or in off-centre wavelengths around 450, 550 and 650 nm (Gausman 1984; Saxena *et al.* 1985; Buschmann and Nagel 1993; Daughtry *et al.* 2000). Table 6.1 lists some examples of remote sensing indices that have been developed for the purpose of detecting stress in plants using only visible wavelengths. Many of these ratios, including some individual visible wavelengths, have been found to correlate strongly with chlorophyll *a* or total chlorophyll content. Unfortunately, the changes in visible reflectance caused by stress may be subtle compared to changes observed at the red edge. For example, Moran *et al.* (2000) found significant relationships between the total chlorophyll concentration of nitrogen stressed spruce needles and reflectance indices built on visible wavelengths; i.e. Photochemical Reflectance Index (PRI; $r^2 = 0.74$) and R_{550} ($r^2 = 0.61$). However, chlorophyll concentration was better correlated with indices or derivatives utilizing NIR wavelengths in their calculation; i.e. REP ($r^2 = 0.89$), R_{698}/R_{760} ($r^2 = 0.91$) and the Structure Independent Pigment Index ($r^2 = 0.76$; SIPI = $(R_{800} - R_{445})/(R_{800} - R_{680})$) designed by Peñuelas *et al.* (1995a) to predict carotenoid:chlorophyll while accounting for the confounding effects of differing leaf structure and surface quality. Since reflectance at 550 nm is often highly correlated with RE reflectance (R_{700}) ($r^2 = 0.997$; Chappelle *et al.* 1992) then the green peak may provide a suitable substitute index wavelength in VI's designed to estimate chlorophyll concentration.

Visible reflectance indices targeting the minor pigments appear to be less reliable, although Peñuelas *et al.* (1993; 1994) have found the Normalised Pigment Chlorophyll Index (NPCI) to be a good indicator of the ratio of total pigments to chlorophyll *a*, which typically increases as plants become

Table 6.1. (A) Some visible reflectance indices developed for the estimation of a selection of physiological stress parameters in vegetation, and (B) correlation coefficients (r) or coefficients of determination (r^2) obtained in studies that have applied these indices to airborne remote sensing (RS) data and/or ground spectrometer data at the leaf or canopy level. Only results $r > 0.6$ (or significant) for leaf or canopy level and $r > 0.4$ (or significant) for RS level have been included. Statistical significance of the result has been included where supplied by the author.

A. VISIBLE REFLECTANCE INDICES

Index	Description	Reference
Anthocyanin content index	$ACI = (\Sigma R_{600} - R_{700})/(\Sigma R_{500} - R_{600})$	Gamon and Surfus 1999
Max G	maximum reflectance at the green peak	Tageeva *et al.* 1961
Yellowness index	YI = the centre divided difference finite approximation of the 2nd derivative of the reflectance spectrum in the green-red region	Adams *et al.* 1999
Photochemical reflectance index	$PRI_1 = (R_{531}-R_{570})/(R_{531}+R_{570})$ $PRI_2 = (R_{570}-R_{539})/(R_{570}+R_{539})$ $PRI_3 = (R_{550}-R_{530})/(R_{550}+R_{530})$ or $PRI_3 = (R_{550}-R_{531})/(R_{550}+R_{531})$	Gamon *et al.* 1992 Peñuelas *et al.* 1995c Peñuelas *et al.* 1994 Zarco-Tejada *et al.* 1999
Lichtenthaler	$Licht. = (R_{440}/R_{690})$	Lichtenthaler *et al.* 1996
Plant stress index	$Carter = (R_{695}/R_{420})$	Carter 1994
Green index	$G = (R_{554}/R_{677})$	Zarco-Tejada *et al.* 1999
Normalised total pigment chlorophyll *a* index	$NPCI = (R_{680}-R_{430})/(R_{680}+R_{430})$	Peñuelas *et al.* 1993
Green first derivative	GF = value of 1st derivative maximum in the green (i.e. green edge)	Peñuelas *et al.* 1993
Green-green 1st derivative normalised difference	$GGFN = (D1_{green\ edge}-D1_{min\ green})/(D1_{green\ edge}+D1_{min\ green})$	Peñuelas *et al.* 1993

Table 6.1. (cont.)

A. VISIBLE REFLECTANCE INDICES (cont.)

Index	Description	Reference
1st derivative of pseudoabsorbance	$D1P_{\lambda}$= 1st derivative of (log 1/R) at specified wavelength	Yoder and Pettigrew-Cosby 1995
2nd derivative of pseudoabsorbance	$D2P_{\lambda}$= 2nd derivative of (log 1/R) at specified wavelength	Yoder and Pettigrew-Cosby 1995
Edge of the red edge	EREP = position of 2nd derivative peak centred at around 692 nm	Demetriades-Shah 1990

*$p<0.05$, **$p < 0.01$, ***$p < 0.001$. R, reflectance; D1, 1st derivative; D2, 2nd derivative

B. APPLICATION OF SOME VISIBLE REFLECTANCE INDICES

Species	Level	Stress parameter investigated	r or r^2	Reference
INDIVIDUAL REFLECTANCE WAVELENGTHS				
R_{480}				
7 tree, 3 shrub and 3 herb spp.	leaf	total chlorophyll content	-0.68*	Saxena *et al.* 1985
R_{526}				
3 Antarctic moss species	canopy	total chlorophyll content	0.06**	Lovelock and Robinson 2002
R_{531}				
sunflower *Helianthus annuus*	canopy	photosynthetic efficiency (F_v:F_m)	0.99	Gamon *et al.* 1990
R_{550}				
beans *Phaseolus vulgaris*	leaf	total chlorophyll content	0.76	Buschmann and Nagel 1993
corn *Zea mays*	leaf	total chlorophyll content	0.77	Daughtry *et al.* 2000
Engelmann spruce *Pinus engelmanii*	leaf	chlorophyll *a* content	0.6	Moran *et al.* 2000
" "	leaf	chlorophyll *b* content	0.60	" "
" "	leaf	total chlorophyll content	0.61	" "
3 Antarctic moss species	canopy	total chlorophyll content	0.08***	Lovelock and Robinson 2002

Table 6.1. (cont.)

B. APPLICATION OF SOME VISIBLE REFLECTANCE INDICES (cont.)

Species	Level	Stress parameter investigated	r or r^2	Reference
R_{570}				
7 tree, 3 shrub and 3 herb spp.	leaf	chlorophyll *a* content	-0.77**	Saxena *et al.* 1985
R_{610}				
7 tree, 3 shrub and 3 herb spp.	leaf	total chlorophyll content	-0.69*	Saxena *et al.* 1985
R_{620}				
7 tree, 3 shrub and 3 herb spp.	leaf	chlorophyll *a* content	-0.74**	Saxena *et al.* 1985
R_{632}				
beans *P. vulgaris*	leaf	total chlorophyll content	0.74	Buschmann and Nagel 1993
Max G				
3 *Eucalyptus* spp.	leaf	total chlorophyll content	0.53*	Stone *et al.* 2001
" "	leaf	photosynthetic efficiency (F_V:F_m)	0.75***	" "
" "	leaf	anthocyanin content	0.60**	" "
INDIVIDUAL DERIVATIVE WAVELENGTHS				
GF				
emergent wetland vegetation	canopy	total chlorophyll content	-0.84*	Peñuelas *et al.* 1993
" "	canopy	total carotenoids:chlorophyll *a* content	-0.85*	" "
" "	canopy	epoxidation state (EPS)	-0.85*	" "
$D1_{636}$				
sugar beet	canopy	total chlorophyll content	0.71	Demetriades-Shah *et al.* 1990
$D1_{656}$				
sugar beet	canopy	total chlorophyll content	0.61	Demetriades-Shah *et al.* 1990
EREP				
sugar beet	canopy	total chlorophyll content	0.75	Demetriades-Shah *et al.* 1990

Table 6.1. (cont.)

B. APPLICATION OF SOME VISIBLE REFLECTANCE INDICES (cont.)

Species	Level	Stress parameter investigated	r or r^2	Reference
D1P$_{452.6}$				
bracken *Pteridium aquilinum*	canopy	total carotenoid content	0.74	Blackburn 1998
D2P$_{664.3}$				
bracken *P. aquilinum*	canopy	chlorophyll *a* content	0.81	Blackburn 1998
" "	canopy	chlorophyll *b* content	0.91	" "
SIMPLE INDICES				
Licht.				
sugar maple *A. saccharum*	leaf	chlorophyll *a* content	0.74	Zarco-Tejada *et al.* 1999
" "	RS	chlorophyll *a* content	0.46	" "
" "	leaf	total chlorophyll content	0.73	" "
" "	RS	total chlorophyll content	0.46	" "
" "	leaf	total carotenoid content	0.74	" "
" "	leaf	photosynthetic efficiency (F_v:F_m)	0.61	" "
G				
sugar maple *Acer saccharum*	leaf	chlorophyll *a* content	-0.74	Zarco-Tejada *et al.* 1999
" "	RS	chlorophyll *a* content	0.46	" "
" "	leaf	total chlorophyll content	-0.72	" "
" "	RS	total chlorophyll content	0.44	" "
" "	leaf	total carotenoid content	-0.71	" "
Carter				
sugar maple *A. saccharum*	leaf	chlorophyll *a* content	-0.77	Zarco-Tejada *et al.* 1999
" "	RS	chlorophyll *a* content	0.46	" "
" "	leaf	total chlorophyll content	-0.75	" "
" "	RS	total chlorophyll content	0.47	" "
" "	leaf	total carotenoid content	-0.74	" "
" "	leaf	photosynthetic efficiency (F_v:F_m)	-0.64	" "

Table 6.1. (cont.)

B. APPLICATION OF SOME VISIBLE REFLECTANCE INDICES (cont.)

Species	Level	Stress parameter investigated	r or r^2	Reference
NORMALISED INDICES				
PRI₁				
14 perennial and 6 herb species	leaf	photosynthetic efficiency (F_v:F_m)	0.75***	Gamon *et al*. 1997
sunflower *H. annuus* on dark-light transition	leaf	epoxidation state (EPS)	0.98	Gamon and Surfus 1999
sugar maple *A. saccharum*	leaf	photosynthetic efficiency (F_v:F_m)	0.72	Zarco-Tejada *et al*. 1999
" "	RS	photosynthetic efficiency (F_v:F_m)	0.59	" "
Engelmann spruce *P. engelmanii*	leaf	chlorophyll *a* content	0.73	Moran *et al*. 2000
" "	leaf	chlorophyll *b* content	0.76	" "
" "	leaf	total chlorophyll content	0.74	" "
3 Antarctic moss species	canopy	total chlorophyll content	0.06**	Lovelock and Robinson 2002
" "	canopy	photosynthetic efficiency (F_v:F_m)	0.43***	" "
PRI₂				
sugar maple *A. saccharum*	leaf	photosynthetic efficiency (F_v:F_m)	-0.68	Zarco-Tejada *et al*. 1999
sugar maple *A. saccharum*	RS	photosynthetic efficiency (F_v:F_m)	0.63	Zarco-Tejada *et al*. 1999
PRI₃				
sunflower *H. annuus* over diurnal cycle	leaf	photosynthetic efficiency (F_v:F_m)	-0.67	Peñuelas *et al*. 1994
sugar maple *A. saccharum*	leaf	photosynthetic efficiency (F_v:F_m)	-0.68	Zarco-Tejada *et al*. 1999
" "	leaf	chlorophyll *a* content	-0.61	" "
" "	RS	chlorophyll *a* content	0.59	" "
" "	leaf	total chlorophyll content	-0.63	" "
" "	RS	total chlorophyll content	0.58	" "
NPCI				
emergent wetland vegetation	canopy	total chlorophyll content	-0.79*	Peñuelas *et al*. 1993
" "	canopy	total carotenoids:chlorophyll *a* content	0.89*	" "
" "	canopy	epoxidation state (EPS)	0.67*	" "
sunflower *H. annuus* across seasons	leaf	total chlorophyll content	-0.64	Peñuelas *et al*. 1994

Table 6.1. (cont.)

B. APPLICATION OF SOME VISIBLE REFLECTANCE INDICES (cont.)

Species	Level	Stress parameter investigated	*r* or r^2	Reference
DERIVATIVE INDICES				
GGFN				
emergent wetland vegetation	canopy	total chlorophyll content	-0.90*	Peñuelas *et al.* 1993
" "	canopy	total carotenoids:chlorophyll *a* content	0.90*	" "
" "	canopy	epoxidation state (EPS)	-0.90	" "
sunflower *H. annuus* over diurnal cycle	leaf	total carotenoids:chlorophyll *a* content	-0.60	Peñuelas *et al.* 1994
" "	leaf	photosynthetic efficiency (F_v:F_m)	-0.63	" "
OTHER INDICES				
ACI				
developing oak leaves *Quercus agrifolia*	leaf	photosynthetic efficiency (F_v:F_m)	0.92	Gamon and Surfus 1999
3 *Eucalyptus* and 5 rainforest tree spp.	leaf	anthocyanin content	0.59**	Coops *et al.* 2004
Eucalyptus saligna	RS	anthocyanin content	-0.76*	" "
YI				
soybean *Glycine max*	canopy	total chlorophyll content	0.97	Adams *et al.* 1999

*p<0.05, **p < 0.01, ***p < 0.001. R, reflectance; D1, 1st derivative; D2, 2nd derivative stressed. This index should be relatively independent of the effects of canopy structure and biomass. Moran *et al.* (2000) on the other hand, observed that the carotenoid concentration of healthy and nitrogen stressed conifer seedlings did not change across a wide enough range when stressed by nutrient deficiency to produce any relationship with the SIPI. Furthermore, individual photosynthetic pigments are highly intercorrelated with each other at the leaf level, and with the LAI in canopy reflectance measurements, and as such the quantification of individual pigments may be a problem for the remote sensing of green plants (Blackburn 1998; Sims and Gamon 2002).

Since normalization has been shown to increase the reliability of vegetation indices, then the NIR reference band might be substituted with a pigment insensitive reference band in the visible instead. It is impossible to select a band where no absorption occurs since at high pigment concentrations absorption occurs to some extent across all visible wavelengths. However, it is possible to select visible regions where absorption does not change measurably across a wide range of pigment concentrations. Datt (1998b) for example, found reflectance around 480 nm and 680 nm to be almost invariant with chlorophyll content while Sims and Gamon (2002) chose 445 nm to represent leaf surface reflectance characteristics rather than pigment content. The 'blue flat area' is a region of absorption by chlorophyll and carotenoid pigments but it is saturated at very low concentrations and rarely changes perceptibly (Sims and Gamon 2002). This region (between 380-480 nm) was also used by Maccioni *et al.* (2001) to measure external reflectance which can be subtracted from measured spectra to give 'internal reflectance' that correlates well with log chlorophyll content, particularly at around 550 nm and 700 nm. The Normalised Phaeophytinization Index (NPQI = $(R_{415}-R_{435})/(R_{415}+R_{435})$) has been used to predict chlorophyll degradation (Peñuelas *et al.* 1995d) and therefore photosynthetic stress. Chlorophyll is converted to phaeophytin when exposed to excessive light, low pH and gases such as SO_2 and NO_2 during senescence (Lichtenthaler 1987). However an index based solely on such short wavelengths may prove unreliable for operational remote sensing since blue wavelengths are scattered strongly in the atmosphere and are also absorbed by dissolved organic substances (gilvin) in estuarine water columns (Kirk 1994).

The use of derivative spectra to predict stress in plants has been dominated by the generally good relationship between REP and chlorophyll content, however, Demetriades-Shah *et al.* (1990) proposed a second derivative index centred at 636 nm that was a much better predictor of canopy chlorosis (based on leaf REP) than either the SVI or canopy REP. In addition, when remote sensing canopies, the position of the 'edge of the red edge', i.e. the 2nd derivative peak centred around 692 nm, may be more sensitive to chlorophyll concentration than canopy REP. This is because increasing LAI due to leaf overlap in a healthy canopy will increase scattering in the NIR and push the RE in the opposite direction to the shift associated with increasing absorption in the red trough caused by higher chlorophyll concentration. Conversely, scattering does not affect the 2nd derivative 'edge of the red edge'. Blackburn (1998) found the best individual correlations with chlorophyll *a* and *b* occurred using the 2nd derivative of pseudoabsorbance (log (1/R)) at 664.3 nm, while carotenoid content was best correlated with the 1st derivative of pseudoabsorbance at 452.6 nm. Peñuelas *et al.* (1993) found 2nd derivative indices too noisy to yield usable information but the normalised difference between the maximum and minimum 1st derivative value in the green (GGFN) predicted total chlorophyll concentration, carotenoids:chlorophyll *a* and EPS equally well and slightly better than the non-normalised value of the maximum 1st derivative in the green region (GF) for wetland emergents.

Derivative indices of the green region were generally good indicators of plant physiological status and acted to highlight chlorosis from background reflectance and the effects of canopy architecture (Peñuelas *et al.* 1993; 1994). Interestingly, Kupeic (1994) found 1st derivatives correlated no better with chlorophyll content than reflectance and indicated that there was no difference between using reflectance or pseudoabsorbance.

Adams *et al.* (1999) were able to predict the chlorophyll content of manganese-deficient soybean leaves very accurately using the Yellowness Index (YI). The YI is a simple, three-point approximation of the spectral second derivative calculated from wavelengths in the visible that are thought to be sensitive to pigment concentration but relatively insensitive to changes in leaf structure or water content. This derivative index provides a measure of the change in shape of the spectral reflectance curve between the maximum visible reflectance near 550 nm and the minimum near 650 nm. As plants become more chlorotic, the shape of the orange shoulder between the green peak and red chlorophyll absorption trough changes from concave up to concave down with increasing stress (Lichtenthaler *et al.* 1996; Adams *et al.* 1999). This characteristic change in reflectance was observed for aging leaves by Gates *et al.* (1965) while Baret *et al.* (1987) subsequently used it to calculate the 'red slope' (between 580 and 660 nm) which they related to NDVI and the proportion of senescing leaves in a wheat canopy without actually relating it to chlorophyll content. Since the magnitude of reflectance at the green peak is a function of both carotenoids:chlorophylls and internal leaf structure, the shifts observed at both shoulders of the peak can provide useful information on various forms of plant stress. For example, a necrotic fungal disease in bean leaves caused a flattening of the green peak that affected reflectance on both shoulders of the peak without a concurrent change in chlorophyll content. Malthus and Madeira (1993) found that

1st derivative reflectance at 525 and 580 nm were highly correlated with infection although stronger correlation did occur at the RE around 720 nm.

The physiological reflectance index (Gamon *et al.* 1992) was developed to take advantage of the green shift in reflectance noted by Bilger *et al.* (1989) and Gamon *et al.* (1990) on the short wavelength shoulder of the green peak. Gamon *et al.* (1992) used it to predict the efficiency with which radiation is used for photosynthesis by normalizing the shoulder reflectance at or around 531 nm to a far green reference wavelength. After further development, the (renamed) photochemical reflectance index (PRI) (Peñuelas *et al.* 1995c) was shown to detect changes in spectral response associated with relative concentrations of xanthophyll cycle pigments within 10 minutes of a change in irradiance (Gamon and Surfus 1999). The PRI (PRI = $R_{ref} - R_{531}/R_{ref} + R_{531}$) therefore acts as an instantaneous measure of photosynthetic efficiency under certain stress conditions as well as a measure of the longer term acclimation of the pigments to the environment. Moran *et al.* (2000) found the PRI a very responsive measure of stress for spruce needles subjected to light and shade treatment in association with nitrogen fertilization treatment while other VI's such as the SIPI were independent of irradiance and responded only to chlorosis. However, this index is not always a successful indicator of plant stress, e.g. the PRI did not detect stress in pine needles infected with Dothistroma fungus (Stone *et al.* 2003). The relationship between PRI and the chlorophyll fluorescence parameters that detect photosynthetic light-use efficiency (F_v/F_m and $\Delta F/Fm'$), and PRI and the total carotenoid to total chlorophyll ratio, is consistent across a range of species, functional types and stress conditions, at least for those upper canopy leaves in full sun that are likely to be sensed by airborne scanners (Peñuelas *et al.* 1994; Gamon *et al.* 1997; Qiu *et al.* 1998; Zarco-Tejada *et al.* 1999; Lovelock and Robinson 2002; Sims and Gamon 2002). While Peñuelas *et al.* (1995b) developed the Structurally Independent Xanthophyll Index (SIXI = $R_{530}-R_{800})/(R_{570}-R_{800})$ as an alternative indicator of photosynthetic efficiency, this index has not been as widely applied by other authors as the PRI. Normalization using an NIR reference wavelength may not actually be an improvement in this case but the SIXI has yet to be tested on remote sensing data.

Increasing anthocyanin concentrations may be a visible symptom of particular types of stress in the leaves of many plant species and for these plants, the Anthocyanin Content Index (Gamon and Surfus 1999) appears to work well as a measure of stress at the leaf scale. However, when applied to high resolution airborne remote sensing data the ACI could not detect anthocyanin increases in the canopies of eucalypts related to insect damage (Coops *et al.* 2004). Researchers have previously observed that anthocyanins have negligible impact on red reflectance but do reduce green reflectance to some extent (Neill and Gould 1999; Sims and Gamon 2002) but in experiments with *Zostera capricorni* (Chapter 5), anthocyanin concentration appeared to have no measurable influence on spectral reflectance changes associated with low and high light stress.

6.1.4 Potential for spectral indices to detect light stress

The studies in this chapter develop and/or test existing and new spectral indices that may be applied to high spectral resolution remote sensing data to predict light stress in seagrass leaves. While red edge position is the most well established remote sensing VI for the prediction of plant stress, the significant attenuation of NIR wavelengths by water means this spectral index cannot be used to detect stress in submerged plants. The spectral reflectance indices investigated in this chapter therefore utilise only the visible wavelengths that do penetrate through a water column. A range of published, modified and newly developed visible reflectance indices were examined to assess their ability to predict stress in seagrass leaves based on the significant and/or dominant pigment changes found to indicate light stress in *Z. capricorni* in laboratory experiments (Chapter 5). The best performing VI was subsequently tested for its effectiveness at predicting light stress using an independent set of field data (Chapter 4). If a visible reflectance index can be found that consistently and accurately estimates the physiological condition of seagrass plants in a range of situations, it could provide estuarine managers with a simple but extremely useful tool for monitoring the health of seagrass meadows by remote sensing.

6.2 Study methods

6.2.1 Testing for intercorrelation

Intercorrelation between the concentrations of individual photosynthetic pigments and the total chlorophyll, total carotenoid and VAZ pool concentrations measured in *Zostera capricorni* samples in the Chapter 5 experiments was determined using a series of simple linear correlations performed in Excel (Microsoft Office 10).

The set of significant pigment ratios found to act as indicators of light stress in *Z. capricorni* in Chapter 5 included the changes in chlorophyll *a*:*b* associated with low light stress, and those changes associated with high light stress, i.e. changes in the absolute and relative size of the xanthophyll cycle pool (VAZ; VAZ:total carotenoids; VAZ:total chlorophyll), in the proportion of the pool present as zeaxanthin (Z:VAZ) and in total carotenoids:total chlorophyll. In order to avoid unnecessary analysis of closely correlated data, a series of simple linear correlations were performed in Excel to assess intercorrelation between these ratios and identify those that may be redundant. Total chlorophyll content, photosynthetic efficiency (F_v:F_m), A+Z:VAZ and the epoxidation state (EPS) were also included for comparison with the stress indicators since they are widely accepted measures of plant stress. In addition, since chlorophyll content was observed to dominate the spectral reflectance response of seagrasses measured in the field and laboratory (Chapters 4 and 5), it was important to assess its relationship with the other stress indicators as well as its influence on the spectral indices.

6.2.2 Selection of wavelengths and derivative wavelengths

The most significantly different and least significantly different visible reflectance wavelengths established in the Chapter 5 experiments for low and high light stressed seagrass were selected by examining the t-values presented in comparisons of control versus light stressed *Z. capricorni* spectra (Figures 5.6 and 5.9). This method was considered superior to ratio analysis of control and treatment spectra for the selection of optimal and invariant wavelengths for use in VI's (Chappelle *et al.* 1992) because, while the latter method graphically highlights the difference between the mean spectra, it ignores variability in the spectral reflectance data and therefore, the significance of differences between spectra.

The visible wavelengths (between 430-700 nm) that were most strongly correlated and least strongly correlated with each of the selected light stress indicators were determined by performing a simple linear correlation at each wavelength across the spectrum using S-Plus software. For each pigment ratio, the correlation coefficient (*r*) generated at each wavelength was plotted against wavelength to produce a correlogram of uncorrected reflectance data and the 1st derivative of uncorrected reflectance data for each of the data sets from the low light and high light stress experiments (Chapter 5), and for the combination of these two data sets. The latter data set, generated by combining the high light – moderate light treated and low light – moderate light treated *Z. capricorni* samples, was assumed to provide a wider range for response and therefore the possibility for stronger linear correlation between the pigment ratio and reflectance. However, since the two experiments were performed at different times and under slightly different environmental conditions, the data sets from the two individual experiments were also retained for individual analysis. The statistical significance of each correlation was assessed in S-Plus using a two-tailed Student's t-test.

All derivative spectra applied in this study were generated and smoothed in PeakFit V.4 (Jandel Scientific) using an improved Savitsky-Golav least squares quartic polynomial fitting function. After preliminary examination of 2nd and 3rd order derivative spectra, and despite attempts at smoothing, higher order spectra were considered too noisy for use in the study.

6.2.3 Candidate vegetation indices

The selected pigment ratios considered to act as indicators of low and high light stress (total chlorophyll content, chlorophyll *a*:*b*, VAZ:total chlorophyll, VAZ:total carotenoids, Z:VAZ) and photosynthetic efficiency (F_v:F_m) were each correlated against a range of previously published narrow band VI's, 'optimised' narrow band NDVI's and several new VI's developed in this study. The aim of the correlations was to assess which spectral index, based solely on visible wavelengths (430-700 nm), would be optimal for the prediction of light stress in seagrass leaves. All correlations were performed by simple linear correlation and the significance of each was subsequently tested with a two-tailed t-test using S-Plus analysis software. Correlations were performed individually on the *Z. capricorni* data sets from the low light stress (n = 40) and high light stress experiments (n = 61) (Chapter 5), and on the combination of these two data sets (n = 101). All spectral reflectance data utilised in the correlations were uncorrected (raw, i.e. not PMSC corrected) reflectance data and the 1st derivatives of uncorrected reflectance data. The VI's applied in the correlations were:

Published visible indices for the prediction of plant stress:

- Physiological Reflectance Index; PRI = $(R_{531}-R_{570})/(R_{531}+R_{570})$ (Gamon *et al.* 1992).
- Normalised total pigment chlorophyll *a* index;
  NPCI = $(R_{680}-R_{430})/(R_{680}+R_{430})$ (Peñuelas *et al.* 1993).
- Green edge position; GEP = wavelength of the maximum 1st derivative value in the green between 500-550 nm (Peñuelas *et al.* 1993).
- Green-green 1st derivative normalised difference;
  GGFN = $(D_{GE}-D_{minG})/(D_{GE}+D_{minG})$ (Peñuelas *et al.* 1993), where D_{GE} = value of the 1st derivative of reflectance at the green edge, D_{minG} = value of the minimum 1st derivative reflectance in green wavelengths.

Modified visible indices for the prediction of plant stress:

- Green NDVI for the estimation of chlorophyll content modified to include a visible reference band; gNDVI = $(R_{495}-R_{550})/(R_{495}+R_{550})$.
- Pigment-specific normalised difference index for estimation of carotenoid content (Blackburn 1998) modified to include a visible reference band; PSNDc = $(R_{495}-R_{470})/(R_{495}+R_{470})$.
- Optimised high light stress NDVI (HL NDVI) and optimised low light stress NDVI (LL NDVI) utilizing the most significantly different wavelengths from Chapter 5 experiments as index wavelengths and the least significantly different wavelengths from experiments as reference wavelengths.
- Optimised pigment ratio NDVI's (e.g. Tchl NDVI) utilizing the wavelengths best-correlated with each pigment ratio as index wavelengths and the least-correlated wavelengths as reference

wavelengths. The best- and least-correlated wavelengths selected were those from the data set (LL, HL or combined) giving the highest overall correlation result.

- Chlorophyll *a:b* index; CABI = (R_{495}/R_{682}) - (R_{495}/R_{643}).
- 1st derivative reflectance wavelength found in Section 6.3.2 to be best-correlated with each of the stress indicator pigment ratios.

<u>New visible indices developed in the current study for the prediction of plant stress:</u>

- Red-orange, red-green; RORG = $(R_{682}/R_{550}) - (R_{682}/R_{635})$, based on the significant orange shoulder and green peak reflectance changes that occur in high and low light stressed *Z. capricorni* (wavelength from the red chlorophyll absorbance trough used as reference band).
- Blue-green, blue-orange; BGBO = $(R_{495}/R_{550}) - (R_{495}/R_{635})$, based on the significant orange shoulder and green peak reflectance changes that occur in high and low light stressed *Z. capricorni* (stable wavelength from the blue flat region used as reference band).
- Orange edge 1 position; OE1P = wavelength of the maximum 1st derivative value on the orange shoulder between 600-650 nm.
- Orange edge 2 position; OE2P = wavelength of the maximum 1st derivative value on the orange shoulder between 651-690 nm.
- Orange edge 2 value; D_{OE2} = value of the maximum 1st derivative of reflectance on the orange shoulder between 651-690 nm.

In addition, each of the pigment ratios was correlated with REP and Datt's (1998a) best normalised index for the prediction of chlorophyll content and plant stress so as to compare the utility of the reflectance and derivative indices based solely on visible wavelengths.

Published visible-NIR indices for the prediction of plant stress:

- Red edge position; REP = wavelength of the maximum 1st derivative value.
- Datt's normalised index = $(R_{850}-R_{710})/(R_{850}-R_{680})$

6.2.4 Developing predictive regression equations

The VI's providing optimal or consistently high correlation with one or more of the pigment ratios, or with photosynthetic efficiency directly, were regressed against these stress indicator/s in order to generate the best possible regression fit. The regression was used to develop a predictive equation for the estimation of the seagrass light stress indicator for use on independent data sets. VI's calculated from uncorrected reflectance or 1st derivative reflectance spectra from both of the experimental data sets in Chapter 5 were plotted against stress indicators using Sigmaplot 8.2 (SPSS Inc.) and fitted with the regression curve affording the highest coefficient of determination. The regression wizard function in Sigmaplot 8.2 allows the user to compare the fit of a large range of regression equations but in most cases, linear least squares and in all other cases, polynomial quadratic regression offered the best, and also the least complex and most valid fit. The statistical significance of the regressions was tested by ANOVA using Sigmaplot 8.2. Despite the fact that regression statistics are robust to many of their underlying assumptions (Zar 1984), the following assumptions of each regression were checked using the statistics reported by the Sigmaplot 8.2 regression function; the assumption that the source population is normally distributed around the regression line (Kolmogorov-Smirnov test), that the variances of the dependent variables in the source population are equal (Levene Median test) and that the residuals are independent of each other (Durbin-Watson statistic). The Root Mean Square Error (RMSE or Standard Error of Estimate) and the power of the model to correctly describe the relationship between the variables were also examined to assess the predictive potential of each regression model.

6.2.5 Testing visible vegetation indices

The BGBO index developed in this study was selected as the best overall predictor of light stress in *Z. capricorni* leaves based on its performance in regression, including criteria such as regression fit, statistical behaviour, simplicity of use and generality in application (see Section 6.3.4). The linear equation generated from the regression of chlorophyll *a:b* against BGBO using data measured from *Z. capricorni* in low and high light stress experiments (Chapter 5) was inverted as follows to allow estimation of chlorophyll *a:b* from the BGBO measured in independent data sets:

$y = 0.121x - 0.506$ where x = chlorophyll *a:b*, y = BGBO (Figure 6.2A)
hence, $x = (y + 0.506)/0.121$, i.e. estimated chlorophyll *a:b* = (measured BGBO + 0.506)/0.121 (Equation 1).

The ability of BGBO to predict the chlorophyll *a:b* content, and hence, light stress in seagrasses, was tested against a data set containing all three of the common south eastern Australian species so as to also test its generality for use on different seagrasses. The test data set consisted of a subset of 52 of the seagrass leaf samples (n = 40 *Z. capricorni*, n = 9 *P. australis* and n = 3 *H. ovalis*) collected during field sampling experiments in Chapter 4. This subset was comprised of field samples for which both spectral reflectance and pigment content had been measured, but contained only those samples in which chlorophyll *b* and lutein had separated completely during HPLC analysis since it was necessary to know the accurate chlorophyll *b* concentration of the samples (see Section 4.4.2). BGBO was calculated for the test data set from uncorrected field-measured spectral reflectance and was used in Equation 1 to estimate the chlorophyll *a*:*b* ratio of the seagrass leaf samples. Estimated chlorophyll *a*:*b* values were then plotted against the measured chlorophyll *a*:*b* of the test data set and a two-tailed paired t-test was used to assess whether predicted and measured values differed from each other.

6.3 Results

6.3.1 Intercorrelation between pigments and between pigment ratios

There was a high degree of intercorrelation between the concentrations of chlorophylls found in *Z. capricorni* leaves whether grown under moderate, high or low light conditions (Table 6.2). In particular, chlorophyll *a* content could be considered indistinguishable from total chlorophyll concentration (r = 0.998 for the high light experiment and r = 0.996 for the low light stress experiment). The concentrations of light harvesting carotenoids such as lutein, neoxanthin and violaxanthin also correlated strongly with that of the chlorophylls. This resulted in a high intercorrelation between chlorophyll and total carotenoid concentrations, particularly in the case of the low light stressed and moderate light grown *Z. capricorni* samples. The lutein concentration of high light treated seagrass leaves was only moderately correlated with chlorophyll concentrations and consequently the correlation between total carotenoid and chlorophyll content was also lower for these samples.

The concentration of α-carotene in the *Z. capricorni* leaves was unique in that this pigment did not appear to be related to the concentration of any other pigment (Table 6.2). The antheraxanthin and zeaxanthin content of seagrass samples also appeared to be quite independent of the concentrations of light harvesting chlorophylls and carotenoids. There was a moderate correlation between antheraxanthin and the total carotenoid content of *Z. capricorni* leaves in both experiments, and between zeaxanthin and the total carotenoid content of *Z. capricorni* in the high

Table 6.2. Intercorrelation between individual pigment concentrations and selected pigment ratios. Values are correlation coefficients (*r*). Data from the high light stress experiment are presented in italics below data from the low light stress experiment (Chapter 5).

	chl *a*	chl *b*	Tchls	Tcars	T β-cars	α-car	VAZ	V	A	Z	L	N
chl *a*	1	0.94	1	0.91	0.90	-0.16	0.78	0.83	0.06	-0.02	0.93	0.85
	1	*0.96*	*1*	*0.75*	*0.93*	*-0.01*	*0.70*	*0.81*	*0.02*	*-0.09*	*0.45*	*0.94*
chl *b*		1	0.96	0.93	0.86	-0.14	0.83	0.85	0.21	0.00	0.97	0.87
		1	*0.97*	*0.78*	*0.90*	*-0.11*	*0.73*	*0.81*	*0.14*	*-0.04*	*0.51*	*0.91*
Tchls			1	0.93	0.90	-0.15	0.80	0.84	0.10	-0.02	0.95	0.87
			1	*0.76*	*0.93*	*-0.03*	*-0.71*	*0.81*	*0.04*	*-0.08*	*0.46*	*0.94*
T cars				1	0.95	-0.02	0.93	0.93	0.35	0.09	0.97	0.97
				1	*0.78*	*0.06*	*0.91*	*0.88*	*0.45*	*0.37*	*0.90*	*0.80*
T β-cars					1	-0.03	0.85	0.84	0.28	0.12	0.90	0.91
					1	*0.19*	*0.74*	*0.83*	*0.10*	*-0.04*	*0.46*	*0.94*
α-car						1	0.00	-0.07	0.21	0.26	-0.07	0.00
						1	*-0.06*	*-0.07*	*-0.12*	*0.19*	*0.03*	*0.02*
VAZ							1	0.97	0.46	0.12	0.87	0.87
							1	*0.95*	*0.58*	*0.36*	*0.73*	*0.79*
V								1	0.24	-0.07	0.88	0.85
								1	*0.32*	*0.10*	*0.65*	*0.85*
A									1	0.62	0.25	0.37
									1	*0.73*	*0.47*	*0.21*
Z										1	0.02	0.10
										1	*0.51*	*0.04*
L											1	0.93
											1	*0.48*

T = total, chl = chlorophyll, car = carotenoid, V = violaxanthin, A = antheraxanthin, Z = zeaxanthin, VAZ = V + A + Z, L = lutein, N = neoxanthin.

light stress experiment. The relationship between the concentrations of antheraxanthin and zeaxanthin and the total VAZ pool followed a similar pattern, while antheraxanthin and zeaxanthin were more strongly correlated with each other.

In general, there was little intercorrelation between the pigment ratios selected in Chapter 5 as indicators of light stress in *Z. capricorni* (Table 6.3). However, EPS could be considered interchangeable with A+Z:VAZ and was also strongly correlated with Z:VAZ. A+Z:VAZ and Z:VAZ

Table 6.3. Intercorrelation between photosynthetic efficiency (F_v:F_m) and selected pigment ratios commonly used as indicators of plant stress. Values are correlation coefficients (r). Data from the high light stress experiment are presented in italics below data from the low light stress experiment (Chapter 5).

	Tchls	chl *a*:*b*	Tcars:Tchls	VAZ:Tchls	VAZ:Tcars	EPS	F_v:F_m	Z:VAZ	A+Z:VAZ
Tchls	1	0.06	-0.20	-0.19	-0.09	0.39	0.39	-0.34	-0.40
	1	*0.04*	*-0.27*	*-0.46*	*-0.45*	*0.48*	*0.02*	*-0.42*	*-0.49*
chl *a*:*b*		1	-0.32	-0.34	-0.26	-0.05	0.04	0.20	-0.00
		1	*-0.07*	*-0.13*	*-0.21*	*0.11*	*-0.00*	*0.09*	*-0.20*
Tcars:Tchls			1	0.89	0.46	-0.30	0.12	0.15	0.34
			1	*0.80*	*0.13*	*-0.38*	*-0.01*	*0.34*	*0.38*
VAZ:Tchls				1	0.82	-0.20	0.24	0.06	0.25
				1	*0.68*	*-0.40*	*-0.01*	*0.27*	*0.44*
VAZ:Tcars					1	-0.03	0.37	-0.04	0.06
					1	*-0.26*	*0.02*	*0.08*	*0.33*
EPS						1	0.26	-0.93	-0.99
						1	*0.43*	*-0.94*	*-0.99*
F_v:F_m							1	-0.23	-0.26
							1	*-0.38*	*-0.43*
Z:VAZ								1	0.87
								1	*0.88*
A+Z:VAZ									1
									1

T = total, chl = chlorophyll, car = carotenoid, V = violaxanthin, A = antheraxanthin, Z = zeaxanthin, VAZ = V + A + Z, EPS = epoxidation state.

showed a relatively high degree of correlation (r = 0.87-0.88). Since Z:VAZ displayed significant stress-related pigment changes in Chapter 5 experiments but A+Z:VAZ did not, Z:VAZ was selected from the two to represent the de-epoxidation state of the xanthophyll cycle pigments in subsequent analyses.

The ratio of VAZ:total chlorophylls was strongly correlated with total carotenoids:total chlorophylls (Table 6.3). Consequently, total carotenoids:total chlorophylls was not used in further analyses because it was considered more likely to be affected by changes in light harvesting carotenoid concentrations that are highly intercorrelated with the chlorophylls. On the other hand, the VAZ pigments offer more information in regard to photoprotection. While VAZ:total chlorophylls and VAZ:total carotenoids are moderately to highly correlated with each other, both were applied in the development of spectral indices since they offered some different information, particular in the case of high light exposed plants. Interestingly, photosynthetic efficiency as a measure of plant stress was not highly- or evenly moderately-well correlated with any of the pigment ratios observed to indicate seagrass stress in the high and low light stress experiments of Chapter 5.

6.3.2 Most- and least-significant wavelengths and best- and least-correlated wavelengths and derivative wavelengths

The most significant visible spectral reflectance changes that occurred in *Z.capricorni* stressed by low light conditions for two weeks were observed on the orange shoulder between 620-650 nm, with important differences also observed around 555 nm and 590 nm (Figure 5.6). However, the strongest spectral reflectance differences between control and treated seagrass leaves were recorded at the wavelength 640 nm (Table 6.4). No significant spectral reflectance change was observed over most of the blue flat region and near-green wavelengths to around 520 nm, but the visible wavelength displaying least difference between control and low light grown *Z.capricorni* was 685 nm in the red chlorophyll absorbance trough. Hence, the optimised low light stress NDVI can be expressed as;

LL NDVI = $(R_{685}-R_{640})/(R_{685}+R_{640})$.

Table 6.4. Reflectance wavelengths at which the most- and least-significant changes in spectral reflectance occurred during paired laboratory stress experiments in which *Z. capricorni* were treated with low or high light (see Chapter 5).

	LOW LIGHT EXPERIMENT		**HIGH LIGHT EXPERIMENT**	
Observed spectral reflectance changes	Most significant	Least significant	Most significant	Least significant
wavelength	640 nm	685 nm	625 nm	515 nm
t	6.446	0.080	4.675	0.202
p	<0.001***	0.938	0.001***	0.842

Values significant at $^{***}p<0.001$

The most significant spectral reflectance differences between control and high light stressed *Z.capricorni* also occurred on the orange shoulder at 625 nm (Table 6.4) although wavelength regions of significant differences also peaked at around 565 nm and 685 nm (Figure 5.9). The wavelength at which high light treatment had the least significant impact was 515 nm, however, the blue-green wavelengths between 480-520 nm were generally invariant to high light stress. The optimised high light stress NDVI was therefore calculated as; HL NDVI = $(R_{515}-R_{625})/(R_{515}+R_{625})$.

The total chlorophyll content of *Z. capricorni* leaves was most significantly correlated with the visible reflectance wavelength 547 nm in the high light stress data set (Table 6.5A; Appendix 2.1B). Both the chlorophyll *a*:*b* and VAZ:total chlorophyll of *Z. capricorni* correlated best with the far-red wavelength 695 nm from the combined data set (Table 6.5B & C; Appendices 2.2C & 2.3C). The ratio of VAZ:total carotenoids in *Z. capricorni* leaves was most strongly correlated with visible reflectance at 544 nm from the high light experiment data set (Table 6.5D; Appendix 2.4B). The proportion of the xanthophyll cycle in the de-epoxidised state

Table 6.5. Reflectance wavelengths and 1st derivative wavelengths best- and least-correlated with indicators of low and high light stress in *Z. capricorni*; (A) total chlorophyll content, (B) chlorophyll *a:b*, (C) VAZ: total chlorophyll content, (D) VAZ:total carotenoid content, (E) Z:VAZ and (F) photosynthetic efficiency (F_v:F_m). Data in bold are the wavelengths selected for use in optimised pigment ratio NDVI's.

	LOW LIGHT EXPERIMENT		**HIGH LIGHT EXPERIMENT**		**COMBINED EXPERIMENTS**	
	best correl.	least correl.	best correl.	least correl.	best correl.	least correl.
A. Total chlorophyll						
REFLECTANCE						
wavelength	700 nm	581 nm	**547 nm**	**643 nm**	544 nm	662 nm
r	-0.164	<0.001	0.374	0.001	0.271	-0.001
p	0.313	0.998	0.003**	0.996	0.006**	0.995
1st DERIVATIVE						
wavelength	667 nm	512 nm	600 nm	679 nm	595 nm	488 nm
r	0.448	<0.001	-0.608	-0.004	-0.383	<0.001
p	0.004**	>0.999	<0.001***	0.978	<0.001***	0.998
B. Chlorophyll *a:b*						
REFLECTANCE						
wavelength	550 nm	697 nm	700 nm	519 nm	**695 nm**	**576 nm**
r	-0.194	0.003	0.235	<0.001	0.420	0.003
p	0.230	0.985	0.028*	>0.999	<0.001***	0.980
1st DERIVATIVE						
wavelength	473 nm	504 nm	550 nm	591 nm	629 nm	643 nm
r	0.485	<0.001	-0.292	<0.001	0.572	-0.006
p	0.002**	0.999	0.022*	0.998	<0.001***	0.952
C. VAZ: total chlorophyll						
REFLECTANCE						
wavelength	697 nm	442 nm	539 nm	625 nm	**695 nm**	**588 nm**
r	0.124	<0.001	-0.251	0.001	0.252	-0.002
p	0.446	>0.999	0.051	0.995	0.011*	0.984
1st DERIVATIVE						
wavelength	434 nm	486 nm	665 nm	490 nm	667 nm	698 nm
r	-0.402	<0.001	-0.557	0.001	-0.561	<0.001
p	0.010**	0.997	<0.001***	0.996	<0.001***	0.995

Table 6.5. **(cont.)**

	LOW LIGHT EXPERIMENT		HIGH LIGHT EXPERIMENT		COMBINED EXPERIMENTS	
	best correl.	least correl.	best correl.	least correl.	best correl.	least correl.
D. VAZ: total carotenoids						
REFLECTANCE						
wavelength	692 nm	441 nm	544 nm	666 nm	547 nm	596 nm
r	0.034	<0.001	-0.353	0.001	-0.254	-0.002
p	0.835	>0.999	0.005**	0.994	0.010**	0.986
1st DERIVATIVE						
wavelength	435 nm	556 nm	561 nm	463 nm	591 nm	548 nm
r	-0.300	<0.001	0.573	-0.002	0.519	<0.001
p	0.060	0.999	<0.001***	0.988	<0.001***	0.997
E. Z:VAZ						
REFLECTANCE						
wavelength	549 nm	611 nm	548 nm	694 nm	**548 nm**	**606 nm**
r	-0.210	<0.001	-0.263	<0.001	-0.261	-0.002
p	0.194	0.996	0.040*	0.997	0.008**	0.987
1st DERIVATIVE						
wavelength	556 nm	460 nm	669 nm	432 nm	595 nm	698 nm
r	0.448	-0.001	-0.428	0.004	0.450	0.004
p	0.044*	0.997	0.001***	0.976	<0.001***	0.969
F. F_v:F_m						
REFLECTANCE						
wavelength	662 nm	576 nm	**446 nm**	**700 nm**	700 nm	505 nm
r	-0.284	-0.002	0.283	0.053	0.168	-0.001
p	0.076	0.989	0.032*	0.687	0.093	0.993
1st DERIVATIVE						
wavelength	675 nm	470 nm	468 nm	638 nm	679 nm	604 nm
r	0.468	<0.001	-0.401	<0.001	0.377	<0.001
p	0.002**	0.999	0.001***	>0.999	<0.001***	0.999

Values significant at the $*p < 0.05$, $**p<0.01$, $***p<0.001$. V = violaxanthin, A = antheraxanthin, Z = zeaxanthin.

(Z:VAZ) was most significantly correlated with the wavelength 548 nm in the combined data set (Table 6.5E; Appendix 2.5C). Photosynthetic efficiency was best correlated with visible reflectance at 446 nm in the high light stress data set (Table 6.5F; Appendix 2.6B). The most significantly correlated wavelengths identified here were subsequently used as index wavelengths, and the corresponding least significantly correlated wavelength from the same data set as reference wavelengths for production of the optimised NDVI specific to each pigment ratio according to the equation;

$$NDVI = (R_{REF} - R_{INDEX})/(R_{REF} + R_{INDEX}).$$

In general, the strength of the correlations between the best-correlated 1st derivative wavelengths and each stress indicator were much higher than for the best-correlated reflectance wavelengths but this is not surprising since differentiation acts to normalise spectral reflectance data for the effects of scattering. Since NDVI's perform a similar normalizing function, the NDVI's constructed from the best-correlated reflectance wavelengths have been compared with the best-correlated 1st derivative wavelengths for each pigment ratio in Table 6.6.

6.3.3 Correlation between visible VI's and seagrass stress indicators

The reflectance and 1st derivative VI's that displayed the best overall degree of correlation with a number of the seagrass stress indicators were the BGBO, GGFN, RORG and GEP spectral indices (Table 6.6). These indices consistently appeared in the 10 most highly correlated VI's for each of the pigment ratios tested. The correlation between VI's and stress indicator pigment ratios or F_v:F_m was not, however, particularly strong for any of the data sets despite the correlation being highly significant in many cases. The maximum r value obtained was -0.655 which was the result of the correlation between Datt's normalised index and VAZ:total chlorophyll. While a VI based on NIR wavelengths was responsible for this result, the NIR VI's did not actually perform better overall.

Table 6.6. Correlation between indicators of low and high light stress in *Z. capricorni*; (A) total chlorophyll content, (B) chlorophyll *a:b*, (C) VAZ: total chlorophyll content, (D) VAZ: total carotenoid content, (E) Z:VAZ and (F) photosynthetic efficiency, and selected hyperspectral vegetation indices for the estimation of light stress in seagrasses. See Section 6.2.3 for calculation of the VI's. Strongest correlations for each indicator are shown in bold.

	LOW LIGHT EXPERIMENT		**HIGH LIGHT EXPERIMENT**		**COMBINED EXPERIMENTS**	
	r	*p*	*r*	*p*	*r*	*p*
A. Total chlorophyll						
REFLECTANCE INDICES						
PRI	0.343	0.030*	0.374	0.003**	0.290	0.003**
NPCI	-0.009	0.956	-0.104	0.423	-0.065	0.517
gNDVI	-0.049	0.764	-0.307	0.016*	-0.193	0.054
PSNDc	-0.102	0.532	0.115	0.376	0.024	0.811
LL NDVI	0.465	0.003**	0.117	0.370	0.227	0.022*
HL NDVI	0.384	0.014*	0.469	<0.001***	0.307	0.002**
Tchl NDVI	-0.233	0.148	-0.550	<0.001***	-0.324	<0.001***
RORG	-0.370	0.019*	-0.553	<0.001***	-0.357	<0.001***
BGBO	-0.388	0.013*	-0.554	<0.001***	-0.351	<0.001***
CABI	**-0.471**	0.002**	-0.188	0.147	-0.268	0.007**
Datt's normalised	0.082	0.614	**0.577**	<0.001***	**0.361**	<0.001***
1st DERIVATIVE INDICES						
GEP	**0.467**	0.002**	0.548	<0.001***	0.445	<0.001***
GGFN	0.467	0.002**	0.551	<0.001***	0.359	<0.001***
OEP1	-0.375	0.017*	-0.588	<0.001***	-0.405	<0.001***
OEP2	-0.332	0.036*	-0.213	0.100	-0.224	0.024*
D_{OE2}	0.353	0.025*	0.347	0.006**	0.278	0.005**
best 1st derivative	0.448	0.004**	**-0.608**	<0.001***	-0.383	<0.001***
REP	0.284	0.076	0.588	<0.001***	**0.471**	<0.001***
B. Chlorophyll *a:b*						
REFLECTANCE INDICES						
PRI	-0.160	0.323	**0.241**	0.062	-0.381	<0.001***
NPCI	0.209	0.196	-0.056	0.670	0.246	0.013*
gNDVI	-0.019	0.908	0.116	0.372	-0.324	<0.001***
PSNDc	0.218	0.177	-0.027	0.837	0.026	0.794
LL NDVI	-0.033	0.838	0.090	0.491	-0.219	0.028*
HL NDVI	-0.272	0.090	0.130	0.319	-0.516	<0.001***
chl *a:b* NDVI	**-0.346**	0.029*	0.034	0.794	-0.535	<0.001***
RORG	0.136	0.404	0.032	0.805	0.543	<0.001***
BGBO	0.200	0.216	0.024	0.858	**0.560**	<0.001***
CABI	0.063	0.700	-0.092	0.482	0.285	0.004**
Datt's normalised	0.196	0.226	0.010	0.942	-0.437	<0.001***

Table 6.6. (cont.)

	LOW LIGHT EXPERIMENT		**HIGH LIGHT EXPERIMENT**		**COMBINED EXPERIMENTS**	
	r	*p*	*r*	*p*	*r*	*p*
B. Chlorophyll *a*:*b* (cont.)						
1st DERIVATIVE INDICES						
GEP	-0.239	0.137	0.002	0.987	-0.402	<0.001***
GGFN	-0.453	0.003**	0.043	0.744	**-0.581**	<0.001***
OEP1	0.157	0.334	0.015	0.907	0.500	<0.001***
OEP2	0.046	0.779	-0.015	0.906	0.375	<0.001***
D_{OE2}	-0.132	0.416	0.155	0.232	-0.416	<0.001***
best 1st derivative	**0.485**	0.002**	**-0.292**	0.022*	0.572	<0.001***
REP	0.089	0.585	0.042	0.746	-0.251	0.011*
C. VAZ:total chlorophyll						
REFLECTANCE INDICES						
PRI	-0.189	0.243	-0.453	<0.001***	-0.482	<0.001***
NPCI	-0.060	0.713	0.079	0.544	0.121	0.228
gNDVI	0.025	0.876	0.149	0.253	0.212	0.034*
PSNDc	-0.120	0.462	-0.020	0.880	-0.056	0.580
LL NDVI	-0.284	0.075	-0.354	0.005**	-0.397	<0.001***
HL NDVI	-0.286	0 .074	-0.439	<0.001***	-0.485	<0.001***
VAZ:Tchl NDVI	-0.127	0.435	-0.258	0.045*	-0.366	<0.001***
RORG	0.222	0.168	0.399	0.001**	0.457	<0.001***
BGBO	0.210	0.194	0.405	0.001**	0.457	<0.001***
CABI	0.285	0.074	0.428	<0.001***	0.449	<0.001***
Datt's normalised	**-0.438**	0.005**	**-0.655**	<0.001***	**-0.646**	<0.001***
1st DERIVATIVE INDICES						
GEP	-0.074	0.648	-0.470	<0.001***	-0.493	<0.001***
GGFN	-0.125	0.441	-0.541	<0.001***	-0.473	<0.001***
OEP1	0.228	0.157	0.496	<0.001***	0.515	<0.001***
OEP2	0.071	0.662	0.301	0.018*	0.313	0.001**
DOE2	-0.241	0.134	-0.546	<0.001***	-0.534	<0.001***
best 1st derivative	**-0.402**	0.010**	-0.557	<0.001***	-0.561	<0.001***
REP	-0.214	0.186	**-0.587**	<0.001***	**-0.570**	<0.001***
D. VAZ:total carotenoids						
REFLECTANCE INDICES						
PRI	-0.071	0.664	-0.387	0.002**	-0.451	<0.001***
NPCI	-0.001	0.997	-0.054	0.682	0.086	0.390
gNDVI	-0.010	0.952	0.378	0.003**	0.339	<0.001***
PSNDc	-0.130	0.425	-0.230	0.074	-0.179	0.073
LL NDVI	0.004	0.981	-0.066	0.614	-0.168	0.094
HL NDVI	-0.116	0.474	-0.352	0.005**	-0.456	<0.001***
VAZ:Tcar NDVI	0.037	0.822	0.474	<0.001***	0.463	<0.001***
RORG	0.018	0.911	0.521	<0.001***	0.515	<0.001***
BGBO	0.010	0.949	**0.524**	<0.001***	0.515	<0.001***
CABI	-0.008	0.959	0.157	0.227	0.241	0.015*
Datt's normalised	**-0.327**	0.039*	-0.490	<0.001***	**-0.558**	<0.001***

Table 6.6. (cont.)

	LOW LIGHT EXPERIMENT		HIGH LIGHT EXPERIMENT		COMBINED EXPERIMENTS	
	r	*p*	*r*	*p*	*r*	*p*
D. VAZ:total carotenoids (cont.)						
1st DERIVATIVE INDICES						
GEP	0.057	0.725	-0.552	<0.001***	**-0.535**	<0.001***
GGFN	0.053	0.746	**-0.602**	<0.001***	-0.502	<0.001***
OEP1	-0.005	0.976	0.567	<0.001***	0.534	<0.001***
OEP2	-0.131	0.421	0.165	0.205	0.215	0.030*
DOE2	-0.034	0.835	-0.255	0.047*	-0.371	<0.001***
best 1st derivative	**-0.300**	0.060	0.573	<0.001***	0.519	<0.001***
REP	-0.052	0.751	-0.486	<0.001***	-0.471	<0.001***
E. Z:VAZ						
REFLECTANCE INDICES						
PRI	-0.333	0.036*	-0.209	0.106	-0.342	<0.001***
NPCI	-0.202	0.212	0.096	0.462	0.061	0.542
gNDVI	0.417	0.007**	0.134	0.304	0.305	0.002**
PSNDc	-0.234	0.146	0.031	0.814	-0.075	0.459
LL NDVI	-0.035	0.830	-0.335	0.008**	-0.302	0.002**
HL NDVI	-0.148	0.364	-0.372	0.003**	-0.393	<0.001***
Z:VAZ NDVI	**0.470**	0.002**	0.372	0.003**	**0.450**	<0.001***
RORG	0.457	0.003**	0.375	0.003**	0.448	<0.001***
BGBO	0.443	0.004**	**0.381**	0.002**	0.446	<0.001***
CABI	0.034	0.833	0.373	0.003**	0.328	<0.001***
Datt's normalised	-0.024	0.882	-0.340	0.007**	-0.336	<0.001***
1st DERIVATIVE INDICES						
GEP	-0.192	0.235	**-0.442**	<0.001***	**-0.456**	<0.001***
GGFN	-0.263	0.101	-0.377	0.003**	-0.404	<0.001***
OEP1	0.149	0.360	0.307	0.016*	0.362	<0.001***
OEP2	-0.028	0.862	0.353	0.005**	0.275	0.005**
D_{OE2}	0.148	0.362	-0.352	0.005**	-0.314	0.001**
best 1st derivative	**0.448**	0.004**	-0.428	<0.001***	0.450	<0.001***
REP	-0.282	0.077	-0.358	0.005**	-0.403	<0.001***
F. F_v:F_m						
REFLECTANCE INDICES						
PRI	0.157	0.334	**0.361**	0.004**	0.039	0.701
NPCI	0.228	0.157	-0.300	0.019*	0.104	0.299
gNDVI	-0.428	0.006**	0.194	0.134	-0.098	0.330
PSNDc	0.210	0.194	-0.331	0.009**	0.024	0.812
LL NDVI	0.166	0.305	0.300	0.019*	0.117	0.244
HL NDVI	0.064	0.695	0.329	0.010**	-0.039	0.695
F_v:F_m NDVI	0.298	0.062	-0.283	0.027*	**0.143**	0.153
RORG	**0.482**	0.002**	-0.054	0.680	-0.018	0.857
BGBO	-0.466	0.002**	-0.067	0.606	-0.007	0.943
CABI	-0.180	0.267	-0.296	0.021*	-0.100	0.318
Datt's normalised	-0.102	0.530	0.054	0.680	-0.136	0.175

Table 6.6. (cont.)

	LOW LIGHT EXPERIMENT		HIGH LIGHT EXPERIMENT		COMBINED EXPERIMENTS	
	r	*p*	*r*	*p*	*r*	*p*
F. F_v:F_m (cont.)						
1st DERIVATIVE INDICES						
GEP	**0.521**	<0.001***	<0.001	0.999	0.019	0.849
GGFN	0.263	0.101	0.082	0.532	-0.025	0.802
OEP1	-0.217	0.179	-0.013	0.922	0.058	0.563
OEP2	-0.253	0.115	-0.338	0.008**	-0.128	0.201
D_{OE2}	-0.098	0.547	0.157	0.228	-0.117	0.244
best 1st derivative	0.468	0.002**	**-0.401**	0.001**	**0.377**	<0.001***
REP	0.473	0.002**	0.140	0.280	0.147	0.143

Values significant at the *$p < 0.05$, **$p<0.01$, ***$p<0.001$. V = violaxanthin, A = antheraxanthin, Z = zeaxanthin.

VI's calculated from only visible wavelengths were more strongly correlated with all stress indicators other than VAZ:total chlorophyll. In terms of overall correlation with the whole range of pigment ratios and F_v:F_m tested, REP and Datt's normalised index ranked only 6th and 7th best behind the visible VI's; BGBO, GGFN, RORG, GEP and the best-correlated 1st derivative wavelength. The previously published visible reflectance indices; PRI, NPCI and the modified PSNDc ranked very poorly in correlations with seagrass stress indicators as did the proposed chlorophyll *a*:*b* index, CABI. 'Optimised' VI's were not well correlated with any of the pigment ratios or F_v:F_m except for Z:VAZ which displayed highest correlation with the optimised Z:VAZ NDVI index.

In all cases, significant and relatively high correlation coefficients ($r > 0.5$) were obtained from the best-correlated 1st derivative wavelength (by nature of their being the most highly correlated wavelength for each individual data set) (Table 6.6). However, the best-correlated 1st derivative wavelength for each pigment ratio varied for the three data sets tested (Table 6.5) and in most cases one single wavelength or a particular narrow wavelength region could not be considered to be strongly associated with the stress indicator in question. Where this did occur, the 1st derivative wavelength was considered to display some generalised relationship with a seagrass stress indicator and was selected for later use in regression in Section 6.3.4. The 1st derivative of reflectance at (or near) 667 nm was significantly correlated with total chlorophyll and VAZ:total chlorophyll in the high light, low light and combined experiment data sets. Similarly, D_{675} was the only VI significantly correlated with F_v:F_m for all three data sets tested.

First derivative indices usually displayed higher r values than those observed for reflectance indices but this was not always the case. Reflectance indices performed better on data from the low light stress experiment, and in most correlations of VI's with VAZ:total chlorophyll. Higher correlation coefficients were generally observed for the high light stress experiment and the combined data sets than for correlations involving data from the low light stress experiment. The exception was for correlations between VI's and F_v:F_m in which the low light stress data set produced the most significant results.

6.3.4 Regression of the seagrass stress indicators against visible VI's

Figure 6.1 displays graphs of the total chlorophyll content of *Z. capricorni* leaves plotted against each of the selected visible VI's (BGBO, RORG, GEP, GGFN and the value of the best 1st derivative, in this case D_{667}). In each of these scatterplots, the spectral reflectances of samples from the high light stress experiment are offset to those from the low light stress experiment. Since the range of total chlorophyll values for the low light stress data lies within that of the high light stress data, it is clear that some factor other than total chlorophyll content had an important effect on spectral reflectance. If total chlorophyll content was dependent only on the

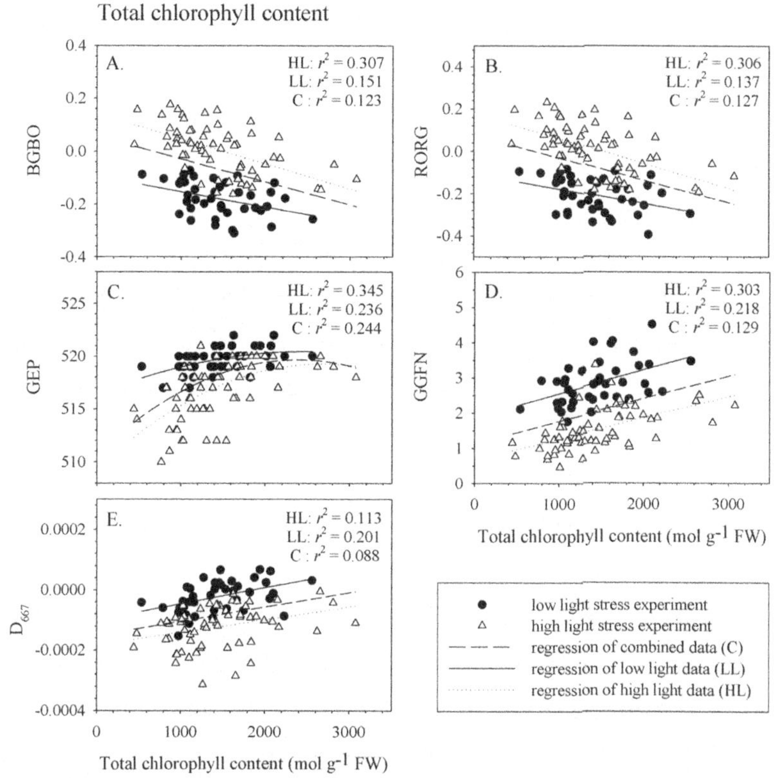

Tchl vs.	Regression equation	*F*	*p*	RMSE	Power
BGBO (HL)	$y = 9.458\text{x}10^{-5}x + 0.141$	26.073	<0.0001***	0.077	0.9974
RORG (HL)	$y = -1.174\text{x}10^{-4}x + 0.177$	25.953	<0.0001***	0.095	0.9973
GEP# (HL)	$y = 509.248 + 7.480\text{x}10^{-3}x - 1.397\text{x}10^{-6}x^2$	15.254	<0.0001***	2.251	0.9992
GGFN (HL)	$y = 5.986\text{x}10^{-4}x + 0.661$	25.668	<0.0001***	0.489	0.9971
D_{667} (LL)	$y = 5.344\text{x}10^{-8}x - 9.937\text{x}10^{-5}$	9.538	0.004**	0.000	0.8346

failed Levene Median test: variances are not homogeneous.

Figure 6.1. Regression of total chlorophyll content of *Z. capricorni* leaves with the visible spectral indices; (A) BGBO, (B) RORG, (C) green edge position, (D) GGFN and (E) the value of 1st derivative reflectance at wavelength 667 nm. Regressions were performed separately on data from the low light and high light stress experiments, and the combined data from both experiments. Regression results are tabulated only for the data set producing the highest r^2 in each case.

PAR level at which seagrass samples were grown then the scatterplot would take the form of a continuous line or curve grading from low light stressed samples, through control samples from both experiments, into high light stressed seagrass samples. Regressions were therefore performed on each individual data set as well as the combined data because data from the high and low light stress experiments did not behave this way. Relatively high coefficients of determination were achieved only from the high light stress data (Figure 6.1). In general, however, none of the VI's tested could be considered to show a strong and unambiguous relationship with total chlorophyll content.

There was a clear gradational increase in the chlorophyll *a*:*b* of low light to high light stressed *Z. capricorni* samples which may suggest that the value of this stress indicator is primarily influenced by the irradiance conditions in which seagrass samples were grown (Figure 6.2). Both BGBO and GGFN demonstrated highly significant linear regressions with chlorophyll *a*:*b*, and although the chlorophyll *a*:*b* content of the seagrass leaves explained only 31% of the variation observed in the BGBO index and 34% of the variation in the GGFN index, these two relationships are the most likely of the regressions to yield equations capable of predicting light stress.

The regression of VAZ:total chlorophyll against GGFN using data from the high light stress experiment resulted in the highest coefficient of determination achieved for any of the regressions performed (r^2 = 0.377; Figure 6.3B). However, each of the regressions involving VAZ:total chlorophyll suffered data offset similar to, but not as extreme as, that observed for total chlorophyll content (Figure 6.3). The wide range of BGBO and GGFN values recorded for the same VAZ:total chlorophyll value, at least for lower values of VAZ:total chlorophyll, suggested that these indices are not likely to be of use in the prediction of VAZ:total chlorophyll content. The relationship between D_{667} and VAZ:total chlorophyll content may prove somewhat better for estimating light stress.

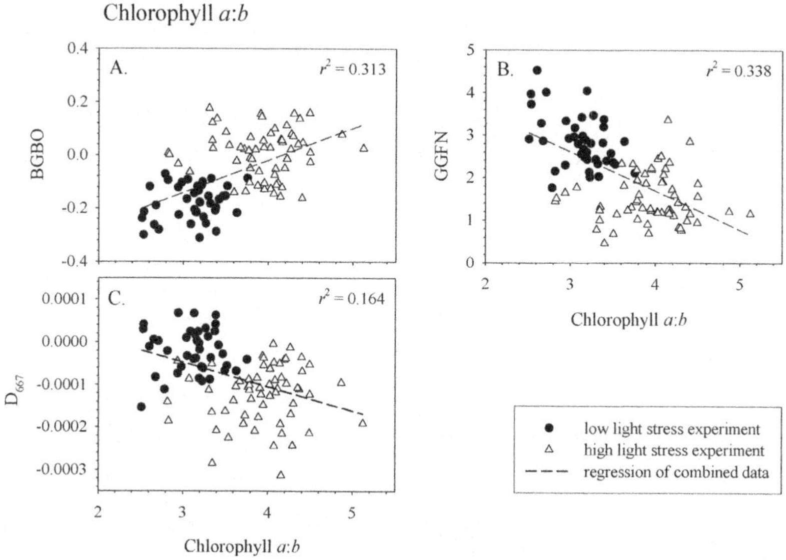

Chl *a:b* vs.	Regression equation	*F*	*p*	RMSE	Power
BGBO§	$y = 0.121x - 0.506$	45.147	<0.0001***	0.102	1.0000
GGFN§	$y = -0.915x + 5.350$	50.515	<0.0001***	0.724	1.0000
D_{667}§	$y = -5.718\text{x}10^{-5}x + 1.238\text{x}10^{-4}$	19.483	<0.0001***	0.000	0.9892

§ Durbin-Watson < 1.5: residuals are not independent.

Figure 6.2. Regression of chlorophyll *a:b* content in *Z. capricorni* leaves with the visible spectral indices; (A) BGBO, (B) GGFN and (C) the value of 1st derivative reflectance at wavelength 667 nm. Regression lines and tabulated results are shown for the combined low and high light stress data sets which produced the highest r^2 in each case.

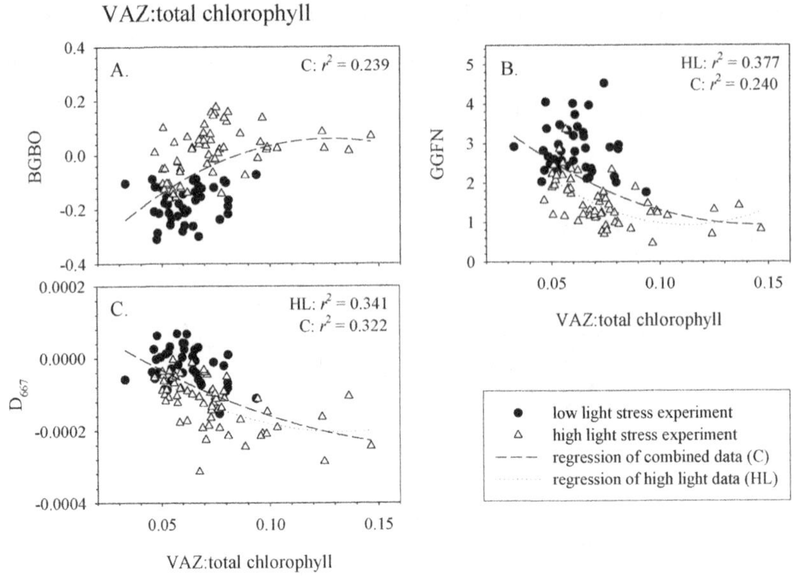

VAZ:Tchl vs.	Regression equation	*F*	*p*	RMSE	Power
BGBO§ (C)	$y = -0.482 + 8.472x - 33.108x^2$	15.405	<0.0001***	0.108	0.9996
GGFN (HL)	$y = 4.731 - 67.524x + 299.758x^2$	17.512	<0.0001***	0.467	0.9998
D_{667} (HL)	$y = 1.535\text{x}10^{-4} - 5.352\text{x}10^{-3} + 2.007\text{x}10^{-2}x^2$	15.026	<0.0001***	0.000	0.9991

§ Durbin-Watson < 1.5: residuals are not independent.

Figure 6.3. Regression of VAZ:total chlorophyll content in *Z. capricorni* leaves with the visible spectral indices; (A) BGBO, (B) GGFN and (C) the value of 1st derivative reflectance at wavelength 667 nm. Regressions were performed separately on data from the low light and high light stress experiments, and the combined data from both experiments. Regression results are tabulated only for the data set producing the highest r^2 in each case.

VAZ:totals carotenoid content only accounted for 27% and 25% respectively of the variation in BGBO and GGFN respectively (Figure 6.4). The strength of the relationship appeared to be in the spread of points from

VAZ:total carotenoids

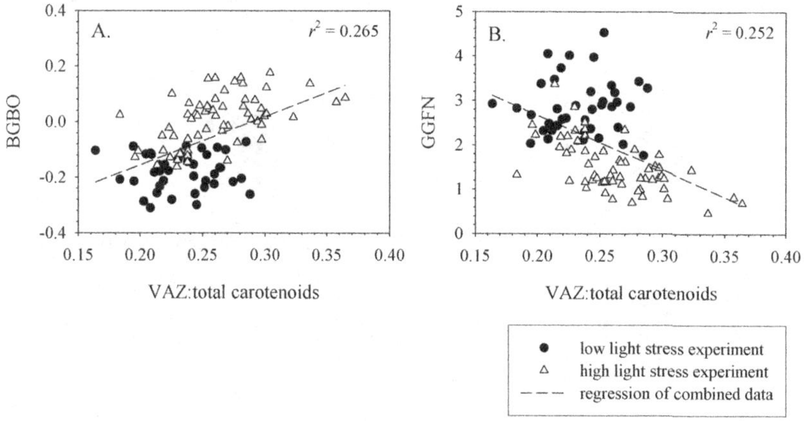

VAZ:Tcars vs.	Regression equation	*F*	*p*	RMSE	Power
BGBO§	$y = 1.735x - 0.502$	35.687	<0.0001***	0.105	0.9999
GGFN§	$y = -12.265x + 5.120$	33.283	<0.0001***	0.770	0.9998

§ Durbin-Watson < 1.5: residuals are not independent.

Figure 6.4. Regression of VAZ:total carotenoid content of *Z. capricorni* leaves with the visible spectral indices; (A) BGBO and (B) GGFN. Regression lines and tabulated results are shown for the combined low and high light stress data sets which produced the highest r^2 in each case.

the high light experiment and the relationship does not follow through into the low light stress data set.

The regressions of Z:VAZ against the RORG, optimised Z:VAZ NDVI and GEP indices (Figure 6.5) also displayed considerable offset of the two experimental data sets and were not considered to offer any useful predictive relationship.

Unfortunately the F_v:F_m values recorded in the experiments did not extend over a large enough range to produce a valid regression with a spectral

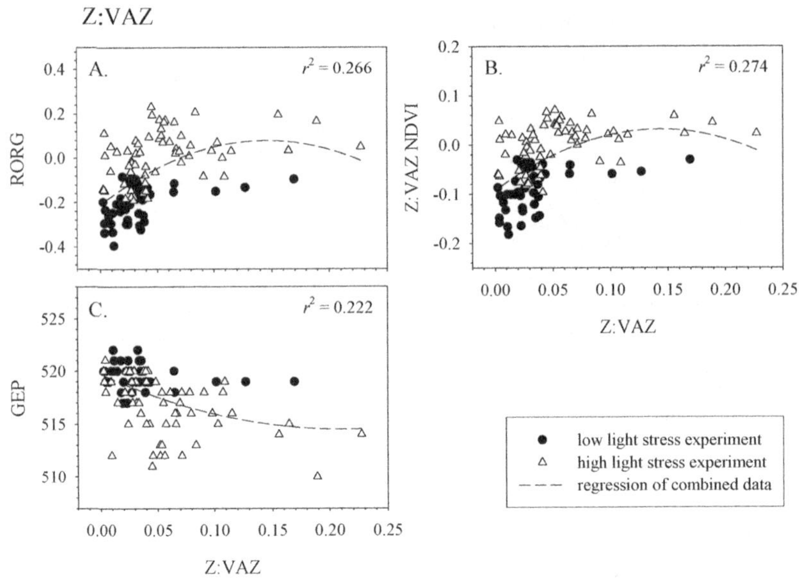

Z:VAZ vs.	Regression equation	*F*	*p*	RMSE	Power
RORG§	$y = -0.210 + 3.965x - 13.600x^2$	17.708	<0.0001***	0.128	0.9999
Z:VAZ NDVI§	$y = -0.093 + 1.735x - 6.047x^2$	18.470	<0.0001***	0.054	0.9999
GEP#	$y = 519.711 - 49.297x + 116.563x^2$	14.015	<0.0001***	2.360	0.9991

§ Durbin-Watson < 1.5: residuals are not independent. # failed Levene Median test: variances are not homogeneous.

Figure 6.5. Regression of Z:VAZ content of *Z. capricorni* leaves with the visible spectral indices; (A) RORG, (B) optimised Z:VAZ NDVI and (C) green edge position. Regression lines and tabulated results are shown for the combined low and high light stress data sets which produced the highest r^2 in each case.

reflectance index. Each of the regressions involving photosynthetic efficiency were forced by one or two low values from the low light experiment and cannot be considered to reliably represent the relationship between photosynthetic stress and the VI's tested (Figure 6.6). In fact, the wide range of values recorded for RORG, GEP and D_{675} at an F_v:F_m of around 0.8, particularly in the case of the high light stress data, throws doubt on the value of any of these spectral indices to directly predict photosynthetic efficiency in these plants.

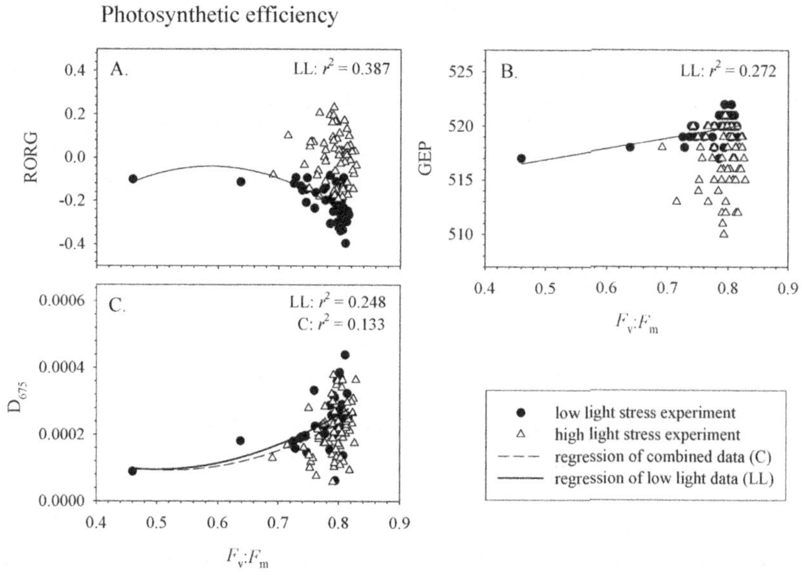

F_v:F_m vs.	Regression equation	F	p	RMSE	Power
RORG (LL)	$y = -1.568 + 5.191x - 4.413x^2$	11.684	0.0001***	0.067	0.9933
GEP (LL)	$y = 9.938x + 511.916$	14.168	0.0006***	1.014	0.9401
D_{675}# (LL)	$y = 5.397\times10^{-4} - 1.791\times10^{-3} + 1.802\times10^{-3}x^2$	6.092	0.0052**	0.000	0.9135

failed Levene Median test: variances are not homogeneous.

Figure 6.6. Regression of the photosynthetic efficiency (F_v:F_m) of *Z. capricorni* leaves with the visible spectral indices; (A) RORG, (B) green edge position and (C) the value of 1st derivative reflectance at wavelength 675 nm. Regressions were performed separately on data from the low light and high light stress experiments, and the combined data from both experiments. Regression results are tabulated only for the data set producing the highest r^2 in each case.

6.3.5 Estimating light stress in field-grown seagrass

From the patterns in the data observed in regression scatterplots (Figures 6.1-6.6), chlorophyll *a:b* was considered to be the most unambiguous and sensitive pigment-based indicator of light stress in *Z. capricorni*. The BGBO index was chosen as the most appropriate VI to test for its ability to estimate the level of light stress (chlorophyll *a:b*) in an independent field-measured seagrass data set. GGFN did perform slightly better than BGBO in regression (Figure 6.2), however, a VI based on 1st derivative reflectance data will not be as generally applicable as one based on reflectance because continuous spectral reflectance data must be available for the derivatives to be calculated and such data is not always available.

The measured chlorophyll *a:b* content of field seagrass samples explained 42 % of the variation in chlorophyll *a:b* values predicted by the BGBO. The value of r^2 increased to 0.476 (RMSE = 0.917) after two outlying data points were removed from the regression (Figure 6.7). These data points were not statistical outliers but, with chlorophyll *a:b* of 8.3 and 7.6, they were considered to represent unusually high ratios for seagrass since chlorophyll *a:b* is usually around 3:1 for higher plants (Kirk 1994). With the outliers removed from the relationship, the regression line lay very close to the straight line that would be generated if predicted values were equal to measured values. However, the considerable variation around the regression line (21.5% of the mean chlorophyll *a:b*) suggested that the predicted chlorophyll *a:b* value at each data point could not be considered statistically equivalent to the measured value ($t = 5.513$, $p < 0.0001$, df = 49).

Since BGBO also performed well in regressions with total chlorophyll content and VAZ:total carotenoids, the capability of BGBO to estimate these stress indicators in the test data set was also investigated. Predicted

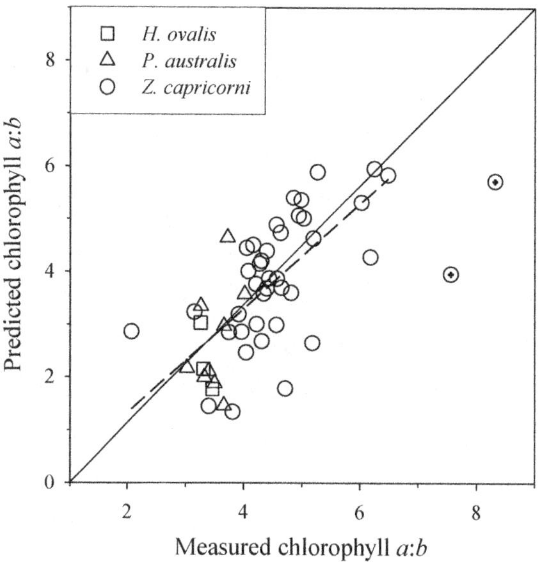

Figure 6.7. Relationship between the chlorophyll *a*:*b* content of seagrass leaves predicted using the BGBO and the chlorophyll *a*:*b* content measured in field samples of three seagrass species. Data outliers are marked with a cross. Dashed line is the regression of measured versus predicted chlorophyll *a*:*b* content.

total chlorophyll content (range -800 to 5100 nmol gFW^{-1}) extended into negative values and well out of the range of measured chlorophyll content (400 to 2400 nmol gFW^{-1}) and hence the predictive value of the equation produced from the regression of laboratory data was extremely poor for this stress indicator. The VAZ:total carotenoids values predicted by BGBO did generally fall within the range of real values but measured VAZ:total carotenoids accounted for only 19% of the variation observed in predicted VAZ: total carotenoids.

6.4 Discussion

6.4.1 Visible vegetation indices with potential to monitor light stress in seagrasses

The BGBO index shows some potential for application to high spectral resolution remote sensing of light stress in seagrass meadows. BGBO measures the difference in reflectance of wavelengths centred at the green peak and on the orange shoulder, with both of these regions referenced to 'blue flat' wavelengths. The wavelengths used in the index (495, 550 and 635 nm) occur in the portion of the spectrum that is least affected by attenuation in a typical estuarine water body. The BGBO index is simple to calculate and can either be applied to single wavelengths drawn from hyperspectral data or to narrow wavebands (with possibly 5-15 nm bandwidths) collected by high resolution optical sensors. The RORG performed nearly as well as the BGBO in many of the correlations with stress indicators, but it is considered to be a less reliable index for general application than the BGBO. The reference band for the RORG lies in the far red wavelengths (682 nm) where chlorophyll absorption was saturated for the data in these experiments, but it may not offer a stable reference for other plant species in various situations where the chlorophyll content is not consistently high.

BGBO displayed a relatively strong relationship with all the stress indicators tested but was most useful in the prediction of chlorophyll *a:b*. The chlorophyll *a:b* content of seagrass leaves is quite sensitive to the recent PAR level to which the plants were exposed and it appears to offer a consistent indicator of light stress across a wide range of field and laboratory conditions (see Figure 5.12). The equation developed from laboratory experiments to estimate light stress (Equation 1) modelled the relationship between BGBO and chlorophyll *a:b* reasonably well for the test field data set but the level of error in the result throws some doubt on the ability of the BGBO to predict light stress in seagrass leaves and canopies to any degree of accuracy. However, none of the other spectral indices tested on any of the pigment ratios performed any better at this task, including previously well-established relationships such as between REP and chlorophyll content (e.g. Horler *et al.* 1983; Curran *et al.* 1990; Curran *et al.* 1995) and between PRI and various plant stress indicators (e.g. Gamon *et al.* 1997; Peñuelas *et al.* 1997, Gamon and Surfus 1999; Sims and Gamon 2002; Stylinski *et al.* 2002). For this reason, it is worth pursuing the BGBO as an index of plant stress in future research involving both ground spectroradiometer and remote sensing data sets.

The principle behind the BGBO and the RORG is that, for light-stressed seagrasses, green reflectance and red reflectance move in opposite directions dependant on the physiological status of the plant. This is not a new concept but has been observed and utilised previously in VI's such as the CARI (Kim *et al.* 1994), ACI (Gamon and Surfus 1999), MCARI (Daughtry *et al.* 2000) and Zarco-Tejada *et al.*'s (1999) 'G' index. Adams *et al.*'s (1999) Yellowness Index, which numerically represents the shape of the orange shoulder (concave up or concave down), is possibly the best of these although it appears only to be applicable when chlorosis is a clear symptom of stress in a plant. The advantage of the BGBO (and RORG) over previous VI's is the use of 635 nm instead of a far red wavelength (660-690 nm) because the former is sensitive to chlorophyll content without the risk of saturation experienced by wavelengths in the chlorophyll absorbance trough. The shorter red wavelengths provide further advantages over the far red wavelengths in that they will also respond to orange-red reflectance from unmasked concentrations of carotenoids and anthocyanins and are less affected by attenuation of the signal by an overlying water column. Reflectance in the green peak wavelengths not only responds to pigment content but carries important information on the internal structure of seagrass leaves which also appears to change in response to light stress in seagrasses.

The green shift in reflectance has previously been utilised to assess the photophysiological status of plants. For example, Malthus and Madeira (1993) and Peñuelas *et al.* (1993) both correlated the value of the 1st derivative at the green edge with plant stress parameters but this current study appears to be the first to apply the position of the green edge to quantify the green shift. GEP does appear to be linked to the de-epoxidation state of the xanthophyll cycle pigments, and to a lesser extent with photosynthetic efficiency, although in the current study these measures of stress did not extend across a wide enough range of values to produce strong relationships and were not tested further. Productive species with high photosynthetic capacity like seagrasses do not invest as much in photoprotective processes as slower growing vegetation types (Gamon *et al.* 1995a). Nonetheless, there is further potential for this index to be applied to the detection of high light stress and other stressors that induce a decline in the photosynthetic function of seagrasses, e.g. toxins or raised water temperatures resulting from industrial discharge into estuaries. The GGFN, calculated as the normalised difference between the 1st derivative reflectance at the green edge and the minimum 1st derivative value in green wavelengths (Peñuelas *et al.* 1993), is also based on the principle of the green shift in reflectance but this index was significantly associated with quite different seagrass stress indicators than GEP; in particular, chlorophyll *a:b* and the size of the VAZ pool. GGFN could potentially provide a somewhat more accurate spectral reflectance measure of light stress in seagrasses than BGBO since the regression of chlorophyll *a:b* with GGFN was actually stronger than with BGBO. However, the GGFN (like the GEP) can only be calculated from continuous hyperspectral data after differentiation of each spectrum and was not tested on an independent data set in this study. There is certainly potential for further testing of this index as an indicator of aquatic plant physiological status. In Peñuelas *et al.*'s (1993) study of the spectral reflectance characteristics of emergent aquatic plants, the GGFN was highly correlated with a number of plant physiological parameters including chlorophyll content, EPS and the carotenoid:chlorophyll ratio (with r values around 0.9) although, in that study, indices incorporating REP were somewhat more successful ($r > 0.9$).

6.4.2 Comparison with published plant stress VI's

The PRI has probably received the most attention in recent literature as a robust measure of physiological status of terrestrial and emergent aquatic plants (e.g. Gamon *et al.* 1997; Peñuelas *et al.* 1997, Sims and Gamon 2002; Stylinski *et al.* 2002) but this VI performed relatively weakly in correlations with the indicators of plant stress measured in the current study, including those associated with photosynthetic function and the de-epoxidation state of the xanthophyll cycle for which the index was specifically intended. The poor performance of the PRI in this study contrasts with results reported by other authors. For example, Gamon and Surfus (1999) achieved an r^2 of 0.98 when regressing the concentration of the xanthophyll cycle carotenoids in the de-epoxidised form (Z + 0.5A) against PRI for sunflowers during a dark-light transition. Similarly, the VAZ pool content of sunflower leaves explained 93% of the variation in PRI observed in this same study. In the current study, PRI was best correlated with VAZ:total chlorophyll using the combined data set ($r = -0.48 \sim r^2 = 0.23$) and although the index was significantly correlated with each of the stress indicators for at least one of the data sets in each case, several other VI's provided much stronger correlations. Plant stressors that induced xanthophyll cycle pigment changes in the seagrass *Z. capricorni* did produce a green shift in the short wavelength shoulder of the green peak but these changes were of small magnitude compared to the green peak – red reflectance offset detected by the BGBO index. In addition, no green peak shift occurred in the reflectance of seagrass leaves stressed by low light availability. The PRI may be particularly useful for detecting high light and water stress in plant species with very responsive xanthophyll pigments (e.g. Gamon *et al.* 1997; Lovelock and Robinson 2002; Stylinski *et al.* 2002) but it falls short as a measure of the potential for plant death in seagrass meadows.

The vegetation indices incorporating NIR wavelengths that were included in this study did not outperform the visible VI's as expected on the basis of the bulk of research published on the spectral detection of stress in vegetation. REP measured from dense, homogeneous plant canopies (or samples in this case) should be strongly associated with total chlorophyll content (e.g. Curran *et al.* 1995; Peñuelas *et al.* 1995b; Curran *et al.* 1998). Since Datt's (1998a) normalised index is also fundamentally based on detection of the blue shift in the RE induced by a narrowing of the chlorophyll absorption trough, it was not surprising that these two indices provided the best correlations with total chlorophyll content and with VAZ:total chlorophyll, particularly for those seagrass leaf samples stressed by high light conditions. However, the coefficients achieved from the correlation of REP and Datt's normalised index with total chlorophyll content were not of the magnitude previously reported for a 'strong association' (r typically > 0.90) even for the high light data set, and they were generally not much higher than for correlations achieved using several of the visible indices. In part this may be due to the presence of red pigments such as anthocyanins in the leaves of *Z. capricorni* which are known to have a detrimental affect on the relationship between chlorophyll concentration and REP (Curran *et al.* 1991; Stone *et al.* 2001). Chlorophyll content is not necessarily a good indicator of plant physiological status when plant stress does not lead to significant chlorosis (e.g. Stylinski *et al.* 2002; Stone *et al.* 2003) and this was certainly the case for *Z. capricorni* which responded more consistently to light stress by changes in chlorophyll *a*:*b* content and leaf structure, factors which did not induce a shift in red edge reflectance. Since chlorophyll absorption appeared to be already saturated by the high total pigment concentration of *Z. capricorni* leaves, low light stress in particular did not result in a shift in REP. Furthermore, the chlorophyll concentration of seagrass leaves is influenced by many factors (e.g. plant nutrition; Alcoverro *et al.* 2001) and can not be considered an absolute indicator of the level of light stress to which the seagrasses were previously exposed, although it may act as a relative indicator in some situations.

6.4.3 Predictive strength of the regressions

Despite the fact that the linear regression was highly significant, the strength of association between BGBO and chlorophyll *a*:*b* using the combined stress experiment data set was relatively poor ($r^2 = 0.31$). This was the case for all VI's and stress indicators tested on the experimental data, including the relationship between total chlorophyll content and the REP for which r^2 values as high as 0.91 have previously been reported in similar leaf level spectral reflectance studies (Curran *et al.* 1990). Lovelock and Robinson (2002) reported similarly low r and r^2 values in the prediction of pigment content from moss canopies using a selection of VI's. It is difficult to explain why the results of the current study and those of Lovelock and Robinson (2002) were so much lower than generally reported in the literature since both of these studies were carried out at the leaf level with measurements made under controlled laboratory conditions using precision techniques, i.e. HPLC for pigment analysis and a very high resolution spectroradiometer for the measurement of spectral reflectance. The methodology used in these studies was similar in two respects; first, spectral reflectance measurements were collected from many layers of leaves in order to saturate pigment absorbance rather than from a single leaf, and second, pigment concentrations were calculated per unit of leaf mass rather than leaf area because of the size and shape of moss and *Z. capricorni* leaves. Typically, vegetation indices are better correlated with pigment content when it is expressed as concentration per unit area of leaf or canopy rather than per unit weight (e.g. Blackburn 1998), at least in studies where reflectance was collected from a single layer of leaves. However, where biomass and LAI vary little spatially and the canopy is optically thick then these measures become functionally equivalent (Pinar and Curran 1996; Curran *et al.* 1998) and this applies to the current study since spectra were collected from a multilayered sample of leaves. Poorer coefficients are generally obtained when leaf biochemistry is correlated with spectral reflectance indices measured from plant canopies rather than from single leaves (e.g. Daughtry *et al.* 2000; Coops *et al.* 2003) because canopy measurements introduce spectral variability from a range of sources which are not present in a single leaf or ordered stack of leaves; e.g. background reflectance, scattering from leaf surfaces at varying angles, variable LAI, etc. Nonetheless, Peñuelas *et al.* (1993) were able to achieve coefficients much higher than recorded in the current study (r = 0.70 - 0.95) when relating the canopy reflectance of wetland species to plant stress parameters.

Poor correlation may be due to high levels of variability in the spectral reflectance of samples, in pigment content, or both. Spectral reflectance in this study was collected from handfuls of seagrass leaves and therefore the spectral response represents an average of many leaves grown under the same conditions. Some variability in the reflectance of samples may be introduced through differences in sample illumination and geometry, even under controlled laboratory conditions, since the data were not PMSC corrected prior to use in correlation and regression. However, most of the inconsistencies stem from measurement of the pigment content of samples. Variability will be intrinsically high when the pigment content of a whole sample is represented by small pieces of 1-3 of the leaves taken from within that sample. Due to water movement and self-shading, seagrass leaves within a single handful, and even different sections of the same leaves, may have been exposed to quite different irradiance conditions for varying periods of time and their biochemistry could vary considerably. In addition, adjacent leaves will vary in age and structure, which will also affect their spectral reflectance response. This point in fact reinforces one advantage of synoptic remote sensing methods for assessing the physiological status of vegetation over point methods which, unless replicated extravagantly, will not provide a true estimate of the condition of a whole meadow.

6.4.4 Applying vegetation indices to remote sensing

Vegetation indices are a simple, convenient and non-intrusive tool for rapidly inferring a number of functionally important leaf and canopy properties from spectral reflectance measured at ground level, or from satellite and aircraft platforms (Peñuelas and Filella 1998; Gamon and Surfus 1999). If a VI such as the BGBO proves robust in application across a wide range of species and under different conditions then monitoring can be simplified and made more cost-effective using instruments purpose built or programmed to measure only the wavebands of interest, e.g. the portable 'leaf reflectometer' developed for field measurement of PRI in agricultural crops (Gamon and Surfus 1999; Méthy 2000). There is some concern, however, that relationships established between spectral indices and plant physiological parameters at the leaf level will not be applicable at the landscape scale and therefore will not be relevant to operational remote sensing. Researchers so far have had mixed success in applying the diagnostic spectral reflectance features identified at leaf scale to airborne remote sensing (e.g. Carter and Miller 1994; Yoder and Pettigrew-Cosby 1995; Pinar and Curran 1996; Curran *et al.* 1998; Zarco-Tejada *et al.* 1999: 2002; Coops *et al.* 2003; 2004). Many factors confound the interpretation of subtle physiological signals in remote sensing data (Gamon *et al.* 1995b). For example, the PRI response mapped for different vegetation types and different successional stages with CASI imagery could be explained by differences in canopy structure, phenology or atmospheric effects as well as by differences in photosynthetic function (Gamon *et al.* 1995b; Qiu *et al.* 1998). The quantification of canopy damage is more difficult for mixed species forests than forests of low diversity due to inherent variability in the chlorophyll content of the species (Carter *et al.* 1998). Canopy bi-directional reflectance imposes a further dimension of variability over that of 'pure' leaf reflectance. LAI and leaf angle control the magnitude of spectral reflectance from dense canopies while background features may dominate reflectance from sparse canopies (Asner 1998). Modelling of canopy structure and atmospheric effects can significantly improve the estimation of physiologically based stress indicators from airborne imagery (Zarco-Tejado *et al.* 1999; 2002) and the knowledge of leaf structure and optical properties has advanced to the point where leaf biochemical content and canopy structure can be reliably modelled (e.g. Jacquemoud *et al.* 1996). Canopy gaps can be masked out using a variety of techniques (Blackburn 2002). However, where canopies are spatially homogeneous and optically thick there is no need for canopy modelling and strong correlations between REP and chlorophyll concentration have been recorded from airborne CASI data (Pinar and Curran 1996; Curran *et al.* 1998; Blackburn 2002). Zarco-Tejada *et al.* (1999) and Curran *et al.* (1998) concluded that leaf level measurements of plant physiological and spectral parameters could be used to produce algorithms that will estimate vegetation condition from above-canopy spectral reflectance.

In the past the operational use of narrow band reflectance indices with airborne or satellite remote sensing data were limited by the capabilities of the remote sensing instruments but recent technology has provided far more potential for their application in monitoring. The technological advances in instrumentation and correction procedures that have occurred since the introduction of hyperspectral sensors may overcome many of the calibration and correction problems that limited Qiu *et al.*'s (1998) use of the PRI for monitoring photosynthetic function with AVIRIS imagery. The PRI predicted relative photosynthetic function quite successfully within a single AVIRIS image but Qiu *et al.* (1998) considered PRI inappropriate for multitemporal vegetation monitoring by remote sensing. This index is very sensitive to calibration error and inadequate atmospheric correction because of the green wavebands used which are too specific and narrow for the half-band-widths of most sensors. Zarco-Tejada *et al.* (1999) subsequently achieved better results with the PRI using CASI imagery because it offers significantly better resolution and greater SNR than AVIRIS, and can be radiometrically corrected using at–sensor radiance and concurrent atmospheric optical depth data.

The BGBO as a plant stress index is more versatile than the PRI because it does not rely on such specific narrow wavelengths and utilises regions of the visible spectrum that are not so sensitive to atmospheric effects. The BGBO should perform well with data obtained from a programmable high resolution sensor such as the CASI, though it is yet to be tested in this capacity. A further problem with the application of the PRI in remote sensing to infer the health of seagrass is that daily and seasonal variation in photosynthetic efficiency may mask any real decline in stressed plants (e.g.

Moran *et al.* 2000). Chlorophyll *a:b* content measured by the BGBO would be less sensitive to rapid changes in environmental factors (particularly irradiance and temperature) than a light stress indicator based on the xanthophyll cycle and instantaneous photosynthetic function, and would therefore provide a better estimate of the seagrass meadow's response to average daily irradiance over the previous weeks.

While VI's can provide a simple, rapid and effective remote sensing procedure for assessing the physiological status of vegetation, they are not the only method available for monitoring the condition of seagrass meadows. There is a good deal of spectral information pertaining to the health of the plants that is not utilised in a VI produced from two to three narrow waveband regions. Modern instruments, computers and analysis methods allow comparisons of whole spectra or large numbers of bands to be made and there is no computational requirement for data reduction, though it may prove beneficial for certain applications. There are many alternative methods that can be applied to remote sensing imagery to detect a change in the spectral reflectance of a seagrass meadow resulting from light stress. Multivariate classification procedures include spectral feature-fitting, spectral mixture analysis, multidimensional scaling and modelling approaches (Gamon and Surfus 1999) which are sensitive to change across the whole visible spectrum. The objective of remote sensing monitoring is not to quantify the exact pigment composition of seagrass meadows but to detect changes in spectral reflectance consistent with a decline in seagrass condition. Monitoring provides management authorities with the capacity to identify problems in an estuary in a timely manner so that measures can be taken to rectify the problem before more permanent damage occurs.

Chapter 7
Implications for the Remote Sensing of Seagrasses

S.K. Fyfe

7.1 Key findings of the spectral studies

The results of the set of spectral studies examined in this book provide essential background information and baseline data, i.e. the tools required to optimise remote sensing methods and procedures for the mapping and monitoring of seagrasses. . The resultant knowledge can also be applied to the remote sensing of other submerged aquatic vegetation, and to improve the remote detection of stress in plants in general. Two key datasets were generated from this study:

- a comprehensive spectral library of the three common seagrass species that dominate meadows in south eastern Australia (*Zostera capricorni*, *Posidonia australis* and *Halophila ovalis*), characterising the spectral variability inherent in the reflectance of these seagrasses growing in their natural environment across different estuaries, habitats, seasons and years (Chapter 4).
- a spectral library characterising the spectral reflectance changes that occurred in *Zostera capricorni* leaves as this seagrass declined in response to conditions of low and high light stress imposed on laboratory grown plants (Chapter 5).

The following conclusions were drawn from analyses performed on these datasets:

1. The seagrass species *Z. capricorni*, *P. australis* and *H. ovalis* were shown to be spectrally distinct across wide regions of the visible wavelengths, despite small but significant levels of spatial and temporal variability within the spectral signature of each species, and regardless of the level of epibiont fouling of seagrass leaves. The spectral reflectance differences between species were consistent and of a magnitude that could be detected by a high resolution imaging spectrometer. In addition, spectral reflectance differences in the visible wavelengths occurred predominantly in regions that are least affected by attenuation in an estuarine water column. Therefore, the mapping of these three seagrasses to species level using remote sensing should be possible in all but the most turbid of estuaries using sensors of appropriate resolution, at least in the estuaries of south eastern NSW where seagrass meadows are typically shallow, dense and monospecific (Chapter 4).
2. Differences in the spectral signatures of the three seagrass species could be attributed more to differences in leaf morphology than to differences in photosynthetic and accessory pigment content, although pigment concentrations did play an important role in determining reflectance. Leaf thickness, surface texture, cell size and shape and in particular, the proportion and arrangement of lacunal air spaces in the leaf had predominant control over reflectance of the weakly absorbed green and NIR wavelengths. *Zostera capricorni* lacked a noticeable green peak and reflected significantly less light than *H. ovalis* and *P. australis* across all visible wavelengths because its leaves contained higher concentrations of both chlorophyll and anthocyanin pigments. The spectral signature of *P. australis* was more typical of terrestrial plants, displaying lower anthocyanin-induced red reflectance, a more enhanced green peak and strong NIR reflectance that could be related to its thicker leaves containing numerous, well-dispersed lacunae (Chapter 4).

3. Epibiont fouling of seagrass leaves did not diminish the significant spectral reflectance differences apparent between species but did influence the shape of their spectral reflectance curves. Fouling tended to increase reflectance at the orange shoulder, reduce green peak reflectance, increase spectral variability in green wavelengths and deepen the red chlorophyll absorption trough (Chapter 4).
4. Within-species differences in the spectral reflectance of seagrass leaves collected from different estuaries, habitats, seasons and years were related to differences in pigment composition as well as structural changes in the internal anatomy of the leaves. Spatial and temporal differences in the visible spectral reflectance response were dominated by differences in seagrass leaf chlorophyll concentration. Differences in the species composition and cover of epibiont foulers also made an important contribution to observed intraspecific spectral reflectance differences (Chapter 4).
5. Piece-wise multiplicative scatter correction (PMSC) is a useful tool for the compilation of spectral libraries since it removes external spectral variation associated with sample illumination and geometry from the reflectance signatures while highlighting the spectral variability inherent in the plant species themselves. PMSC correction achieves this by forcing each of the sample spectra to a common but realistic baseline value, in this case, the mean signature of all samples measured in the field. The signatures retained in the spectral library after PMSC correction are in a format that can be directly applied to image processing and modelling procedures (Chapter 4).
6. Eelgrass *Zostera capricorni* had the ability to photoacclimate and chromatically acclimate to short term changes in ambient irradiance levels. Photosynthetic rate was reduced and efficiency of light use was maximised in light deprived *Z. capricorni* leaves compared with control grown plants. Low light stressed leaves displayed lower chlorophyll

a:b, zeaxanthin and antheraxanthin concentrations than control grown plants within two weeks of exposure, but over the longer term showed a general increase in photosynthetic pigment concentration to maximise light harvesting. Significant spectral reflectance changes occurred in concert with these pigment changes although changes in green peak and NIR reflectance were considered to represent changes in leaf internal structure rather than pigment content. The green peak and NIR reflectance from eelgrass leaves increased while orange shoulder reflectance decreased in response to shade treatment. Despite the efforts of the plant to maximise light-harvesting and efficiency of light use, *Z. capricorni* samples subjected to low light treatment received irradiance levels below their compensation point and within one to three months these samples were dead or had lost most of their biomass (Chapter 5).

7. Exposure to high irradiance levels led to a decrease in the efficiency of photosynthetic energy conversion in *Z. capricorni* leaves but an overall increase in the photosynthetic rate compared to samples grown at lower light levels. High light treatment significantly increased the concentration of xanthophyll cycle carotenoids relative to other *Z. capricorni* leaf pigments and increased the proportion of VAZ pool occurring in the de-epoxidised state compared to controls. Hence, *Z. capricorni* displayed light-dependent down-regulation of photosynthesis in response to high light stress and did not suffer photo-oxidative damage or reductions in growth or biomass during high light experiments. High light treated *Z. capricorni* leaves displayed significantly lower green and far red reflectance and significantly higher orange shoulder reflectance concurrent with pigment changes and probable changes in leaf structure. In addition, a blue shift of the red edge and a red shift of the green edge were detected in the spectral reflectance of high light stressed seagrass leaves (Chapter 5).

8. No relationship was found between the anthocyanin pigment content of *Z. capricorni* leaves and ambient irradiance conditions. In addition, anthocyanin content appeared to have no influence on changes in the spectral reflectance of *Z. capricorni* leaves during laboratory experiments despite the contribution these accessory pigments apparently make to the characteristically flat spectral signature of this seagrass species (Chapter 5).
9. When the data for laboratory and field grown *Z. capricorni* were combined (Chapters 4 and 5), significant linear relationships were observed in the concentration of chlorophyll *a*:*b*, VAZ:total carotenoids, VAZ:total chlorophyll, and A+Z:VAZ concentrations as a function of average daily irradiance. In general, the concentration of light-harvesting pigments increased while photoprotective pigments decreased in response to decreasing daily irradiance. Significant linear trends were also recorded between the spectral reflectance response of *Z. capricorni* leaves sampled from both the field and laboratory at wavelengths 550, 630 and 685 nm and daily irradiance levels (once a field data outlier was removed from regressions; see Section 7.2 for congruence in the laboratory and field spectral reflectance results).
10. Several reflectance indices and derivative indices were significantly correlated with pigment-based indicators of light stress in *Z. capricorni* including the GGFN, GEP, BGBO and RORG, although correlation coefficients were typically low ($r < 0.66$). The BGBO index, which utilises the wavelengths that displayed significant spectral reflectance changes in laboratory light stress experiments on *Z. capricorni*, showed the best potential as a remote sensing index for monitoring light stress in seagrass meadows. There were significant linear relationships between the BGBO and consistent pigment indicators of light stress in *Z. capricorni* including chlorophyll *a*:*b* content and the relative proportion of VAZ pool pigments to total chlorophyll and carotenoid

concentrations. For a test data set consisting of field-measured spectral reflectance for three seagrass species, measured chlorophyll *a:b* content explained 48% of the variation in chlorophyll *a:b* predicted by the BGBO index. The ability of the BGBO to accurately estimate the level of light stress in seagrass leaf samples from this data set was questioned, however, since there was considerable variation around the regression line (21.5% of the mean). No other spectral index performed as well as the BGBO at this task, including well-established reflectance indices such as the REP and PRI that are widely applied in remote sensing projects to predict stress in vegetation (Chapter 6).

11. While VI's offer a simple, rapid and generally effective method for remote sensing the physiological status of vegetation, they should not be considered as the only technique for detecting stress-induced changes in the spectral reflectance response of vegetation. Modern instruments, computers and procedures allow for multivariate analysis of hyperspectral datasets and may offer more sensitive methods for the detection of the spectral reflectance changes in light-stressed seagrass meadows characterised in this study.

7.2 Applying spectral detection of light stress in the field

The spectral reflectance changes associated with light stress in seagrass leaves identified in this study should be considered *relative* changes that can be applied to detect stress by monitoring the spectral response of any particular meadow over time. It is not likely that the library of spectral signatures can be used as *absolute* values to compare the relative levels of light stress in seagrass meadows growing in different estuaries or in different sites within an estuary (e.g. marine versus brackish habitats), unless other environmental factors that may affect seagrass leaf biochemistry, morphology and therefore reflectance, are the same for each meadow. Factors such as the depth at which the seagrass are growing and variations in daily irradiance due to tidal immersion and exposure will influence seagrass physiology and hence, affect the spectral signatures recorded at different sites. For example, *Z. capricorni* meadows in Lake Illawarra might be expected to display the spectral reflectance symptoms of low light stress (i.e. higher green peak reflectance and lower orange shoulder reflectance) compared with meadows in Port Hacking or Sussex Inlet, because of the poorer water quality in this estuary. In addition, the brackish backwaters of most estuaries are usually more turbid than the well-flushed marine waters of the entrance because basin sites accumulate fine particles and organics that become stirred up by wind and waves during the day. There is more potential for seagrasses to be subjected to low light stress at brackish sites assuming the seagrasses are growing at the same depth and under the same tidal regime. Analysis of the effects of estuary and habitat on the spectral reflectance of winter sampled *Z. capricorni* at 530, 550, 635 and 682 nm using two-way ANOVA tests gave inconclusive results because sample variances were heterogeneous for the first three analyses (despite attempts to transform the data) and because of significant interactions between estuary and habitat in each case (Table 7.1). The interactions between habitat and estuary demonstrate that site specific environmental conditions do have an important effect on spectral reflectance. The site where *Z. capricorni* spectra showed the most obvious symptoms of low light stress was not the brackish habitat in the more turbid Lake Illawarra but the brackish Port Hacking site (Figure 7.1). There is a strong tidal influence in the estuary at Gray's Point and although the water is less turbid than in Koona Bay, the *Z. capricorni* are immersed in up to a metre of water for a large portion of each day. Koona Bay is not influenced by tides so *Z. capricorni* grows in very shallow water with the leaves exposed at or near the water surface all day. Seagrass leaves in Koona Bay therefore receive a higher dose of daily irradiance and do not

Table 7.1. Two-way analysis of variance to test for differences in the spectral reflectance of unfouled *Z. capricorni* leaves at wavelengths 530, 550, 635 and 682 nm as a function of the estuary (Lake Illawarra, Port Hacking or St Georges Basin) and the habitat (marine or brackish) from which the leaves were sampled. All samples were collected during winter 1999 and 2000 (n = 10).

Source	df	MS	F	*p*	Signif. pairwise
530 nm#					
Estuary	2	0.000021	13.33	<0.001***	LI-SG, PH-SG
Habitat	1	0.000059	36.44	<0.001***	
Estuary*Habitat	2	0.000006	3.43	0.040*	
Error	54	0.000002			
550 nm#					
Estuary	2	0.000032	6.87	0.002**	LI-SG, PH-SG
Habitat	1	0.000010	2.09	0.154	
Estuary*Habitat	2	0.000041	8.63	<0.001***	
Error	54	0.000005			
635 nm§					
Estuary	2	0.000024	3.24	0.047*	PH-SG
Habitat	1	0.000022	2.98	0.090	
Estuary*Habitat	2	0.000030	3.97	0.025*	
Error	54	0.000007			
682 nm					
Estuary	2	0.000222	17.02	<0.001***	LI-SG, LI-PH
Habitat	1	0.000807	61.90	<0.001***	
Estuary*Habitat	2	0.000076	5.84	0.005**	
Error	54	0.000013			

Cochrane's C value significantly heteroscedastic; # at $p < 0.05$, § at $p < 0.01$. Values significant at $^{*}p < 0.05$, $^{**}p < 0.01$, $^{***}p < 0.001$.

suffer the level of low light stress shown by seagrasses at Gray's Point. Similarly, one could conclude that the *Z. capricorni* leaves most affected by high light stress (i.e. having lowest reflectance at 530, 550 and 682 nm and highest reflectance at 635 nm) were those sampled from the brackish backwater of St Georges Basin (Figure 7.1). This may indeed be the case since the brackish waters at the Island Point site were unusually clear (at

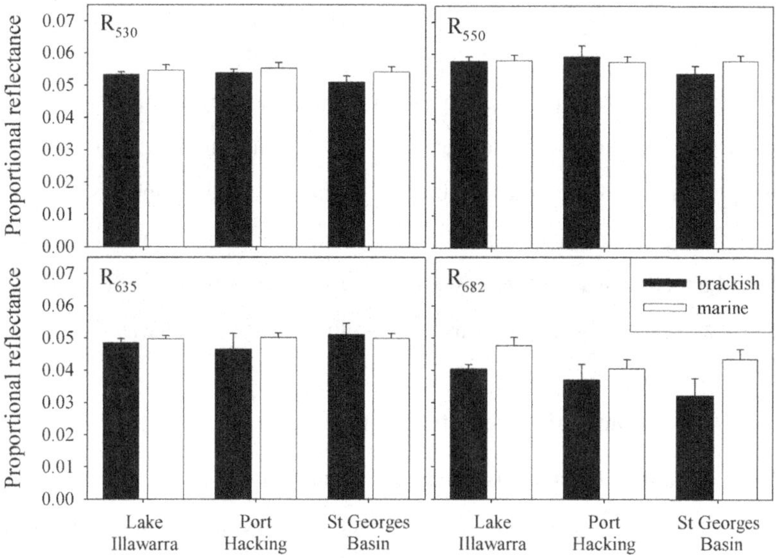

Figure 7.1. Mean + SD spectral reflectance at wavelengths 530, 550, 635 and 682 nm of unfouled *Zostera capricorni* leaves collected from marine and brackish habitats in three NSW estuaries during winter 1999 and 2000 (n = 10).

least as clear as in the tidal entrance channel) and *Z. capricorni* leaves were sampled from shallow water consistently less than 300 mm deep. At any one location, however, relative changes in the spectral reflectance could be used to monitor for symptoms of light stress in the seagrass leaves and hence, for environmental change detrimental to the health of the seagrass meadows.

The spectral reflectance and physiological changes identified in the laboratory do not represent the full extent of changes in the field, at least those recorded for the effects of high light stress on *Z. capricorni*. The laboratory environment could not simulate the high light conditions experienced by plants growing in the field in summer nor the low-high light changes that a seagrass can experience over the course of a day due to changing tides and weather conditions. However, the high light conditions to which seagrasses are typically exposed in south eastern Australia may affect their photosynthetic performance but do not appear to impact on their growth or survival in the longer term. Monitoring high light stress as an early warning of dieback is therefore of far less management value than monitoring for low light stress, although the need for photoprotection will be enhanced in seagrass leaves suffering from the combined effects of light stress and another stress such as toxic pollution in the waterway.

7.3 Application to remote sensing

7.3.1 Potential for mapping seagrass meadows

There is great potential for mapping seagrass meadows to species level. Species discrimination in the remote sensing of vegetation is achievable so long as the species under study are spectrally distinct over space and time. The fact that the three common south eastern Australian species are spectrally distinct at the level of leaf reflectance provides a solid baseline from which remote sensing of these species can begin. The methodology employed in this study provides a repeatable approach for assessing the capability for spectral discrimination of aquatic or terrestrial vegetation species.

The accuracy of benthic vegetation maps will improve with the use of spectral libraries that describe the natural variability in the reflectance of the species involved. This study has produced the first spectral library of aquatic vegetation to characterise the natural levels of both spatial and temporal variation in the regional spectral reflectance response of the plant species involved. Comprehensive spectral libraries are particularly effective when used for hyperspectral classification procedures and when applied in radiative transfer modelling (e.g. Jupp *et al.* 1996; Anstee *et al.* 2000).

7.3.2 Potential for monitoring the condition of seagrass meadows

Intraspecific differences in seagrass reflectance may be much more difficult to detect by remote sensing because these differences are very small. The fact that there were significant differences in laboratory-measured spectra of low and high light stressed *Z. capricorni* samples compared with control grown samples, and in field-measured spectra associated with the estuary, habitat, season and year of sampling, does suggest some potential for the monitoring of seagrass condition. However, the conclusions drawn from this study are yet to be tested on real image data sets. Whether such small reflectance differences will be detected by remote sensing after attenuation through an atmosphere and water column will depend on a wide range of factors including the SNR and the spatial, spectral and radiometric resolution of the sensor, the clarity, depth and constituents of the water column and atmosphere, and the density and homogeneity of the seagrass meadow itself. Phytoplankton and benthic microalgae may have quite an impact on the ability of remote sensing to detect light stress in the seagrass leaves, for example using the BGBO index, since marine green algae tend to have low chlorophyll *a*:*b* ratios. Although the results reported here are promising, they are based on leaf-level observations and must be verified at the canopy level for remote sensing applications (Adams *et al.* 1999). Work in terrestrial ecosystems has established some basis for scaling physiological processes up from leaf to landscape level (reviewed in Field and Ehrlinger 1993).

If the short term spectral reflectance changes indicative of light stress in seagrasses prove too small to be detected by even the most advanced remote sensing scanners then the next best option would be to monitor longer term structural changes in the meadows. Seagrasses respond to changing irradiance conditions with a range of morphological adaptations that occur over a longer time scale than the physiological adaptations examined in this study. In particular, shoot density and meadow biomass are sensitive indicators of seagrass stress (Bulthuis 1983; Neverauskus 1988; Dennison 1990; Abal *et al.* 1994; Abal and Dennison 1996; Dixon 2000; Neely 2000; Alcoverro *et al.* 2001; Meehan 2001). Shoot density varies inversely and predictably with light levels for a wide range of species and is independent of other environmental factors such as nutrient availability and water temperature (e.g. Agawin *et al.* 2001). Meadow extent, location and patchiness (e.g. Hastings *et al.* 1995) offer further information about the condition of the seagrasses in an estuary. These structural parameters have been accurately measured for seagrasses by remote sensing (see Table 1.1) and could be readily monitored over time. Alternately, monitoring of estuarine water quality can provide an early warning of potential threats to seagrass meadows (e.g. Dennison *et al.* 1993; Gallegos and Kenworthy 1996; Moore *et al.* 1996). It should be possible to monitor water clarity regularly over large scale areas using satellite imagery to infer the likelihood of seagrass dieback, based on both the severity of turbidity and the length of time that turbid conditions persist (Stumpf and Frayer 1997). Airborne remote sensing methods for monitoring water quality have been developed to a high level of sophistication and would be appropriate for smaller estuaries or wherever more precise or accurate measurements are required (e.g. Dekker *et al.* 1992; 1995).

Excess seagrass epiphyte growth, or blooms of certain phytoplankton species in the water column, may provide another means for early detection of water quality problems in an estuary (May *et al.* 1978; Kirkman 1996; 1997). Algal epiphytes can be quite sensitive indicators of environmental conditions (May *et al.* 1978) because different species have different life histories and preferences for nutrients, light and temperature (Kirkman

1997). Different classes of algae have been discriminated in high spectral resolution remote sensing data (Richardson *et al.*1994; Richardson 1996; Malthus *et al.* 1997b; Aguirre-Gómez *et al.* 2001a) and would be particularly easy to detect when blooms form on the upper canopy of the seagrass and along the surface of the water as they often do during eutrophication events.

7.3.3 Suitable sensors and platforms

The potential for mapping and monitoring of seagrass meadows will be maximised by using appropriate sensors with the appropriate spectral, radiometric and spatial resolution to detect the spectral features of interest. For example, Lubin *et al.* (2001) modelled upwelling spectral radiance at the air-water interface and the top-of-atmosphere near-nadir radiance from *in situ* reflectance spectra for coral species, sand and algae over a range of water depths to test the utility of Landsat TM satellite data for coral reef mapping. Satellite imagery was not able to detect the spectral features that were successfully used to distinguish coral species in ground or aircraft data (Lubin *et al.* 2001). In contrast, Jupp *et al.* (1996) and Anstee *et al.* (2000) were able to map seagrasses and other benthic plant species using radiative transfer models in conjunction with airborne high resolution imagery.

A suitable scanner for detailed seagrass species mapping and monitoring of seagrass condition would record a spectrum with a resolution of 5-15 nm in the bands suggested in Table 7.2. The level of noise in the spectral data should be about an order of magnitude smaller than the depth of the absorption feature of interest (Goetz and Calvin 1987). Very high spectral resolution is a feature of airborne imaging spectrometers and hyperspectral satellite scanners, and although high signal-to-noise ratio may be harder to

Table 7.2. Suggested bandsets for application to high spectral resolution remote sensing scanners for the detailed mapping of seagrasses and for monitoring light stress in seagrass meadows.

SPECIES MAPPING		**MONITORING LIGHT STRESS**	
band centre	bandwidth	band centre	bandwidth
major seagrass peaks and troughs		**green peak reflectance**	
550 nm	5-15 nm	550 nm	5-15 nm
620 nm	5-15 nm		
675 nm	5-15 nm	**orange shoulder reflectance**	
		635 nm	5-10 nm
and/or the shoulders of the major seagrass peaks and troughs		**chlorophyll absorption trough**	
530 and 570 nm	5 nm	682 nm	5 nm
660 and 682 nm	5 nm		
		the green edge	
major epiphyte peaks		530 nm	5 nm
590 nm	5-15 nm		
640 nm	5-15 nm	**a spectrally 'invariant' region**	
		e.g. 495 nm	5-15 nm
a spectrally 'invariant' region where separation between species is poor			
e.g. 440 or 495 nm	5-15 nm		
NIR waveband for emersed or floating seagrass			
e.g. 750 nm	5-15 nm		

achieve (Curran 1989), sensor technology has advanced significantly in this regard in recent years.

Optimal results will be achieved from an airborne sensor if image data collection missions are planned carefully and this is particularly the case when working in the estuarine environment. For example, data should be collected on calm days when the atmosphere is clear and when winds do not generate waves or stirring of the sediments in shallow water.

Collecting image data during low tide will minimise the attenuating affects of water depth on the signal, but consideration must also be given to the tidal period when best water clarity is achieved, which may not be during low tide. Shadowing increases with low sun angles (Jupp *et al.* 1986) so it is best to collect remote sensing imagery when the sun angle is high. However, flight paths must be carefully planned in relation to the zenith and azimuth angle of the sun so as to maximise signal while minimising hot spot effects and sun glint from waves and ripples.

7.3.4 Maximising the potential for remote sensing of seagrass

In south eastern New South Wales, the three common seagrass species (*P. australis*, *Z. capricorni* and *H. ovalis*) tend to form discrete meadows of characteristic colour and tone, although *Halophila* spp. will also form mixed and low density beds that may be harder to detect. It should therefore be easy to discriminate seagrass meadows and even the different species by optical remote sensing but other information will be of benefit to mapping in deeper, more turbid waters or where seagrass meadows are not dense or monospecific. The effects of the atmosphere and an estuarine water column on spectral reflectance must be accounted for when mapping submerged plants. Empirical or model-based atmospheric and radiometric corrections have become standard pre-classification processing procedures in modern remote sensing but correcting for variable water depths will be crucial image processing step in any benthic mapping exercise (Mumby *et al.* 1997b; Maritorena *et al.* 1994). Benthic discrimination has been improved by the use of algorithms (e.g. Bierwirth *et al.* 1993; Hick 1997) or simple water depth masking (Zainal *et al.* 1993; Bosma 1998) to reduce the effects of depth and water column components. Lyzenga (1981) demonstrated how the apparent reflectance spectrum of a submerged object was modified by the water column and proposed techniques for removing these effects. Lyzenga's method is commonly applied in the correction of imagery from regions of very clear water (optically deep) but it has been less successful for optically shallow water (turbid or coloured waters). Radiative transfer theory was used to model upwelling radiance in an attempt to map estuarine vegetation (using a parametric approach) as early as 1983 (Ackleson and Klemas 1983). A number of radiative transfer models have been applied to model the effects of a turbid or coloured water column on the reflectance of benthic vegetation, for example Tassan's (1996) modified Lyzenga's method for macroalgal detection in turbid waters. Optical water models have been developed by Jupp *et al.* (1996), Malthus *et al.* (1997a) and Plummer *et al.* (1997) using spectral libraries of seagrasses, macroalgae and substrates, and using water depth and water quality parameters as inputs to the models.

Classical methods for image analysis do not take full advantage of the spectral dimensionality of hyperspectral data. Methodologies for hyperspectral processing that Kruse *et al.* (1997) identified as being relevant for coastal mapping applications include:

- the use of linear transformation to minimise noise and determine or reduce data dimensionality (e.g. minimum noise fraction (MNF) transform)
- locating the most spectrally pure pixels which often correspond to the spectral endmembers (e.g. pixel purity index (PPI))
- extracting endmember spectra by n-dimensional visualization
- spatial mapping of endmembers (e.g. the Spectral Angle Mapper (SAM) is an automated method for comparing the similarity of the spectrum in each image pixel to endmember spectra or reference spectra in a spectral library).
- spectral unmixing of the materials or endmembers of interest that are typically mixed in image pixels. A simple additive linear model (linear

spectral unmixing) can be used to estimate the fractional abundances of the 'pure' spectra or endmembers in each pixel. Alternately, matched filtering maximises the response of a known endmember and suppresses the response of the composite unknown background. It may therefore provide a rapid and computationally more efficient procedure than unmixing for detecting specific target spectra from a spectral library or from image endmembers.

Benthic classification accuracy may be improved by incorporating spectral reflectance data with contextual information, bathymetry and other knowledge based attributes in a GIS. For example, information on bottom type, water movement, salinity, the habitat preferences of certain species and the typical size shape of meadows will aid in the discrimination of seagrass species. New procedures, e.g. neural networks that can integrate knowledge based attributes with spectral reflectance data in classifications may also improve the accuracy of mapping and may allow for decision making about the type of change that has occurred within an image pixel during monitoring (Malthus *et al.* 1997b). The advantages of applying human knowledge and textural recognition to map interpretation have been widely recognised in API but contextual editing and textural filters can both be used to improve the classification accuracy of airborne and satellite remote sensing as well (e.g. Mumby *et al.* 1997a; Mumby and Edwards 2002).

The data sets generated by this study have so far been applied in the development of two other rapid and useful techniques specifically aimed at identifying stress in seagrass meadows using hyperspectral remote sensing; i.e. a narrow band vegetation index, the BGBO and neural networks used to estimate photosynthetic efficiency of seagrass (Ressom *et al.* 2003; Sriranganam *et al.* 2003).

7.4 Management implications

Butler and Jernakoff (1999) identified areas where there is a lack of knowledge on seagrass monitoring, assessment and management. Such areas included resource assessment, impact assessment, assessment of the effectiveness of management, plus information on spatial and temporal variability in meadows, the spatial scales that are biologically appropriate for monitoring, and in general, how and what to monitor. Many of these areas could be addressed using a carefully planned and implemented remote sensing monitoring program that baseline maps and utilises repeatable methods for monitoring.

7.4.1 Seagrass dynamics

Seagrass managers, e.g. government agencies and coastal zone managers, need to know the extent of natural changes in seagrass meadows so that human impacts can be separated from normal background variation (Lee Long *et al.* 1996a). Changes can occur in the location, areal extent, shape or depth of a meadow, but changes in biomass, species composition, productivity, and the flora and fauna associated with a meadow may also occur with or without a distributional change. Tropical seagrass meadows are usually more dynamic than temperate meadows and may change shape markedly between years without necessarily changing in abundance (Lee Long *et al.* 1996a). However, abundance can also be highly variable (e.g. Mellors *et al.* 1993; McKenzie 1994). Large annual species distribution and biomass changes have been recorded for some temperate species without a significant change in the total area of the seagrass within the estuary (King and Hodgson, 1986). A baseline study may be required to establish the level of seasonal and between-year variation against which impacts can judged (Kirkman 1996). CCL (1998) and Meehan (2001) attempted to define the scale of natural temporal variation in seagrass cover for certain species to provide meaningful baseline information for impact assessment. Long *et al.* (1996) considered a 50-70% change in above ground biomass to be the trigger point to prompt environmental management action for tropical seagrasses. Meehan (2001), however, found that natural variation in the shoot density of the temperate seagrass *P. australis* lies generally within 20% of the mean and therefore considered shoot losses of >30% of the mean to be the trigger level for management intervention. BACI (before/after, control/impact) style experimental designs that have been applied in the field monitoring of seagrass parameters (e.g. shoot density, Fyfe and Davis 2007; leaf production and rhizome elongation rates, Guidetti 2001) will also be very useful for separating natural dynamics from impacts. These statistically valid approaches could readily be applied to detection of change by remote sensing and thus obviate the need for long term data on the dynamics of each individual seagrass species. One problem facing this method, however, may be that appropriate (unimpacted) control sites may not be available.

Successful management of seagrass meadows represents a significant challenge to estuary managers given the high levels of natural spatial and temporal variability encountered in these systems. Long term permanent transects have been used to establish temporal dynamics at the small scale (Kirkman and Kirkman 2000; Morris *et al.* 2000) but remote sensing could offer much more information on seagrass dynamics and landscape ecology across the larger scale. Field and Ehrlinger (1993) noted that prediction and analysis of the structure and function of ecological systems over large spatial and temporal scales may be a difficult research challenge but the repeatable and synoptic methods offered by remote sensing appear to be the logical approach to answering such questions.

7.4.2 Physiological responses of different seagrass species

Seagrass managers must also consider the particular characteristics of the seagrass species that occur within their region in their planning and decision making. *Zostera capricorni* grows both intertidally and subtidally and is highly tolerant of adverse environmental conditions (Dawson and Dennison 1996) but not all species have the same capacity for photoadaptation and chromatic acclimation as *Z. capricorni*. For example, many tropical seagrasses, and particularly the subtidal species, do not display much capacity for phenotypic plasticity (e.g. Neely 2000; Major and Dunton 2000) and therefore do not survive well under persistent stress in polluted estuaries (Agawin *et al.* 2001). Large, robust species such as *P. sinuosa* and *P. australis* will display symptoms of stress and a photoadaptive response only after several months of shading (Neverauskus 1988; Gordon *et al.* 1994; Masini and Manning 1997). It may be more difficult to apply photophysiological parameters as indicators of environmental stress in these species because their ability for photoacclimation is limited (Major and Dunton 2000) and the early symptoms of stress may not be as apparent as they would be in an adaptable species like *Z. capricorni*. For slow-growing species that are poor colonisers such as *P. australis*, however, it will be even less acceptable to wait until meadow dieback indicates that the seagrasses are under stress because the plants will then have reached a critical state.

The ability of seagrasses to survive and recover from a shading event depends on the timing and duration of the event as well as the species involved. Seagrasses evolved in coastal estuaries and bays and therefore have some ability to deal with erosion, sedimentation and pulsed events of light deprivation associated with periodic floods and storms. However, anthropogenic changes to the rate, timing and duration of sedimentation or turbidity events may be fatal (e.g. Dennison and Alberte, 1982, 1985;

Zimmerman *et al.* 1995; Dixon 2000). Seagrass species have different levels of tolerance linked to factors such as their growth rate and capacity for rhizome storage of carbohydrates (Bulthuis 1983; Longstaff and Dennison 1999; Longstaff *et al.* 1999). For example, *Z. capricorni* survives for approximately 2-3 months at around 3% of the summer daily surface irradiance typical for south eastern Australia. The tropical seagrasses *H. wrightii* and *T. testudinum* last between 9-11 months under similar conditions of light deprivation (Czerny and Dunton 1995). *Posidonia sinuosa* can survive up to 2 years at light levels below their minimum requirement (Gordon *et al.* 1994) but complete plant death occurs in *H. ovalis* after only 30 days of low light stress (e.g. Longstaff *et al.* 1999). The limited ability of temperate plants to grow in winter may limit their responsiveness to sedimentation at this time of year although lower respiration rates may also reduce photosynthetic light demand (Vermaat *et al.* 1996). Summer bouts of turbidity may be more catastrophic for permanently submerged seagrass than for intertidal populations. In general, even small increases in turbidity will have greater impact on subtidal seagrass species that normally grow in clear waters.

Catchment development usually leads to permanent changes to the runoff of nutrients and sediments into estuaries and coastal waters. If managers wish to realise the concept of ecologically sustainable development around coastal waterways then difficult decisions must be made about the type, extent and timing of development, human activity or resource use based on a clear understanding of the characteristics of the seagrass meadows under their jurisdiction.

7.5 Future research directions

These ground-based spectral reflectance studies have generated baseline data and laid the groundwork for the mapping and monitoring of seagrass meadow condition using hyperspectral remote sensing. The next step in applying these results to operational remote sensing would be to use radiative transfer modelling to predict whether the spectral signatures characteristic of the different species, and those characteristic of high and low light-stressed and non-stressed seagrass plants, could still be discriminated after a water column and atmosphere have attenuated the signals. The spectral libraries generated in this study should subsequently be applied to high resolution image data sets to test their utility for the detailed mapping and monitoring of seagrass meadows. It would be ideal to use high resolution image data to map seagrass meadows in an estuary prior to the onset of some human impact, e.g. a new residential development, and then to collect and analyse imagery in a temporal sequence following the impact. Different hyperspectral processing procedures, e.g. the BGBO, neural network models and hyperspectral classification and change detection techniques could be tested to determine their effectiveness at this task.

Further research might also be carried out to assess the influence of other types of physiological stress on the spectral reflectance of seagrasses including the impacts of nutrients and pollutants associated with the eutrophication of estuaries.

7.6 Conclusions and recommendations

Seagrass monitoring has, up until now, been performed as a management response to observed problems of seagrass loss and degeneration rather than as a trigger for management intervention which may halt or abate the cause of the problem before significant damage occurs (Thomas *et al.* 1999). Ideally baseline mapping and subsequent monitoring of seagrass beds should be undertaken in all estuaries, particularly those located in relatively pristine areas where new developments of any type are planned in the catchment, since seagrass management and catchment management are inextricably linked. However, mapping of seagrasses as a prelude to protection may be a drain on available resources (Leadbitter *et al.* 1999) and therefore it is crucial that the most appropriate and cost-effective methods are chosen for the particular objective (Phinn *et al.* 2000). Managers need to have a clear idea of the type and scale of information, and the degree of precision they require from monitoring (Phinn *et al.* 2000).

At this stage, airborne imaging spectrometers generally offer the best solution for many seagrass monitoring applications. These sensors offer flexibility in spatial resolution, temporal resolution and provide high and/or programmable spectral resolution (see Section 1.4.2.2). Thomas *et al.* (1999) suggested that airborne remote sensing might still be of low value in turbid water because it requires extensive ground truthing and the use of site specific algorithms, yet there is no better alternative for synoptic mapping of seagrass in turbid water (e.g. see Rollings *et al.* 1998). As for other mapping methods, it is important to correct airborne scanner imagery for the effect of variable water depths on the signatures received at the sensor, however modeling of high spectral resolution data can also account for the effects of variable water quality and substrates as well as water depths (e.g. Jupp *et al.* 1996; Malthus *et al.* 1997a; Plummer *et al.* 1997). Accurate seagrass maps are currently being produced for use in environmental studies (Bajjouk *et al.* 1996; Clark *et al.* 1997; Held *et al.* 1997) but they are expensive and difficult to obtain because of the specialist processing techniques used in their production. It would be unlikely that a local council, for example, could afford to regularly obtain new seagrass maps produced in this manner. However, for the purposes of monitoring seagrass health, it should only be necessary to produce one accurate baseline benthic map of an estuary and then apply simpler, semi-automatic procedures, using the software they are already familiar with, for subsequent detection and interpretation of change.

The cost of obtaining airborne images and the initial set up costs involved in purchasing hardware and software, training, etc, may be high in comparison to equivalent costs for satellite imagery and API, particularly if the area to mapped is large (Mumby *et al.* 1999). Once a monitoring programme has been established, however, airborne remote sensing becomes more cost effective because image acquisition costs will be offset against the huge investment of time required to create maps from API (O'Neill *et al.* 1997; Mumby *et al.* 1999). If colour aerial photographs are purpose flown for a particular project then the cost of image acquisition may actually be comparable to that for airborne scanner imagery (O'Neill *et al.* 1997). Due to the patchy and dynamic nature of many seagrass beds and the low resolution of the less expensive methods of remote sensing, it has been considered 'difficult' (i.e. too expensive) to map seagrass at the resolution required to detect short term or small scale losses within individual meadows (Watford and Williams 1998). However, new legislation requires management authorities to monitor seagrass meadows to a high degree of accuracy and at a scale that can be used to detect anthropogenic impacts and any of the appropriate high resolution methods for fine scale habitat mapping will be equally expensive (Mumby *et al.* 1999).

Seagrass meadows have economical importance and are of high conservation significance. Seagrass plants respond rapidly to their external environment; some species have great capacity for physiological adaptation to changing conditions while others do not cope well with environmental stress. In either way they are sensitive bioindicators of the condition of estuaries. Since dieback events are almost always linked to human impacts, estuary managers have a moral and legal responsibility to monitor waterways to protect this important resource and the ecosystems that depend on it. Successful management depends on appropriate monitoring of the health and extent of seagrass meadows. Current field monitoring methods are limited spatially and temporally, and while colour aerial photography is flown routinely over some estuaries, API will not detect meadow stress until losses have already occurred. Estuary managers need to look toward the future by establishing high spectral and spatial resolution remote sensing baseline mapping and monitoring campaigns for the early and accurate detection of change or stress in seagrass meadows.

References

Abal, E.G., and W.C. Dennison, 1996. Seagrass depth range and water quality in southern Moreton Bay, Queensland, Australia, *Marine Freshwater Res.*, **47**:763-771.

Abal, E.G., N. Loneragan, P. Bowen, C.J. Perry, J.W. Udy, and W.C. Dennison, 1994. Physiological and morphological responses of the seagrass *Zostera capricorni* Aschers to light intensity, *J. Exp. Mar. Biol. Ecol.*, **178**:113-129.

Ackleson, S.G., and V. Klemas, 1983. Remote reconnaissance of submerged aquatic vegetation: a radiative transfer approach, *3rd Landsat-4 Workshop*, NASA Goddard Space Flight Center, Greenbelt, MD, 6-7 December, 1983, 17 pp.

Ackleson, S.G., and V. Klemas, 1987. Remote sensing of submerged aquatic vegetation in Lower Chesapeake Bay: a comparison of Landsat MSS and TM imagery, *Remote Sens. Environ.*, **22**:235-248.

Adams, M.L., W.D. Philpot, and W.A. Norvell, 1999. Yellowness index: an application of spectral second derivatives to estimate chlorosis of leaves in stressed vegetation, *Int. J. Remote Sens.*, **20**(18):3663-3675.

Adams, III W.W., and B. Demmig-Adams, 1992. Operation of the xanthophyll cycle in higher plants in response to diurnal changes in incident sunlight. *Planta*, **186**:390-398.

Adams, III W.W., B. Demmig-Adams, A.S. Verhoeven, and D.H. Barker, 1995. Photoinhibition during winter stress: involvement of sustained xanthophyll cycle-dependent energy dissipation. *Aust. J. Plant Physiol.*, **22**:261-276.

Agawin, N.S.R., C.M. Duarte, M.D. Fortes, J.S. Uri, and J.E. Vermaat, 2001. Temporal changes in the abundance, leaf growth, and photosynthesis of three co-occurring Philippine seagrasses, *J. Exp. Mar. Biol. Ecol.*, **260**:217-239.

Aguirre-Gómez, R., S.R. Boxall, and A.R. Weeks, 2001a. Detecting photosynthetic algal pigments in natural populations using a high-spectral-resolution spectroradiometer, *Int. J. Remote Sens.*, **22**(15):2867-2884.

Aguirre-Gómez, R., A.R. Weeks, and S.R. Boxall, 2001b. The identification of phytoplankton pigments from absorption spectra, *Int. J. Remote Sens.*, **22**(2&3):315-338.

Alberotanza, L., V.E. Brando, G. Ravagnan, and A. Zandonella, 1999. Hyperspectral aerial images. A valuable tool for submerged vegetation recognition in the Ortebello Lagoons, Italy. *Int. J. Remote Sens.*, **20**(3):523-533.

Alcoverro, T., E. Cerbiān, and E. Ballesteros, 2001. The photosynthetic capacity of the seagrass *Posidonia oceanica*: influence of nitrogen and light. *J. Exp. Mar. Biol. Ecol.*, **261**:107-120.

Anderson, M.J., 1999. Multivariate analysis for biology and ecology, Marine Ecology Laboratories, University of Sydney, Australia.

Anstee, J.M., D.L.B. Jupp, and G.T. Byrne, 1997. The shallow benthic cover map and optical water quality of Port Phillip Bay, *Proc. 4th International Conference on Remote Sensing for Marine and Coastal Environments*, Orlando, Florida, 17-19 March 1997, 19 pp.

Anstee, J., A.G. Dekker, G. Byrne, P. Daniel, A. Held, and J. Miller, 2000. Use of hyperspectral imaging for benthic species mapping in South Australian coastal waters. *Proc. 10th Australasian Remote Sens. Photogramm.* Conf., Adelaide, Australia, pp. 1051-1061.

Armstrong, R.A., 1993. Remote sensing of submerged vegetation canopies for biomass estimation, *Int. J. Remote Sens.*, **14**(3):621-627.

Ashton, H.I., 1973. Aquatic plants of Australia, Melbourne University Press, Melbourne.

Asner, G.P., 1998. Biophysical and biochemical sources of variability in canopy reflectance, *Remote Sens. Environ.*, **64**:234-253.

Bajjouk, T., B. Guillaumont, and J. Populus, 1996. Application of airborne imaging spectrometry system data to intertidal seaweed classification and mapping. *Hydrobiologia*, **326/327**:463-471.

Bajjouk, T., J. Populus, and B. Guillaumont, 1998. Quantification of subpixel cover fractions using principal component analysis and a linear programming method: Application to the coastal zone of Roscoff (France), *Remote Sens. Environ.*, **64**:153-165.

Baret, F., and G. Guyot, 1991. Potentials and limits of vegetation indices for LAI and APAR assessment, *Remote Sens. Environ.*, **35**:161-173.

Baret, F., I. Champion, G. Guyot, and A. Podaire, 1987. Monitoring wheat canopies with a high spectral resolution radiometer, *Remote Sens. Environ.*, **22**:367-378.

Barker, D.H., G.G.R. Seaton, S.A. Robinson, 1997. Internal and external photoprotection in developing leaves of the CAM plant *Cotyledon orbiculata*. *Plant Cell Physiol.*, **20**:617-624.

Bearlin, A.R., M.A. Burgman, and H.M. Regan, 1999. A stochastic model for seagrass (*Zostera muelleri*) in Port Phillip Bay, Victoria, Australia, *Ecological Modelling*, **118**:131-148.

Beer, S., 1996. Inorganic carbon transport in seagrasses, pp. 43-47. *In*, Kuo, J., R.C. Phillips, D.I. Walker, and H. Kirkman (eds), Seagrass Biology: proceedings of an international workshop, Rottnest Island, Western Australia, January 1996.

Beer, S., and M. Björk, 2000. Measuring rates of photosynthesis of two tropical seasgrasses by pulse amplitude modulated (PAM) fluorometry, *Aquat. Bot.*, **66**:69-76.

Beer, S., M. Björk, R. Gademan, and P. Ralph, 2001. Measurements of photosynthesis in seagrass, pp. 183-198, *In*, Short, F.T., and R. Coles (eds), Global seagrass research methods, Elsevier, The Netherlands.

Beer, S., B. Vilenkin, A. Weil, M. Veste, L. Susel, and A. Eshel, 1998. Measuring photosynthetic rates in seagrasses by pulse amplitude modulated (PAM) fluorometry, *Mar. Ecol. Prog. Ser.*, **174**:293-300.

Belay, A., 1981. An experimental investigation of inhibition of phytoplankton photosynthesis at lake surfaces. *New Phytol.*, **89**:61-74.

Bell, J.D., and D.A. Pollard, 1989. Ecology of fish assemblages and fisheries associated with seagrass beds, pp. 565-609, *In* A.W.D. Larkum, A.J. McComb, and S.A. Shepherd (eds), Biology of seagrasses: a treatise on the biology of seagrasses with special reference to the Australian region, Elsevier.

Ben Moussa, H., M. Viollier, and T. Belsher, 1989. Télédétection des algues macrophytes de l'Archipel de Moléne (France) radiométrie de terrain et application aux données du satellite SPOT, *Int. J. Remote Sens.*, **10**(1):53-69.

Bierwirth, P.N., T.J. Lee, and R.V. Burne, 1993. Shallow sea-floor reflectance and water depth derived by unmixing multispectral imagery, *Photogramm. Eng. Remote Sens.*, **59**:331-338.

Bilger, W., O. Björkman, and S.S. Thayer, 1989. Light-induced spectral absorbance changes in relation to photosynthesis and the epoxidation state of xanthophyll cycle components in cotton leaves, *Plant Physiol.*, **91**(2):542-551.

Billings, W.D., and R.J. Morris, 1951. Reflection of visible and infrared radiation from leaves of different ecological groups, *Am. J. Bot.*, **38**:327-331.

Biswal, B., 1997. Chloroplasts, pigments, and molecular responses of photosynthesis under stress, pp. 877-885, *In*, Pessarakli, M. (ed.), Handbook of Photosynthesis, Marcel Dekker Inc, New York.

Björkman, O., 1981. Responses to different quantum flux densities, pp. 57-107, *In*, Lange, O.L., P.S. Nobel, C.B. Osmond, and H. Ziegler (eds), Physiological plant ecology I: Responses to the physical environment, Encyclopedia of Plant Physiology New Series, Vol. 12A, Springer-Verlag, Berlin.

Björkman, O., and B. Demmig, 1987. Photon yield of O_2-evolution and chlorophyll fluorescence characteristics at 77K among vascular plants of diverse origins, *Planta*, **170**:489-504.

Björkman, O., and B. Demmig-Adams, 1995. Regulation of photosynthetic light energy capture, conversion, and dissipation in leaves of higher plants, pp.17-47, *In*, Schulze, E.-D., and M.M. Caldwell (eds), Ecophysiology of photosynthesis, Springer-Verlag, Berlin.

Blackburn, G.A., 1998. Quantifying chlorophylls and carotenoids from leaf to canopy scales: an evaluation of some hyperspectral approaches. *Remote Sens. Environ.*, **66**:273–285.

Blackburn, G.A., 2002. Remote sensing of forest pigments using airborne imaging spectrometer and LIDAR imagery *Remote Sens. Environ.*, **82**:311-321.

Blackburn, G.A. and J.I. Pitman, 1999. Biophysical controls on the directional spectral reflectance properties of bracken (*Pteridium aquilinum*) canopies: results of a field experiment, *Int. J. Remote Sens.*, **20**(11):2265-2282.

Blackburn, G.A., and C.M. Steele, 1999. Towards the remote sensing of matorral vegetation physiology: relationships between spectral reflectance, pigment, and biophysical characteristics of semiarid bushland canopies, *Remote Sens. Environ.*, **70**:282-292.

Boardman, N.K., 1977. Comparative photosynthesis of sun and shade plants, *Ann. Rev. Plant Physiol.*, **28**:355-377.

Boochs, F., G. Kupfer, K. Dockter, and W. Kuhbauch, 1990. Shape of the red edge as vitality indicator for plants, *Int. J. Remote Sens.*, **11**(10):1741-1753.

Borowitzka, M.A., and R.C. Lethbridge. 1989. Seagrass epiphytes, p. 458-499. *In* A.W.D. Larkum, A.J. McComb, and S.A. Shepherd (eds), Biology of seagrasses: a treatise on the biology of seagrasses with special reference to the Australian region, Elsevier.

Borregaard, T., H. Nielsen, L. Nørgaard, and H. Have, 2000. Crop-weed discrimination by line imaging spectroscopy, *J. Agric. Engng Res.*, **75**:389-400.

Bosma, D., 1998. Mapping of an invasive alga on the NSW coast using remote sensing techniques, Honours Thesis (unpub.), University of Wollongong, Australia.

Boyer, M., J. Miller, M. Belanger, and E. Hare, 1988. Senescence and spectral reflectance in leaves of Northern Pin Oak (*Quercus palustris* Muenchh.), *Remote Sens. Environ.*, **25**:71-87.

Brakke, T.W., J.A. Smith, and J.M. Harnden, 1989. Bidirectional scattering of light from tree leaves, *Remote Sens. Environ.*, **29**:175-183.

Brix, H., and J.E. Lyngby, 1985. Uptake and translocation of phosphorus in eelgrass (*Zostera marina*), *Mar. Biol.*, **90**:111-116.

Bulthuis, D.A., 1983. Effects of in situ light reduction on density and growth of the seagrass *Heterozostera tasmanica* (Martens ex Aschers.) den Hartog in Western Port, Victoria, Australia, *J. Exp. Mar. Biol. Ecol.*, **67**:91-103.

Bulthius, D.A., 1984. Control of the seagrass *Heterozostera tasmanica* by benthic screens, *J. Aquat. Plant Manag.*, **22**:41–43.

Bulthuis, D.A., 1987. Effects of temperature on the photosynthesis and growth of seagrasses, *Aquat. Bot.*, **27**:27-40.

Bulthuis, D.A., 1995. Distribution of seagrasses in a North Puget Sound estuary: Padilla Bay, Washington, USA, *Aquat. Bot.*, **50**:99-105.

Burkholder, J.M., K.M. Mason, and H.B. Glasgow, 1992. Water-column nitrate enrichment promotes decline of eelgrass *Zostera marina*: evidence from seasonal mesocosm experiments, *Mar. Ecol. Prog. Ser.*, **81**:163-178.

Buschmann, C., and E. Nagel, 1991. Reflection spectra of terrestrial vegetation as influenced by pigment-protein complexes and the internal optics of the leaf tissue, *Proc. IEEE Int. Geosc. Remote Sens. Symp., IGARSS 1991*, June 3-6, Vol 4, pp. 1909-1912.

Buschmann, C., and E. Nagel, 1993. *In vivo* spectroscopy and internal optics of leaves as basis for remote sensing of vegetation, *Int. J. Remote Sens.*, **14**(4):711-722.

Butler, A., and P. Jernakoff, 1999. Appendix 6.1: Research needs identified by working groups, pp.199-207, *In*, Butler, A., and P. Jernakoff (eds), FRDC Project 98/223: Seagrass in Australia: strategic review and development of an R & D plan, CSIRO Publishing, Victoria.

Cambridge, M.L., and A.J. McComb, 1984. The loss of seagrasses in Cockburn Sound, Western Australia. I. The time course and magnitude of seagrass decline in relation to industrial development, *Aquat. Bot.*, **24**:229-243.

Cambridge, M.L., A.W. Chiffings, C. Brittan, L. Moore, and A.J. McComb, 1986. The loss of seagrass from Cockburn Sound, Western Australia. II. Possible causes of seagrass decline, *Aquat. Bot.* **24**:269-285.

Carter, G.A., 1993. Responses of leaf spectral reflectance to plant stress, *Am. J. Bot.*, **80**(3):239-243.

Carter, G.A., 1994. Ratios of leaf reflectances in narrow wavebands as indicators of plant stress, *Int. J. Remote Sens.*, **15**(3):697-703.

Carter, G.A., 1998. Reflectance wavebands and indices for remote estimation of photosynthesis and stomatal conductance in pine canopies, *Remote Sens. Environ.*, **63**(1):61-72.

Carter, G.A., and A.K. Knapp, 2001. Leaf optical properties in higher plants: Linking spectral characteristics to stress and chlorophyll concentration, *Am. J. Bot.*, **88**(4):677-684.

Carter, G.A., and R.L. Miller, 1994. Early detection of plant stress by digital imaging within narrow stress-sensitive wavebands, *Remote Sens. Environ.*, **50**(3):295-302.

Carter, G.A., M.R. Seal, and T. Haley, 1998. Airborne detection of southern pine beetle damage using key spectral bands, *Can. J. Forest Res.*, **28**(7):1040-1045.

Carter, G.A., K. Paliwal, U. Pathre, T.H. Green, R. J. Mitchell, and D.H. Gjerstad, 1989. Effect of competition and leaf age on visible and infrared reflectance in pine foliage, *Plant Cell Environ.*, **12**:309-315.

CCL (Cockburn Cement Limited), 1998. Changes in seagrass coverage on Success and Parmelia Banks between 1965 and 1995. Report to CCL prepared by NGIS, The University of Western Australia and D.A. Lord and Assoc., April 1998, CCL, Success, Western Australia.

Chalker-Scott, L., 1999. Environmental significance of anthocyanins in plant stress responses. *Photochem. Photobiol.*, **70(1):1-9.**

Chappelle, E.W., M.S. Kim, and J.E. McMurtrey III, 1992. Ratio analysis of reflectance spectra (RARS): An algorithm for the remote estimation of the concentrations of chlorophyll a, chlorophyll b, and carotenoids in soybean leaves, *Remote Sens. Environ.*, **39**:239-247.

Chauvaud, S., C. Bouchon, and R. Manière, 2001. Cartographie des biocénoses marines de Guadeloupe à partir de données SPOT (récifs coralliens, phanérogames marines, mangroves), *Oceanologica Acta,* **24 (Supplement)**: S3–S16.

Claasen, D. van R., R.A. Kenchington, and T.S. Shearn, 1986. Managing coral reefs: Benefits of remote sensing in marine park planning, *Proc. 10th Canadian Symp. Remote Sens.*, Edmonton, Canada, 5-8 May 1986, 11 pp.

Claasen, D. van R., D.L.B. Jupp, J. Bolton, and L.D. Zell, 1984. An initial investigation into the mapping of seagrass and water colour with CZCS and Landsat in North Queensland, Australia, *Proc. 10th Int. Symp. Machine Proc. Remotely Sensed Data*, Indiana, pp. 190-201.

Clark, C.D., H.T. Ripley, E.P. Green, A.J. Edwards, and P.J. Mumby, 1997. Mapping and measurement of tropical coastal environments with hyperspectral and high spatial resolution data, *Int. J. Remote Sens.*, **18**(2):237-242.

Clarke, K.R., 1993. Non-parametric multivariate analyses of changes in community structure. *Aust. J. Ecol.* **18**:117-143.

Clarke, S.M., and H. Kirkman, 1989. Seagrass dynamics, pp. 304-345, *In*, A.W.D. Larkum, A.J. McComb, and S.A. Shepherd (eds), Biology of Seagrasses: A treatise on the biology of seagrasses with special reference to the Australian region, Elsevier, New York.

Clarke, K.R., and R.M. Warwick, 1994. Change in marine communities: an approach to statistical analysis and interpretation. Natural Environment Research Council, UK.

Clevers, J.G.P.W., 1988. The derivation of a simplified reflectance model for the estimation of leaf-area index, *Remote Sens. Environ.*, **25**(1):53-69.

Clough, B.F, and P.M. Attiwell, 1980. Primary productivity of *Zostera muelleri* Irmisch ex Aschers. in Westernport Bay (Victoria, Australia), *Aquat. Bot.*, **9**:1-13.

Coles, R.G., W.J. Lee Long, R.A. Watson, and K.J. Derbyshire, 1993. Distribution of seagrasses, and their fish and penaeid prawn communities, in Cairns Harbour, a tropical estuary, northern Queensland, Australia, *Aust. J. Mar. Freshwater Res.*, **44**:193-210.

Congalton, R.G., 1991. A review of assessing the accuracy of classifications of remotely sensed data, *Remote Sens. Environ.*,**37**:35-46.

Coops, N.C., Stone, C., Culvenor, D.S., Chisholm, L. 2004. Assessment of crown condition in eucalypt vegetation by remotely sensed optical indices, *J. Environ.Qual.*, **33**(3):956-964.

Coops, N., C. Stone, D.S. Culvenor, L.A. Chisholm, and R.N. Merton, 2003. Chlorophyll content in eucalypt vegetation at the leaf and canopy scales as derived from high resolution spectral data, *Tree Physiol.*, **23**(1):23-31.

Cracknell, A.P., 1999. Remote sensing techniques in estuaries and coastal zones - an update, *Int. J. Remote Sens.*, **20**(3):485-496.

Critchley, C., 1981. Studies on the mechanism of photoinhibition in higher plants. *Plant Physiol.*, **67**:1161-1165.

Critchley, C., 1988. The molecular mechanism of photoinhibition – facts and fiction. *Aust. J. Plant Physiol.*, **15**:27-41.

Curran, P.J., 1989. Remote sensing of foliar chemistry, *Remote Sens. Environ.*, **30**:271-278.

Curran, P.J., and N.W. Wardley, 1988. Radiometric leaf-area index, *Int. J. Remote Sens.*, **9**(2):259-274.

Curran, P.J., J.L. Dungan, and D.L. Peterson, 2001. Estimating the foliar biochemical concentration of leaves with reflectance spectrometry: Testing the Kokaly and Clark methodologies, *Remote Sens. Environ.*, **76**:349-359.

Curran, P.J., R.A. Jago, and M.E. Cutler, 1998. Red-edge research: Estimating canopy chlorophyll concentration form both field and airborne spectra, *Proc. 9th Australasian Remote Sens. Photogramm. Conf.*, Sydney, Australia, 20 pp., CD-Rom.

Curran, P.J., J.L. Dungan, and H.L. Gholz, 1990. Exploring the relationship between reflectance red edge and chlorophyll content in slash pine, *Tree Physiol.*, **7**:33-48.

Curran, P.J., W.R. Windham, and H.L. Gholz, 1995. Exploring the relationship between reflectance red edge and chlorophyll concentration in slash pine leaves, *Tree Physiol.*, **15**:203-206.

Curran, P.J., J.L. Dungan, B.A. Macler, and S.E. Plummer, 1991. The effect of a red leaf pigment on the relationship between red edge and chlorophyll concentration. *Remote Sens. Environ.*, **35**:69-76.

Czerny, A.B., and K.H. Dunton, 1995. The effects of *in situ* light reduction on the growth of two subtropical seagrasses, *Thalassia testudinum* and *Halodule wrightii*, *Estuaries*, **18**:418-427.

Datt, B., 1998a. Remote sensing of chlorophyll *a*, chlorophyll *b*, chloropyll *a+b*, and total carotenoid content in *Eucalyptus* leaves, *Remote Sens. Environ.*, **66**:111-121.

Datt, B., 1998b. Remote sensing of leaf chlorophyll content in Eucalyptus species. *Proc. 9th Australasian Remote Sens. Photogramm. Conf.*, Sydney, Australia, 16 pp., CD-Rom.

Datt, B., 2000. Recognition of Eucalyptus forest species using hyperspectral reflectance data. *Proc. IEEE Int. Geosc. Remote Sens. Symp.*, Honolulu, Hawaii, **4**:1405-1407.

Daughtry, C.S.T., C.L. Walthall, M.S. Kim, E.B. de Colstoun, and J.E. McMurtrey, 2000. Estimating corn leaf chlorophyll concentration from leaf and canopy reflectance, *Remote Sens. Environ.*, **74**(2):229-239.

Dawes, C.J. 1998. Marine Botany, 2nd ed., J. Wiley and Sons.

Dawes, C.J., and D.A. Tomasko, 1988. Depth distribution in two *Thalassia testudinum* meadows on the west coasts of Florida: a difference in effect of light availability, *Mar. Ecol.*, **9**:123-230.

Dawson, S.P., and W.C. Dennison, 1996. Effects of ultraviolet and photosynthetically active radiation on five seagrass species. *Mar. Biol.*, **125**:629-638.

Dekker, A.G., J.M. Anstee, and V.E. Brando, 2003. Seagrass change assessment using satellite data for Wallis Lake, NSW, CSIRO Land and Water, Technical Report 13/03, April 2003, 64pp.

Dekker, A.G., T.J. Malthus, and H.J. Hoogenboom, 1995. The remote sensing of inland water quality, pp. 123- , *In*, Danson, F.M., and S.E. Plummer (eds), Advances in Environmental Remote Sensing, John Wiley and Sons, UK.

Dekker, A.G., T.J. Malthus, M.M. Wijnen, and E. Seyhan, 1992. Remote sensing as a tool for assessing water quality in Loosdrecht lakes. *Hydrobiologia*, **233**: 137-159.

Demetriades-Shah, T.H., M.D. Steven, and J.A. Clark, 1990. High resolution derivative spectra in remote sensing, *Remote Sens. Environ.*, **33**:55-64.

Demmig, B., K. Winter, A. Kruger, and F. Czygan, 1987. Photoinhibition and zeaxanthin formation in intact leaves: a possible role of the xanthophyll cycle in the dissipation of excess light energy, *Plant Physiol.*, **84**:218-224.

Demmig-Adams, B., 1990. Carotenoids and photoprotection in plants: a role for the xanthophyll zeaxanthin, *Biochim. Biophys. Acta*, **1020**:1-24.

Demmig-Adams, B., and W.W. Adams III, 1992a. Photoprotection and other responses of plants to high light stress, *Ann. Rev. Plant Physiol. Plant Mol. Biol.*, **43**:599-626.

Demmig-Adams, B., and W.W. Adams III, 1992b. Carotenoid composition in sun and shade leaves of plants with different life forms, *Plant, Cell and Environ.*, **15**:411-419.

Demmig, B., and O. Björkman, 1987. Comparison of the effect of excessive light on chlorophyll fluorescence (77K) and photon yield of oxygen evolution in leaves of higher plants, *Planta*, **171**:171-184.

Demmig-Adams, B., W.W. Adams III, V. Ebbert, and B.A. Logan, 1999. Ecophysiology of the xanthophyll cycle, pp. 245-269, *In*, Frank, H.A., A.J. Young, B.G. Britton, and R.J. Cogdell (eds), The photochemistry of carotenoids, Kluwer Academic, Dordrecht, The Netherlands.

den Hartog, C., 1970. The seagrasses of the world, *Verh. Kon. Ned. Akad. Wetensch. Natuurk Reeks 2*, **59**:1-275.

den Hartog, C., 1996. Sudden decline of seagrass beds: "Wasting disease" and other disasters, pp. 307-314, *In*, Kuo, J., R.C. Phillips, D.I. Walker, and H. Kirkman (eds), Seagrass Biology: Proceedings of an International Workshop, 25-29 January, Rottnest Island, Western Australia.

Dennison, W.C., 1987. Effects of light on seagrass photosynthesis, growth and depth distribution, *Aquat. Bot.*, **27**:15-26.

Dennison, W.C., 1990. Shoot density, pp. 61-63, *In*, Phillips, R.C., and C.P. McRoy (eds), Seagrass Research Methods, UNESCO, Paris.

Dennison, W.C., and R.S. Alberte, 1982. Photosynthetic responses of *Zostera marina* L. (eelgrass) to in situ manipulations of light intensity. *Oecologia*, **55**:137-144.

Dennison, W.C., and R.S. Alberte, 1985. Role of daily light period in the depth distribution of *Zostera marina* (eelgrass). *Mar. Ecol. Prog. Ser.*, **25**:51-61.

Dennison, W.C., and H. Kirkman, 1996. Seagrass survival model, pp.341-344, *In*, Kuo, J., R.C. Phillips, D.I. Walker, and H. Kirkman (eds), Seagrass Biology: Proceedings of an International Workshop, Rottnest Island, Western Australia, 25-29 January 1996.

Dennison, W.C., R.J. Orth, K.A. Moore, J.C. Stevenson, V. Carter, S. Kollar, P.W. Bergstrom, and R.A. Batiuk, 1993. Assessing water quality with submersed aquatic vegetation: habitat requirements as barometers of Chesapeake Bay health, *BioScience*, **43**:86-94.

Deysher, L.E., 1993. Evaluation of remote-sensing techniques for monitoring giant-kelp populations, *Hydrobiologia*, **26**:307-312.

Dierrsen, H.M., R.C. Zimmerman, R.A. Leathers, T.V. Downes, and C.O. Davis, 2003. Ocean colour remote sensing of seagrass and bathymetry in the Bahamas Banks by high-resolution airborne imagery, *Limnol. Oceanogr.*, **48**(1 part 2):444-455.

Dixon, L.K., 2000. Establishing light requirements for the seagrass *Thalassia testudinum*: An example from Tampa Bay, Florida, pp. 9-31, *In*, Bortone, S.A. (ed), Seagrasses: Monitoring, ecology, physiology, and management, CRC Press, London.

Drew, E.A., 1979. Physiological aspects of primary production in seagrasses. *Aquat. Bot.*, **7**:139-150.

Duarte, C.M., 1990. Seagrass nutrient content, *Mar. Ecol. Prog. Ser.*, **67**:201-207.

Duarte, C.M., 1991. Seagrass depth limits. *Aquat. Bot.*, **40**:363-377.

Duarte, C.M., 1999. Seagrass ecology at the turn of the millennium: challenges for the new century, *Aquat. Bot.*, **65**:7-20.

Duarte, C.M., and C.L. Chiscano, 1999. Seagrass biomass and production: a reassessment, *Aquat. Bot.*, **65**:159-174.

Duggin, M.J., 1980. The field measurement of reflectance factors, *Photogramm. Eng. Remote Sens.*, **46**(5):643-647.

Duggin, M.J., 1987. Impact of radiance variations on satellite sensor calibration, *Appl. Optics*, **26**(7):1264-1271.

Dunk, I., and M. Lewis, 2000. Seagrass and shallow water feature discrimination using HyMap imagery, *Proc. 10th Australasian Remote Sens. Photogramm. Conf.*, Adelaide, Australia, pp. 1092-1108.

Dunn, J.L., J.D. Turnbull, and S.A. Robinson, 2004. Comparison of solvent regimes for the extraction of photosynthetic pigments from leaves of higher plants, *Functional Plant Biol.*, **31**: 195-202.

Dunton, K.H., 1994. Seasonal growth and biomass of the subtropical seagrass *Halodule wrightii* in relation to continuous measurements of underwater irradiance, *Mar. Biol.*, **120**(3):479-489.

Dunton, K.H., and D.A. Tomasko, 1994. In-situ photosynthesis in the seagrass *Halodule wrightii* in a hypersaline subtropical lagoon, *Mar. Ecol. Prog. Ser.*, **107**(3):281-293.

Enríquez, S., M. Merino, and R. Iglesias-Prieto, 2002. Variations in the photosynthetic performance along the leaves of the tropical seagrass *Thalassia testudinum, Mar. Biol.*,**140**(5):891-900.

Falkowski, P.G., and J.A. Raven, 1997. Aquatic Photosynthesis. Blackwell Science, Carlton.

Falkowski, P.G., and J. LaRoche, 1991. Acclimation to spectral irradiance in algae, *J. Phycol.*, **27**:8-14.

Ferguson, R.L., and K. Korfmacher, 1997. Remote sensing and GIS analysis of seagrass meadows in North Carolina, USA, *Aquat. Bot.*, **58**:241-258.

Field, C.B., and J.R. Ehrlinger, 1993. Introduction: Questions of scale, pp. 1-4, *In*, Ehrlinger, J.R., and C.B. Field (eds), Scaling ecological processes: Leaf to globe, Academic Pres, Sydney.

Field, C.B., J.A. Gamon, and J. Penuelas, 1995. Remote sensing of terrestrial photosynthesis, pp. 511-527, *In*, Schulze, E.- D., and M.M. Caldwell (eds), Ecophysiology of Photosynthesis, Springer-Verlag, Berlin.

Flanigan, Y.S., and C. Critchley, 1996. Light response of D1 turnover and photosystem II efficiency in the seagrass *Zostera capricorni*, *Planta*, **198**:319-323.

Fong, C.W., S.Y. Lee, and R.S.S. Wu., 2000. The effects of epiphytic algae and their grazers on the intertidal seagrass *Zostera japonica*. *Aquat. Bot.*, **67**:251-261.

Fonseca, M.S., 1996. Scale dependence in the study of seagrass systems, pp. 95-104, *In*, Kuo, J. R.C. Phillips, D.I. Walker, and H. Kirkman (eds), Seagrass Biology: Proceedings of an International Workshop, Rottnest Island, Western Australia, 25-29 January 1996.

Fonseca M.S., and W.J. Kenworthy, 1987. Effects of current on photosynthesis and distribution of seagrasses, *Aquat. Bot.* **27**:59-78.

Foppen, J.H., 1971. Tables for the identification of carotenoid pigments. *Chromatog. Rev.*, **14**:133-298.

Francis, F.J., 1982. Analysis of anthocyanins, pp. 181-207, *In*, P. Markakis (ed.), *Anthocyanins as Food Colours*, Academic Press, New York.

Fuleki, T., and F.J. Francis, 1968. Quantitative methods for anthocyanins. 2. Determination of total anthocyanin and degradation index for cranberry juice, *J. Food Science*, **33**:78-83.

Fyfe, S.K., 2003. Spatial and temporal variation in spectral reflectance: are seagrass species spectrally distinct?, *Limnol. Oceanogr.*, Coastal Optics Special Issue, 48(1):464-479.

Fyfe, S.K., and Davis, A.R., 2007. Spatial scale and the detection of impacts on the seagrass Posidonia australis following pier construction in an embayment in southeastern Australia, *Estuarine, Coastal and Shelf Science*, **74**:297-305.

Gallegos, C.L., and W.J. Kenworthy, 1996. Seagrass depth limits in the Indian River Lagoon (Florida, U.S.A.): Application of an optical water quality model, *Estuarine, Coastal and Shelf Science*, **42**:267-288.

Gamon, J.A., and J.S. Surfus, 1999. Assessing leaf pigment content and activity with a reflectometer, *New Phytol.*, **143**:105-117.

Gamon, J.A., J. Peñuelas, and C.B. Field, 1992. A narrow-waveband spectral index that tracks diurnal changes in photosynthetic efficiency, *Remote Sens. Environ.*, **41**:35-44.

Gamon, J.A., C.B. Field, M.L. Goulden, K.L. Griffin, A.E. Hartley, G. Joel, J. Peñuelas, and R. Valentini, 1995a. Relationship between NDVI, canopy structure and photosynthesis in three Californian vegetation types, *Ecol. Appl.*, **5**(1):28-41.

Gamon, J.A., D.A. Roberts, and R.O. Green, 1995b. Evaluation of the photochemical reflectance index in AVIRIS imagery, Summaries of the 5th Annual JPL Airborne Earth Science Workshop, Vol. 1 AVIRIS Workshop (R.O. Green, Ed.), Pasadena, USA, 23-27 Jan 1995, JPL Publication 95-1, pp. 55-58.

Gamon, J.A., L. Serrano, and J.S. Surfus, 1997. The photochemical reflectance index: an optical indicator of photosynthetic radiation use efficiency across species, functional types, and nutrient levels, *Oecologia*, **112**:492-501.

Gamon, J.A., C.B. Field, W. Bilger, O. Björkman, A.L. Fredeen, and J. Peñuelas, 1990. Remote sensing of the xanthophyll cycle and chlorophyll fluorescence in sunflower leaves and canopies, *Oecologia*, **85**:1-7.

García-Plazaola, J.I., S. Matsubara, and C.B. Osmond. 2007. The lutein epoxide cycle in higher plants: its relationships to other xanthophyll cycles and and possible functions, *Functional Plant Biol.*, **34**:759-773.

Gates, D.M., H.J. Keegan, J.C. Schleter, and V.R. Weidner, 1965. Spectral properties of plants. *App. Optics*, **4**(1):11-20.

Gausman, H.W., 1977. Reflectance of leaf components, *Remote Sens. Environ.*, **6**:1-9.

Gausman, H.W., 1982. Visible light reflectance, transmittance, and absorptance of differently pigmented cotton leaves. *Remote Sens. Environ.*, **13**:233-238.

Gausman, H.W., 1984. Evaluation of factors causing reflectance differences between sun and shade leaves, *Remote Sens. Environ.*, **15**:177-181.

Gausman, H.W., and W.A. Allen, 1973. Optical parameters of leaves of 30 plant species, *Plant Physiol.*, **52**:57-62.

Gausman, H.W., W.A. Allen, V.I. Myers, and R. Cardenas, 1969. Reflectance and internal structure of cotton leaves, *Gossypium hirsutum* L., *Agron. J.*, **61**:374-376.

Gausman, H.W., J.E. Quisenberry, J.J. Burke, and C.W. Wendt, 1984. Leaf spectral measurements to screen cotton strains for characters affected by stress, *Field Crops Res.*, **9**:373-381.

Gausman, H.W., J.R. Thomas, D.E. Escobar, and A. Berumen, 1975. Cotton leaf air volume and chlorophyll concentration affect reflectance of visible light, *J. Rio Grande Valley Hort. Soc.*, **29**:109-114.

Gausman, H.W., W.A. Allen, D.E. Escobar, R.R. Rodriguez, and R. Cardenas, 1971. Age effects of cotton leaves on light reflectance, transmittance, and absorptance and on water content and thickness, *Agron. J.*, **63**:465-469.

Gilmore, A.M., and H.Y. Yamamoto, 1991. Resolution of lutein and zeaxanthin using a non-endcapped, lightly carbon-loaded C_{18} high-performance liquid chromatography column, *J. Chromatography,* **543**:137-145.

Ginsburg, R.N., and H.A. Lowenstam, 1958. The influence of marine bottom communities on the depositional environment of sediments, *J. Geology*, **66**:318-319.

Gitelson, A.A., and M.N. Merzlyak, 1996. Signature analysis of leaf reflectance spectra: algorithm development for remote sensing of chlorophyll, *J. Plant Physiol.*, **148**:494-500.

Goetz, A.F.H., and W.M. Calvin, 1987. Imaging spectrometry: Spectral resolution and analytical identification of spectral features, Imaging Spectros. II, Vane, G. (ed.), *Proc. SPIE,* **384**:158-165.

Gong, P., R. Pu and B. Yu, 1997. Conifer species recognition: An exploratory analysis of in situ hyperspectral data. *Remote Sens. Environ.*, **62**:189-200.

Goodman, J.L., K.A. Moore, and W.C. Dennison, 1995. Photosynthetic responses of eelgrass (*Zostera marina* L.) to light and sediment sulfide in a shallow barrier island lagoon, *Aquat. Bot.*, **50**:37-47.

Gordon, D.M., K.A. Grey, S.C. Chase, and C.J. Simpson, 1994. Changes to the structure and productivity of a *Posidonia sinuosa* meadow during and after imposed shading, *Aquat. Bot.*, **47**:265-275.

Gordon, D.M., P. Collins, I.N. Baxter, and I. LeProvost, 1996. Regression of seagrass meadows, changes in seabed profiles and seagrass composition at dredged and undredged sites in the Owen Anchorage region of south-western Australia, pp. 323-332. *In*, Kuo, J., R.C. Phillips, D.I. Walker, and H. Kirkman (eds), Seagrass Biology: proceedings of an international workshop, Rottnest Island, Western Australia, January 1996.

Goward, S.N., B. Markham, D.G. Dye, W. Dulaney, and J. Yang, 1991. Normalized difference vegetation index measurements from the Advanced Very High Resolution Radiometer, *Remote Sens. Environ.*, **35**:257-277.

Grice, A.M., N.R. Loneragan, and W.C. Dennison, 1996. Light intensity and the interactions between physiology, morphology and stable isotope ratios in five species of seagrass, *J. Exp. Mar. Biol. Ecol.*, **195**:91-110.

Guidetti, P., 2001. Detecting environmental impacts on the Mediterranean seagrass *Posidonia oceanica* (L.) Delile: the use of reconstructive methods in combination with 'beyond BACI' designs, *J. Exp. Mar. Biol. Ecol.*, **260**:27-39.

Guillaumont, B., L. Callens, and P. Dion, 1993. Spatial distribution and quantification of *Fucus* species and *Ascophyllum nodosum* beds in intertidal zones using spot imagery, *Hydrobiologia*, **260/261**:297-305.

Guillaumont, B., S.B. Ben Mustapha, H. Ben Moussa, J. Zaouali, N. Soussi, A.B. Mammou, and C. Ganov, 1995. Pollution impact study in Gabes Gulf (Tunisia) using remote sensing data, *Mar. Tech. Soc. J.*, **29**(2):46-58.

Guyot, G., 1990. Optical properties of vegetation canopies, pp. 19-43, *In*, Clark, J.A., and M.D. Steven (eds), Applications of remote sensing in agriculture, University Press, Cambridge.

Hale, M.G., and D.M. Orcutt, 1987. The Physiology of Plants under Stress, John Wiley and Sons, Brisbane.

Hancock, W.S., and J.T. Sparrow, 1984. HPLC analysis of biological compounds – A laboratory guide, Marcel Dekker Inc., New York.

Harden, G. J. (ed.), 1993. Flora of New South Wales, Vol. 4, NSW University Press, Sydney.

Harris, M.McD., 1977. Ecological studies on Illawarra Lake with special reference to *Zostera capricorni* Ascherson, M.Sc. Thesis (unpub.), School of Botany, University of NSW, Sydney.

Harris, M.McD., R.J. King, and J. Ellis, 1980. The eelgrass *Zostera capricorni* in Lake Illawarra, New South Wales, *Proc. Linn. Soc. NSW*, **104**:23-33.

Harrison, B.A., and D.L.B. Jupp, 1990. Introduction to Image Processing, CSIRO Division of Water Resources, Canberra.

Hart, D., 1997. Near-shore seagrass change between 1949 and 1996 mapped using digital aerial orthophotography: Metropolitan Adelaide area Largs Bay – Aldinga South Australia, Image Data Services, Resource Information Group, South Australian Department of Environment and Natural Resources.

Hastings, K., P. Hesp, and G.A. Kendrick, 1995. Seagrass loss associated with boat moorings at Rottnest Island, Western Australia, *Ocean Coastal Management*, **26**(3):225-246.

Heber, U., R. Bligny, P. Streb, and R. Douce, 1996. Photorespiration is essential for the protection of the photosynthetic apparatus of C_3 plants against photoinactivation under sunlight. *Bot. Acta*, **109**:307-315.

Heilman, J.L., and M.R. Kress 1987. Effects of vegetation on spectral irradiance at the soil surface, *Agron. J.*, **79**(5):765-768.

Heinz Walz GmbH, 1993. Portable fluorometer PAM-2000 and data acquisition software DA-2000, edition 2.1, June 1993, software version 2.00, Effelhrich, Germany.

Held, A., G. Byrne, J. Anstee, N. Williams, and C. Field, 1997. Monitoring of sensitive coastal ecosystems with the *casi*, *3rd Int. Airborne Remote Sens. Conf. Exhibition*, 7-10 July 1997, Copenhagen, Denmark, 8 pp.

Hellblom, F., and M. Björk, 1999. Photosynthetic responses in *Zostera marina* to decreasing salinity, inorganic carbon content and osmolality, *Aquat. Bot.*, **65**:97-104.

Hick, P., 1997. Determination of water column characteristics in coastal environments using remote sensing. Ph.D. Thesis (unpub.), Curtin University, Bently, Western Australia.

Hillman, K., D.I. Walker, A.W.D. Larkum, and A.J. McComb, 1989. Productivity and nutrient limitation, pp. 635-685. *In,* A.W.D. Larkum, A.J. McComb, and S.A. Shepherd (eds), Biology of seagrasses: a treatise on the biology of seagrasses with special reference to the Australian region, Elsevier Science Publishers, Amsterdam.

Hochberg, E.J., and M.J. Atkinson, 2000. Spectral discrimination of coral reef benthic communities, *Coral Reefs*, **19**:164-171.

Holden, H., and E. LeDrew, 1999. Hyperspectral identification of coral reef features, *Int. J. Remote Sens.*, **20**(13):2545-2563.

Hootsmans, M.J.M., L. Santamaría, and J.E. Vermaat, 1995. How to survive darkness? Photosynthetic and other solutions provided by three submerged aquatic macrophytes (*Potamogeton pectinatus* L., *Ruppia drepanensis* Tineo and *Zostera noltii* Hornem.), *Wat. Sci. Tech.*, **32**(4):49-51.

Horler, D.N.H., M. Dockray, and J. Barber, 1983. The red edge of plant leaf reflectance, *Int. J. Remote Sens.*, **4**:272-288.

Horton, P., A.V. Ruban, and M. Wentworth, 2000. Allosteric regulation of the light-harvesting system of photosystem II, *Phil. Trans. R. Soc. Lond.* B, **355**:1361-1370.

Huete, A.R., 1988. A soil-adjusted vegetation index (SAVI), *Remote Sens. Environ.*, **25**:295-309.

Hutchings, P.A. and Recher, H.F., (1974). The fauna of Careel Bay with comments on the ecology of mangrove and seagrass communities, *Aust. Zool.*, **18**:99-127.

Invers, O., J. Romero, and M. Pérez, 1997. Effects of pH on seagrass photosynthesis: a laboratory and field assessment, *Aquat. Bot.*, **59**:185-194.

Irlandi, E.A., W.G. Ambrose Jr, and B.A. Orlando, 1995. Landscape ecology and the marine environment: how spatial configuration of seagrass habitat influences growth and survival of the bay scallop, *Oikos*, **72**:307-313.

Isaksson, T., and B. Kowalski, 1993. Piece-wise multiplicative scatter correction applied to near-infrared diffuse transmittance data from meat products, *Appl. Spectrosc.*, **47**(6):702-709.

Jacquemoud, S., and F. Baret, 1990. PROSPECT: A model of leaf optical properties spectra, *Remote Sens. Environ.*, **34**:75-91.

Jacquemoud, S., F. Baret, B. Andrieu, F.M. Danson, and J.K. Jaggard, 1995. Extraction of vegetation biophysical parameters by inversion of the PROSPECT+SAIL models on sugar beet canopy reflectance data. Application to TM and AVIRIS sensors, *Remote Sens. Environ.*, **52**:163-172.

Jacquemoud, S., S.L. Ustin, J. Verdebout, G. Schmuck, G. Andreoli, and B. Hosgood, 1996. Estimating leaf biochemistry using the PROSPECT leaf optical properties model, *Remote Sens. Environ.*, **56**:163-172.

Jacobs, S.W.L., and A. Williams, 1980. Notes on the genus *Zostera* s. lat. in New South Wales, *Telopea*, **1**:451-455.

Jaubert, J.M., J.R.M. Chisholm, H.T. Ripley, L. Pritchett, and D. Cadot, 1998. Mapping of the spread of the invasive alga *Caulerpa taxifolia* in shallow water habitats of the French Riviera using high resolution multispectral airborne imagery: preliminary results, IEEE Oceans '98 Conference Proceedings, 28 Sep-1 Oct 1998, Vol. 3:1790-1791.

Jeffrey, S.W., 1981. Responses to light in aquatic plants, pp. 249-276, *In*, Lange, O.L., P.S. Nobel, C.B. Osmond, and H. Ziegler (eds), Physiological plant ecology I: Responses to the physical environment, Encyclopedia of Plant Physiology New Series, Vol. 12A, Springer-Verlag, Berlin.

Jensen, J.R., 1996. Introductory Digital Image Processing: a remote sensing perspective, 2nd ed., Prentice Hall.

Jernakoff, P., and P. Hick, 1994. Spectral measurements of marine habitat: simultaneous field measurement and CASI data. *Proc. 7th Australasian Remote Sens. Conf.*, pp. 706-713.

Jiménez, C., F.X. Niell, and P. Algarra, 1987. Photosynthetic adaptation of *Zostera noltii* Hornem., *Aquat. Bot.*, **29**:217-226.

Jupp, D.L.B., J. Walker, and L.K. Penridge, 1986. Interpretation of vegetation structure in Landsat MSS imagery - A case-study in disturbed semiarid eucalypt woodlands. 2. Model-based analysis, *J. Environ. Manage.*, **23**(1):35-57.

Jupp, D.L.B., G.T. Byrne, J.M. Anstee, T.R. McVicar, E.R. McDonald, and D. Parkin, 1995. The use of *casi* spectral data to monitor disturbance in shallow areas of Port Phillip Bay, *Int. Symp. Spectral Remote Sens. Res.*, ISSSR'95, 26-29 November, Melbourne, Australia, 16 pp.

Jupp, D.L.B., G. Byrne, J. Anstee, E. McDonald, T. McVicar, J. Kirk, C. Hurlstone, and S. Chiidgey. 1996. Port Phillip Bay benthic habitat mapping project, Final report 1: project outcomes. Environmental Study Task G2.2, CSIRO Division of Water Resources, Canberra, Australia, 35 pp.

Jupp, D.L.B., J.T.O. Kirk, and G.P. Harris, 1994. Detection, identification and mapping of cyanobacteria- using remote sensing to measure the optical water quality of turbid inland waters, *Aust. J. Mar. Freshwater Res.*, **45**:801-828.

Kamermans, P., Hemminga, and D.J. De Jong, 1999. Significance of salinity and silicon levels for growth of a formerly estuarine eelgrass (*Zostera marina*) population (Lake Grevelingen, the Netherlands), *Mar. Biol.*, **133**(3):527-539.

Kauth, R.J., and G.S. Thomas, 1976. The tasselled cap – a graphic description of the spectral-temporal development of agricultural crops as seen by Landsat, *Proc. Symp. Machine Processing of Remotely Sensed Data*, West Lafayette, USA, 1976, pp.41-51.

Keely, S.J., 1991. Interactive role of stresses on structure and function in aquatic plants, pp. 329-343, *In*, Mooney, H.A., W.E. Winner, E.J. Pell, and E. Chu (eds), Response of plants to multiple stresses, Academic Press, Inc.

Kendrick, G.A., J.Eckersley, and D.I. Walker, 1999. Landscape-scale changes in seagrass distribution over time: a case study from Success Bank, Western Australia, *Aquat. Bot.*, **65**:293-309.

Kendrick, G.A., B.J. Hegge, A. Wyllie, A. Davidson, and D.A. Lord, 2000. Changes in seagrass cover on Success and Parmelia Banks, Western Australia between 1965 and 1995. *Estuarine Coastal Shelf Science*, **50**:341-353.

Kendrick, G. A., M. J. Aylward, B. J. Hegge, M. L. Cambridge, K. Hillman, A. Wyllie, and D. A. Lord, 2002. Changes in seagrass coverage in Cockburn Sound, Western Australia between 1967 and 1999, *Aquat. Bot.*, **73**:75-87.

Kim, M.S., C.S.T. Daughtry, E.W. Chappelle, and J.E. McMurtrey, 1994. The use of high spectral resolution bands for estimating absorbed photosynthetically active radiation (APAR), In, *Proc. ISPRS'94,* Val d'Isere, France, 17-21 January 1994, pp. 299-306.

King, R.J., 1981. Mangroves and saltmarsh plants, pp. 308-328, *In*, Clayton, M.N., and R.J. King (eds), Marine botany: an Australasian perspective, Longman Cheshire.

King, R.J., 1986. Aquatic angiosperms in coastal saline lagoons of New South Wales. I. Vegetation of Lake Macquarie, *Proc. Linn. Soc. NSW*, **109**:11-23.

King, R.J., 1988. The seagrasses of Lake Illawarra, New South Wales, *Wetlands (Australia)*, **8**(1):21-26.

King, R.J., and B.R. Hodgson, 1986. Aquatic angiosperms in coastal saline lagoons of New South Wales. 4. Long term changes, *Proc. Linn. Soc. N.S.W.*,**109**:51-60.

King, R.J., and V.M. Holland, 1986. Aquatic angiosperms in coastal saline lagoons of New South Wales. II. Vegetation of Tuggerah Lakes, *Proc. Linn. Soc. NSW*, **109**:25-39.

Kirk, J.T.O., 1994. Light and photosynthesis in aquatic ecosystems, 2nd ed., Cambridge University Press, Melbourne.

Kirk, J.T.O., and D.J. Goodchild, 1972. Relationship of photosynthetic effectiveness of different kinds of light to chlorophyll content and chloroplast structure in greening wheat and in ivy leaves, *Aust. J. Biol. Sci.*, **25**:215-241.

Kirkman, H., 1978. Declines of seagrass in northern Moreton Bay, Queensland, *Aquat. Bot.*, **5**:63-76.

Kirkman, H., 1989. Growth, density and biomass of *Ecklonia radiata* at different depths and growth under artificial shading off Perth, Western Australia, *Aust. J. Mar. Freshwater* Res., **40**(2):169-177.

Kirkman, H., 1996. Baseline and monitoring methods for seagrass meadows, *J. Environ. Management*, **47**:191-201.

Kirkman, H., 1997. Seagrasses of Australia, Australia: State of the Environment Technical Paper Series (Estuaries and the Sea), Department of the Environment, Canberra, 48 pp.

Kirkman, H., and J. Kirkman, 2000. Long-term seagrass meadow monitoring near Perth, Western Australia, *Aquat. Bot.*, **67**:319-332.

Kirkman, H., and J. Kuo, 1990. Patterns and process in southern Western Australian seagrasses, *Aquat. Bot.*, **37**:367-382.

Kirkman, H. and C. Manning, 1993. A search for eutrophication in the seagrass meadows of Rottnest Island and a baseline study for future monitoring, pp. 471–479, *In*, Wells, F. E., D. I. Walker, H. Kirkman and R. Lethbridge (eds), Proc. 4th Int. Mar. Biol. Workshop: The Marine Flora and Fauna of Rottnest Island, Western Australia, Perth, Western Australian Museum.

Kirkman, H., and D.I. Walker, 1989. Regional studies – Western Australian seagrass, pp. 157-181, *In*, A.W.D. Larkum, A.J. McComb, and S.A. Shepherd (eds), Biology of Seagrasses: A treatise on the biology of seagrasses with special reference to the Australian region, Elsevier, New York.

Kirkman, H., I.H. Cook, and D.D. Reid, 1982. Biomass and growth of *Zostera capricorni* Aschers. in Port Hacking, N.S.W., Australia, *Aquat. Bot.*, **12**:57-67.

Kirkman, H., J. Fitzpatrick, and P.A. Hutchings, 1995. Seagrasses, pp. 137-142, *In* Cho, G., A. Georges, R. Stoutjesdijk, and R. Longmore (eds), Kowari 5- Jervis Bay: a place of cultural, scientific and educational value, Australian Nature Conservation Agency, Canberra.

Kleshnin, A.F. and I.A. Shul'gin. 1959. The optical properties of plant leaves. *Academiya Nauk SSSR, Bot. Sci.*, **125**:108-110.

Klumpp, D.W., R.K. Howard, and D.A. Pollard, 1989. Trophodynamics and nutritional ecology of seagrass communities, pp.394-457, *In*, A.W.D. Larkum, A.J. McComb, and S.A. Shepherd (eds), Biology of seagrasses: a treatise on the biology of seagrasses with special reference to the Australian region, Elsevier Science Publishers, Amsterdam.

Knapp, A.K., and G.A. Carter, 1998. Variability in leaf optical properties among 26 species from a broad range of habitats, *Am. J. Bot.*, **85**(7):940-946.

Koch, E,W., 1996. Hydrodynamics of a shallow *Thalassia testudinum* bed in Florida, USA, pp. 1005-110, *In*, Kuo, J., R.C. Phillips, D.I. Walker, and H. Kirkman (eds), Seagrass Biology: proceedings of an international workshop, Rottnest Island, Western Australia, January 1996.

Komatsu, T., 1996. Influence of a *Zostera* bed on the spatial distribution of water flow over a broad geographic area, pp. 111-116, *In*, Kuo, J., R.C. Phillips, D.I. Walker, and H. Kirkman (eds), Seagrass Biology: proceedings of an international workshop, Rottnest Island, Western Australia, January 1996.

Königer, M., G.C. Harris, A. Virgo, and K. Winter, 1995. Xanythophyll-cycle pigments and photosynthetic capacity in tropical forest species: a comparative field study on canopy gap and understorey plants, *Oecologia*, **104**:280-290.

Korobov, R.M., and V.Ya. Railyan, 1993. Canonical correlation relationships among spectral and phytometric variables for twenty winter wheat fields, *Remote Sens. Environ.*, **43**:1-10.

Kraemer, G.P., and R.S. Alberte, 1995. Impact of daily light period on protein synthesis and carbohydrate stores in *Zostera marina* L. (eelgrass) roots: implications for survival in light limited environments, *J. Exp. Mar. Biol. Ecol.*, **185**:191-202.

Krause, G.H., and E. Weis, 1988. The photosynthetic apparatus and chlorophyll fluorescence: an introduction, pp. 3-11 *In*, Lichtenthaler, H,K. (ed), Applications of chlorophyll fluorescence, Kluwer Academic, Dordrecht.

Krause, G.H., and E. Weis, 1991. Chlorophyll fluorescence and photosynthesis: the basics, *Ann. Rev. Plant Physiol. Plant Mol. Biol.*, **42**:313-349.

Kruse, F.A., L.L. Richardson, and V.G. Ambrosia, 1997. Techniques developed for geologic analysis of hyperspectral data applied to near-shore hyperspectral ocean data, *Proc. 4th International Conference on Remote Sensing for Marine and Coastal Environments*, Orlando, Florida, 17-19 March 1997, Vol I, pp. 233-246.

Kumar, L., and A.K. Skidmore, 1998. Use of derivative spectroscopy to identify regions of differences between some Australian eucalypt species. *Proc. 9th Australasian Remote Sens. Photogramm. Conf.*, Sydney, Australia, p. 3103-3118.

Kuo, J., 1978. Morphology, anatomy and histochemistry of Australian seagrasses of genus *Posidonia* Konig (Posidoniaceae). 1. Leaf blade and leaf sheath of *Posidonia australis* Hook f., *Aquat. Bot.*, **5**(2):171-190.

Kuo, J., 1983. The nacreous walls of sieve elements in seagrasses, *Am. J. Bot.*, **70**(2):159-164.

Kuo, J., and A.J. McComb, 1989. Seagrass taxonomy, structure and development, pp. 6-73, *In*, A.W.D. Larkum, A.J. McComb, and S.A. Shepherd (eds), Biology of Seagrasses: A treatise on the biology of seagrasses with special reference to the Australian region, Elsevier, New York.

Kupeic, J.A., 1994. The remote sensing of foliar chemistry, Ph.D. Thesis (unpub.), University College of Swansea, University of Wales.

Kurz, R.C., D.A. Tomasko, D. Burdlick, T.F. Ries, K. Patterson, and R. Finck, 2000. Recent trends in seagrass distributions in southwest Florida coastal waters, pp. 157-166, *In* S.A. Bortone (ed), Seagrasses: monitoring, ecology, physiology and management, CRC Press, Florida.

Kutser, T., W. Skirving, J. Parslow, L. Clementson, T. Done, M. Wakeford, and I. Miller, 2001. Spectral discrimination of coral reef bottoms, *Proc. Geoscience and Remote Sensing Symp., IGARSS 2001*, 9-13 July, Vol. 6:2872 – 2874.

Larkum, A.W.D., and C. den Hartog, 1989. Evolution and biogeography of seagrasss, pp.112-156. *In,* A.W.D. Larkum, A.J. McComb, and S.A. Shepherd (eds), Biology of seagrasses: a treatise on the biology of seagrasses with special reference to the Australian region, Elsevier Science Publishers, Amsterdam.

Larkum, A.W.D., and P.L. James, 1996. Towards a model for inorganic carbon uptake in seagrasses involving carbonic anhydrase, pp. 191-196, *In*, Kuo, J., R.C. Phillips, D.I. Walker, and H. Kirkman (eds), Seagrass Biology: proceedings of an international workshop, Rottnest Island, Western Australia, January 1996.

Larkum , A.W.D., and R.J. West, 1983. Stability, depletion and restoration of seagrass beds, *Proc. Linn. Soc. NSW*, **106**:201-212.

Larkum , A.W.D., and R.J. West, 1990. Long term changes of seagrass meadows in Botany Bay, Australia, *Aquat. Bot.*, **37**:55-70.

Larkum, A.W.D., and W.F. Wood, 1993. The effect of UV-B radiation on photosynthesis and respiration of phytoplankton, benthic macroalgae and seagrasses, *Photosyn. Res.*, **36**:17-23.

Larkum, A.W.D., G. Roberts, J. Kuo, and S. Strother, 1989. Gaseous movement in seagrasses, pp 686-722, *In,* A.W.D. Larkum, A.J. McComb, and S.A. Shepherd (eds), Biology of seagrasses: a treatise on the biology of seagrasses with special reference to the Australian region, Elsevier Science Publishers, Amsterdam.

Leadbitter, D, W. Lee Long, and P. Dalmazzo, 1999. Seagrasses and their management - implications for research, pp. 140-171, *In*, Butler, A., and P. Jernakoff (eds), FRDC Project 98/223: Seagrass in Australia: strategic review and development of an R & D plan, CSIRO Publishing, Victoria.

Lee, D.W., R.A. Bone, S.L. Tarsis, and D. Storch, 1990. Correlates of leaf optical properties in tropical forest sun and extreme-shade plants, *American J. Bot.*, 77:370-380.

Lee, Z.P., K. Carder, R. Steward, and B. Weigle, 1997. Bottom depth and type for shallow waters: Hyperspectral observations from a blimp, *Proc. 4th International Conference on Remote Sensing for Marine and Coastal Environments*, Orlando, Florida, 17-19 March, Vol. II, pp. 459-467.

Lee Long, W.J., L.J. McKenzie, M.A. Rasheed, and R.G. Coles, 1996a. Monitoring seagrasses in tropical ports and harbours, pp. 345-350, *In*, Kuo, J., R.C. Phillips, D.I. Walker, and H. Kirkman (eds), Seagrass Biology: Proceedings of an International Workshop, Rottnest Island, Western Australia, 25-29 January 1996.

Lee Long, W.J., R.G. Coles, and L.J. McKenzie, 1996b. Deepwater seagrasses in Northeastern Australia – How deep, how meaningful, pp. 41-50, *In*, Kuo, J., R.C. Phillips, D.I. Walker, and H. Kirkman (eds), Seagrass Biology: Proceedings of an International Workshop, Rottnest Island, Western Australia, 25-29 January 1996.

Lee Long, W.J., C.A. Roder, L.J. McKenzie, and A.J. Hundley, 1998. Preliminary evaluation of an acoustic technique for mapping tropical seagrass habitats, Research Publication no. 52, Great Barrier Reef Marine Park Authority, Australia.

Lehmann, A. and J.B. Lachavanne, 1997. Geographic information systems and remote sensing in aquatic botany, *Aquat. Bot.*, **58**(3-4):195-207.

Lelong, C.C.D., P.C. Pinet, and H. Poilve, 1998. Hyperspectral imaging and stress mapping in agriculture – a case study on wheat in Beauce (France), *Remote Sens. Environ.*, **66**(2):179-191.

Lennon, P., and P. Luck, 1990. Seagrass mapping using Landsat TM data: a case study in Southern Queensland, *Asian Pacific Remote Sens. J.*, **2**:1-6.

Les, D.H., M.L. Moody, S.W.L. Jacobs, and R.J. Bayer, 2002. Systematics of seagrasses (Zosteraceae) in Australia and New Zealand, *System. Bot.*, **27**(3):468-484.

Leuschner, C., S. Landwehr, and U. Mehlig, 1998. Limitation of carbon assimilation of intertidal *Zostera noltii* and *Z. marina* by dessication at low tide. *Aquat. Bot.*, **62**:171-176.

Levings, C.D., M.S. North, G.E. Piercey, G. Jamieson, and B. Smiley, 1999. Mapping nearshore and intertidal marine habitats with remote sensing and GPS: The importance of spatial and temporal scales, *Proc. Oceans '99 MTS/IEEE*,13-16 September 1999, Vol. 3, pp. 1249-1255.

Li, X., and A.H. Strahler, 1985. Geometric-optical modeling of a conifer forest canopy, *IEEE Trans. Geosc. Remote Sens.*, **23**(5):705-721.

Lichtenthaler, H.K., 1987. Chlorophyll fluorescence signatures of leaves during the autumnal chlorophyll breakdown, *J. Plant Physiol.*, **131**(1-2):101-110.

Lichtenthaler, H.K., A.A. Gitelson, and M. Lang, 1996. Non destructive determination of chlorophyll content of leaves of a green and an aurea mutant of tobacco by reflectance measurements, *J. Plant Physiol.*, **148**:483-493.

Lillesaeter, O., 1982. Spectral reflectance of partly transmitting leaves: Laboratory measurements and mathematical models, *Remote Sens. Environ.*, **12**:247-254.

Lillesand, T.M., and R.W. Kiefer. 1994. Remote Sensing and Image Interpretation, 3rd ed., J. Wiley and Sons.

Loeblich, L.A., 1982. Photosynthesis and pigments influenced by light intensity and salinity in the halophile *Dunaliella salina* (Chlorophyta), *J. Mar. Biol. Ass. U.K.*, **62**:493-508.

Long, B.G., D.M. Dennis, T.D. Skewes, and I.R Poiner, 1996. Detecting an environmental impact of dredging on seagrass beds with a BACIR sampling design, *Aquat. Bot.*, **53**(3-4):235-243.

Longstaff, B.J., and W.C. Dennison, 1999. Seagrass survival during pulsed turbidity events: the effects of light deprivation on the seagrasses *Halodule pinifolia* and *Halophila ovalis*. *Aquatic Botany*, **65**:105-121.

Longstaff, B.J., N.R. Loneragan, M.J. O'Donohue, and W.C. Dennison, 1999. Effects of light deprivation on the survival and recovery of the seagrass *Halophila ovalis* (RBr.) Hook.*f.*, *J. Exp. Mar. Biol. Ecol.*, **234**:1-27.

Lorenzen, B., and A. Jensen, 1989. Changes in leaf spectral properties induced in barley by cereal powdery mildew, *Remote Sens. Environ.*, **27**:201-209.

Lovelock, C.E., and S.A. Robinson, 2002. Surface reflectance properties of Antarctic moss and their relationship to plant species, pigment composition and photosynthetic function, *Plant Cell Environ.*, **25**(10):1239-1250.

Lovelock, C.E., M. Jebb, and C.B. Osmond, 1994. Photoinhibition and recovery in tropical plant species - response to disturbance, *Oecologia*, **97**(3):297-307.

Lubin, D., W. Li, P. Dustan, C. Mazel, and K. Stamnes, 2001. Spectral signatures of coral reefs: features from space. *Remote Sens. Environ.*, **75**:127-137.

Luther, J.E., and A.L. Carroll, 1999. Development of an index of balsam fir vigor by foliar spectral reflectance, *Remote Sens. Environ.*, **69**:241-252.

Lyzenga, D.R., 1981. Remote sensing of bottom reflectance and water attenuation parameters in shallow water using aircraft and Landsat data, *Int. J. Remote Sens.*, **2**:71-82.

Maccioni, A., G. Agati, and P. Mazzinghi, 2001. New reflectance indices for remote measurement of chlorophylls based on leaf directional reflectance spectra, *J. Photochem. Photobiol. B: Biol.*, **61**(1-2):52-61.

Macinnis-Ng, C.M.O., and P.J. Ralph, 2002. Towards a more ecologically relevant assessment of the impact of heavy metals on the photosynthesis of the seagrass, *Zostera capricorni*. *Mar. Poll. Bull.*, Special Issue, **45**(1-12):100-106.

Macleod, R.D., and R.G. Congalton, 1998. A quantitative comparison of change-detection algorithms for monitoring eelgrass from remote sensing data, *Photogramm. Eng. Remote Sens.*, **64**(3):207-216.

Major, K.M., and K.H. Dunton, 2000. Photosynthetic performance in *Syringodium filiforme*: seasonal variation in light-harvesting characteristics, *Aquat. Bot.*, **68**:249-264.

Major, K.M., and K.H. Dunton, 2002. Variations in light-harvesting characteristics of the seagrass, *Thalassia testudinum*: evidence for photoacclimation, *J. Exp. Mar. Biol. Ecol.*, **275**:173– 189.

Malthus, T.J., and D.G. George, 1997. Airborne remote sensing of macrophytes in Cefni Reservoir, Anglesey, UK, *Aquat. Bot.*, **58**(3-4):317-332.

Malthus, T.J., and A.C. Madeira, 1993. High resolution spectroradiometry: Spectral reflectance of field bean leaves infected by *Botrytis fabae*, *Remote Sens. Environ.*, **45**:107-116.

Malthus, T.J., B. Andrieu, F.M. Danson, K.W. Jaggard, and M.D. Steven, 1993. Candidate high-spectral-resolution infrared indexes from crop cover, *Remote Sens. Environ.*, **46**(2):204-212.

Malthus, T.J., L. Grieve, and M.D. Harwar, 1997b. Spectral modeling for the identification and quantification of algal blooms: A test of approach, *Proc. 4th International Conference on Remote Sensing for Marine and Coastal Environments*, Orlando, Florida, 17-19 March 1997, Vol. I, pp. 223-232.

Malthus, T.J., G. Ciraolo, G. La Loggia, C.D. Clark, S.E. Plummer, S. Calvo, and A. Tomasello, 1997a. Can biophysical properties of submersed macrophytes be determined by remote sensing?, *Proc. 4th International Conference on Remote Sensing for Marine and Coastal Environments*, Orlando, Florida, 17-19 March 1997, Vol. I, pp. 562-571.

Mancinelli, A.L., 1985. Light-dependant anthocyanin synthesis: a model system for the study of plant photomorphogenesis. *The Bot. Rev.*, **51**(1):107-157.

Maritorena, S., A. Morel, and B. Gentili, 1994. Diffuse-reflectance of oceanic shallow waters - influence of water depth and bottom albedo, *Limnol. Oceanogr.*, **39**(7):1689-1703.

Masini, R., and C.R. Manning, 1997. The photosynthetic responses to irradiance and temperature of four meadow-forming seagrasses. *Aquat. Bot.*, **58**:21-36.

Masini, R.J., J.L. Cary, C.J. Simpson, and A.J. McComb, 1995. Effects of light and temperature on the photosynthesis of temperate meadow-forming seagrasses in Western Australia, *Aquat. Bot.*, **49**:239-254.

Maxwel, K. and G.N.Johnson. 2000. Chlorophyll fluorescence – a practical guide, *J. Exper. Botany*, **51**:659-668.

May, V., A.J. Collins, and L.C. Collett. 1978. A comparative study of epiphytic algal communities on two common genera of seagrasses in eastern Australia. *Aust. J. Ecol.*, **3**:91-104.

McCarthy, E.M., and B. Sabol, 2000. Acoustic characterisation of submerged aquatic vegetation: Military and environmental monitoring applications, *Proc. Oceans 2000 MTS/IEEE Conf. Exhibition*, 11-14 September 2000, Vol. 3, pp. 1957-1961.

McComb, A.J., M.L. Cambridge, H. Kirkman, and J. Kuo, 1981. The biology of Australian seagrasses, pp. 258-293, *In*, Pate, J.S., and A.J. McComb (eds), The biology of Australian plants, University of WA Press, Nedlands.

McKenzie, L.J., 1994. Seasonal changes in biomass and shoot characteristics of a *Zostera capricorni* Aschers. dominant meadow in Cairns harbour, northern Queensland, *Aust. J Marine Freshwater Res.*, **45**:1337-1352.

McNeill, S., 1996. The design of marine protected areas to protect seagrass communities: lessons from case studies in New South Wales, *In* Thackaway, R. (ed), Developing Australia's representative system of marine protected areas: criteria and guidelines for identification and selection, *Proceedings of a technical meeting held at the South Australian Aquatic Sciences Centre*, West Beach, Adelaide, 22-23 April 1996, Department of Environment, Sport and Territories, Canberra.

McRee, K.J., 1971. The action spectrum, absorptance and quantum yield of photosynthesis in crop plants, *Agric. Meteorol.*, **9**:191-216.

Meehan, A.J., 2001. Conservation Status of the seagrass *Posidonia australis* Hook *f.* in south east Australia, Ph.D. Thesis (unpub.), University of Wollongong, Australia.

Meehan, A.J., and R.J. West, 2000. Recovery times for a damaged *Posidonia australis* bed in south eastern Australia, *Aquat. Bot.*, **67**:161-167.

Meinesz, A., 1999. Killer Algae, University of Chicago Press, Chicago.

Mellors, J.E., H. Marsh, and R.G. Coles, 1993. Intra-annual changes in seagrass standing crop, Green Island, North Queensland, *Aust. J. Mar. Freshwater Res.*, **44**:33-41.

Méthy, M., 2000. A two-channel hyperspectral radiometer for the assessment of photosynthetic radiation-use efficiency, *J. Agric. Engng Res.*, **75**:107-110.

Meulstee, C. P.H. Nienhuis, and H.T.C. Van Stokkom, 1988. Aerial photography for biomass assessment in the intertidal zone, *Int. J. Remote Sens.*, **9**(10/11):1859-1867.

Mobley, C.D., 1994. Light and Water: Radiative transfer in natural waters, Academic Press, Sydney.

Moore, K., H.A. Neckles, and R.J. Orth, 1996. *Zostera marina* (eelgrass) growth and survival along a gradient of nutrients and turbidity in the lower Chesapeake Bay, *Mar. Ecol. Progr. Ser.*, **142**:247-259.

Moore, K. A., and R.L. Wetzel, 2000. Seasonal variations in eelgrass (*Zostera marina* L.) responses to nutrient enrichment and reduced light availability in experimental ecosystems, *J. Exp. Mar. Biol. Ecol.*, **244**:1-28.

Moran, J.A., A.K. Mitchell, G. Goodmanson, and K.A. Stockburger, 2000. Differentiation among effects of nitrogen fertilization treatments on conifer seedlings by foliar reflectance: a comparison of methods, *Tree Physiol.*, **20**:1113-1120.

Moriarty, D.J.W., and P.I. Boon, 1989. Interactions of seagrasses with sediments and water, pp.500-535. *In,* A.W.D. Larkum, A.J. McComb, and S.A. Shepherd (eds), Biology of seagrasses: a treatise on the biology of seagrasses with special reference to the Australian region, Elsevier Science Publishers, Amsterdam.

Morris, L.J., R.W. Virnstein, J.D. Miller, and L.M. Hall, 2000. Monitoring seagrass changes in Indian River Lagoon, Florida using fixed transects, pp. 167- , *In*, S.A. Bortone (ed), Seagrasses: monitoring, ecology, physiology and management, CRC Press, Florida.

Müller, P., X-P. Lee, and K.K. Niyogi, 2001. Non-photochemical quenching. A response to excess light energy, *Plant Physiol.*, **125**:1558-1566.

Mumby, P.J., and A.J. Edwards, 2002. Mapping marine environments with IKONOS imagery: enhanced spatial resolution can deliver greater thematic accuracy, *Remote Sens. Environ.*, **82**:248-257.

Mumby, P.J., E.P. Green, C.D. Clark, and A.J.Edwards, 1998. Digital analysis of multispectral airborne imagery of coral reefs. *Coral Reefs*, **17**:59– 69.

Mumby, P.J., E.P. Green, A.J. Edwards, and C.D. Clarke, 1997a. Coral reef habitat mapping: how much detail can remote sensing provide?, *Mar. Biol.*, **130**:193-202.

Mumby, P.J., E.P. Green, A.J. Edwards, and C.D. Clarke, 1997b. Measurement of seagrass standing crop using satellite and digital airborne remote sensing. *Mar. Ecol. Prog. Ser.*, **159**:51-60.

Mumby, P.J., E.P. Green, A.J. Edwards, and C.D. Clarke, 1999. The cost-effectiveness of remote sensing for tropical coastal resources assessment and management, *J. Environ. Manag.*, **55**:157-166.

Mumby, P.J., A.J. Edwards, E.P. Green, C.W. Anderson, A.C. Ellis, and C.D. Clark, 1997c. A visual assessment technique for estimating seagrass standing crop. *Aquat. Conserv.: Mar.Freshwater Ecosys.*, 7:239–251.

Murchie, E.H., and K.K. Niyogi. 2011. Manipulation of photoprotection to improve plant photosynthesis, *Plant Physiol.*, **155**:86-92.

Myers, J. and J.R. Graham, 1963. Enhancement in *Chlorella*, *Plant Physiol.*, **38**:105-116.

Myers, J., J.T. Hardy, C.H. Mazel, and P. Dustan, 1999. Optical spectra and pigmentation of Caribbean reef corals and macroalgae, *Coral Reefs*, **18**(2):179-186.

Neely, M.B., 2000. Somatic, respiratory, and photosynthetic responses of the seagrass *Halodule wrightii* to light reduction in Tampa Bay, Florida including a whole plant carbon budget, pp. 33-48, *In*, Bortone, S.A. (ed), Seagrasses: Monitoring, ecology, physiology, and management, CRC Press, London.

Neill, S., and K.S. Gould, 1999. Optical properties of leaves in relation to anthocyanin concentration and distribution, *Canadian J. Botany*, **77**:1777-1782.

Neverauskas, V.P., 1988. Response of a *Posidonia australis* community to prolonged reduction in light, *Aquat. Bot.*, **31**:361-366.

Nishio, J.N., J. Sun, and T.C. Vogelmann, 1994. Photoinhibition and the light environment within leaves, pp. 221-237, *In*, Baker, N.R., and J.R. Bowyer (eds), Photoinhibition of photosynthesis: From molecular mechanisms to the field, Bios. Scientific Publishers, Oxford.

Norris, J.G., and S. Wyllie-Echeverria, 1997. Estimating maximum depth distribution of seagrass using underwater videography, *Proc. 4th International Conference on Remote Sensing for Marine and Coastal Environments*, Orlando, Florida, 17-19 March 1997, Vol. 1, pp. 603-610.

Norris, J.G., S. Wyllie-Echeverria, T. Mumford, A. Bailey, and T. Turner, 1997. Estimating basal area coverage of subtidal seagrass beds using underwater videography, *Aquat. Bot.*, **58**:269-287.

Nyitrai, P., 1997. Development of functional thylakoid membranes: Regulation by light and hormones, *In*, Pessarakli, M. (ed.), Handbook of Photosynthesis, Marcel Dekker Inc, New York.

O'Neill, A.L., S. Hardy, S.J. Fraser, and K.R. McCloy, 1990. Leaf morphology and the spectral reflectance of some Australian plant species. *Proc. 5th Australasian Remote Sens. Conf.*, Perth, Australia, pp. 1096-1105.

O'Neill, A.L., J.K. Marthick, and C. Chafer, 1997. Remote sensing for coastal monitoring. *GIS User*, **23**: 26-28.

Ong, C., A. Wyllie, and P. Hick, 1994. Mapping marine habitats using geoscan mk2 airborne multispectral scanner data, *Proc. 7th Australasian Remote Sens. Conf.*, Melbourne, March 1994.

Onuf, C.P., 1994. Seagrasses, dredging and light in Laguna Madre, TX, USA, *Estuarine Coastal Shelf Sci.*, **39**:75-91.

Orth, R.J., and K.A. Moore, 1983. Chesepeake Bay: An unprecedented decline in submerged aquatic vegetation, *Science*, **222**:51-53.

Osmond, C.B., 1994. What is photoinhibition? Some insights from comparisons of shade and sun plants, pp. 1-24, *In*, Baker, N.R., and J.R. Bowyer (eds), Photoinhibition of photosynthesis: From molecular mechanisms to the field, Bios. Scientific Publishers, Oxford.

Osmond, C.B., and S.C. Grace, 1995. Perspectives on photoinhibition and photorespiration in the field: quintessential inefficiencies of the light and dark reactions of photosynthesis, *J. Exp. Bot.*, **46**:1351-1362.

Owens, T.G., 1994. Excitation energy transfer between chlorophylls and carotenoids. A proposed molecular mechanism for non-photochemical quenching, pp. 95-109, *In*, Baker, N.R., and J.R. Bowyer (eds), Photoinhibition of photosynthesis: From molecular mechanisms to the field, Bios. Scientific Publishers, Oxford.

Pasqualini, V., and C. Pergent-Martini, 1996. Monitoring of *Posidonia oceanica* meadows using image processing, pp. 351-358, *In*, Kuo, J., R.C. Phillips, D.I. Walker, and H. Kirkman (eds), Seagrass Biology: Proceedings of an International Workshop, Rottnest Island, Western Australia, 25-29 January 1996.

Pasqualini, V., C. Pergent-Martini, and G. Pergent, 1999. Environmental impact identification along the Corsican coast (Mediterranean sea) using image processing, *Aquat. Bot.*, **65**:311-320.

Pasqualini, V., C. Pergent-Martini, P. Clabaut and G. Pergent, 1998. Mapping of *Posidonia oceanica* using aerial photographs and side scan sonar: Application off the Island of Corsica (France), *Estuarine, Coastal and Shelf Science*, **47**:359–367

Pasqualini, Y., C. Pergent-Martini, P. Clabaut, H. Morteel, and G. Pergent, 2001. Integration of aerial remote sensing, photogrammetry and GIS technology in seagrass mapping, *Photogramm. Engin. Remote Sens.*, **67**(1):99-105.

Pearcy, R.W., 1994. Photosynthetic response to sunflecks and light gaps: mechanisms and constraints, pp. 255-271, *In*, Baker, N.R., and J.R. Bowyer (eds), Photoinhibition of photosynthesis: From molecular mechanisms to the field, Bios. Scientific Publishers, Oxford.

Penhale, P.A., and W.O. Smith Jr. 1977. Excretion of dissolved organic carbon by eelgrass (*Zostera marina*) and its epiphytes. *Limnol. Oceanogr.*, **22**:400-407.

Peñuelas, J., and I. Filella, 1998. Visible and near-infrared reflectance techniques for diagnosing plant physiological status, *Trends in Plant Science*, **3**:151-156.

Peñuelas, J., F. Baret, and I. Filella, 1995a. Semi-empirical indices to assess carotenoids/chlorophyll *a* ratio from leaf spectral reflectance, *Photosynthetica*, **31**:221-230.

Peñuelas, J., I. Filella, S. Elvira, and R. Inclán, 1995b. Reflectance assessment of summer ozone fumigated Mediterranean White Pine seedlings, *Environ. Exp. Bot.*, **35**:299-307.

Peñuelas, J., I. Filella, and J.A. Gamon, 1995c. Assessment of photosynthetic radiation-use efficiency with spectral reflectance, *New Phytol.*, **131**:291-296.

Peñuelas, J., I. Filella, J.A. Gamon, and C. Field, 1997. Assessing radiation-use efficiency of emergent aquatic vegetation from spectral reflectance, *Aquat. Bot.*, **58**(3-4):307-315.

Peñuelas, J., J.A. Gamon, K.L. Griffith, and C.B. Field, 1993. Assessing community type, plant biomass, pigment composition, and photosynthetic efficiency of aquatic vegetation from spectral reflectance. *Remote Sens. Environ.*, **46**:110-118.

Peñuelas, J., J.A. Gamon, A.L. Fredeen, J. Merino, and C.B. Field. 1994. Reflectance indices associated with physiological changes in nitrogen- and water-limited sunflower leaves, *Remote Sens. Environ.*, **48**(2):135-146.

Pérez M., and J. Romero, 1992. Photosynthetic response to light and temperature of the seagrass *Cymodocea nodosa* and the prediction of its seasonality, *Aquat. Bot.*, **43**(1):51-62.

Pérez-Llorens, J.L., S. Strother, and F.X. Niell. 1994. Species differences in short-term pigment levels in four Australian seagrasses in response to dessication and rehydration. *Botanica Marina*, **37**:91-95.

Pergent-Martini, C., and G. Pergent, 1996. Spatio-temporal dynamics of *Posidonia oceanica* beds near a sewage outfall (Mediterranean – France), pp.299-306, *In*, Kuo, J., R.C. Phillips, D.I. Walker, and H. Kirkman (eds), Seagrass Biology: Proceedings of an International Workshop, Rottnest Island, Western Australia, 25-29 January 1996.

Phillips, R.C., and C.P. McRoy 1980. Handbook of seagrass biology: An ecosystem perspective, Garland STPM Press, New York.

Phinn, S.R., C. Menges, G.J.E.Hill, and M. Stanford, 2000. Optimizing remotely sensed solutions for monitoring, modelling and managing coastal environments, *Remote Sens. Environ.*, **73**:117-132.

Pinar, A., and P.J. Curran, 1996. Grass chlorophyll and the reflectance red edge, *Int. J. Remote Sens.*, **17**(2):351-357.

Pirc, H., 1986. Seasonal aspects of photosynthesis in *Posidonia oceanica*: influence of depth, temperature and light intensity, *Aquat. Bot.*, **26**:203-212.

Plummer, S.E., T.J. Malthus, and C.D. Clark, 1997. Adaptation of a canopy reflectance model for sub-aqueous vegetation: Definition and sensitivity analysis, *Proc. 4th International Conference on Remote Sensing for Marine and Coastal Environments*, Orlando, Florida, 17-19 March 1997, Vol. 1, pp. 149-157.

Poiner, I.R., D.I. Walker, and R.G. Coles, Regional studies – seagrasses of tropical Australia, pp. 279-296, *In*, A.W.D. Larkum, A.J. McComb, and S.A. Shepherd (eds), Biology of Seagrasses: A treatise on the biology of seagrasses with special reference to the Australian region, Elsevier, New York.

Pollard, P.C., and M. Greenway, 1993. Photosynthetic characteristics of seagrasses (*Cymodocea srrulata*, *Thalassia hemprichii*, and *Zostera capricorni*) in a low-light environment, with a comparison of leaf-marking and lacunal-gas measurements of productivity, *Aust. J. Mar. Freshwater Res.*, **44**:127-139.

Ponzoni, GF.J., and J.L. de M. Gonçalves, 1999. Spectral features associated with nitrogen, phosphorus, and potassium deficiencies in *Eucalyptus saligna* seedling leaves, *Int. J. Remote Sens.*, **20**(11):2249-2264.

Powis, B.J. and K.I.M. Robinson, 1980. Benthic macrofaunal communities in the Tuggerah Lakes, NSW, *Aust. J. Mar. Freshwat. Res.*, **31**:803-815.

Prange, J.A., and W.C. Dennison, 2000. Physiological responses of five seagrass species to trace metals, *Mar. Poll. Bull.*, **41**:327-336.

Preen, A., 1995. Impacts of dugong foraging on seagrass habitats; observational and experimental evidence for cultivation grazing, *Mar. Ecol. Progr. Ser.*, **124**:201-213.

Price, J.C., 1994. How unique are spectral signatures?, *Remote Sens. Environ.*, **49**:181-186.

Qiu, H., J.A. Gamon, D.A. Roberts, and M. Luna, 1998. Monitoring post fire succession in the Santa Monica Mountains using hyperspectral imagery, *In*, Green, R.O., and Q. Tong (eds), *Proc. SPIE, Vol. 352, Hyperspectral Remote Sensing and Applications*, September 1998, Beijing, China.

Ralph, P.J., 1996. Diurnal photosynthetic patterns of *Halophila ovalis* (R.Br.) Hook. *f.*, pp.197-202, *In*, Kuo, J., R.C. Phillips, D.I. Walker, and H. Kirkman (eds), Seagrass Biology: Proceedings of an International Workshop, Rottnest Island, Western Australia, 25-29 January 1996.

Ralph, P.J., 1998. Photosynthetic responses of *Halophila ovalis* (R.Br.) Hook. *f.* to osmotic stress. *J. Exp. Mar. Biol. Ecol.*, **227**:203-220.

Ralph, P.J., 1999. Photosynthetic response of *Halophila ovalis* (R.Br.) Hook. *f.* to combined environmental stress, *Aquat. Bot.*, **65**:83-96.

Ralph, P.J., and M.D. Burchett, 1995. Photosynthetic responses of the seagrass *Halophila ovalis* (RBr.) Hook. *f.* to high irradiance stress, using chlorophyll *a* fluorescence. *Aquat. Bot.*, **51**:55-66.

Ralph, P.J., and M.D. Burchett, 1998a. Photosynthetic response of *Halophila ovalis* to heavy metal stress. *Environ. Poll.*, **103**:91-101.

Ralph, P.J., and M.D. Burchett, 1998b. Impact of petrochemicals on the photosynthesis of *Halophila ovalis* using chlorophyll fluorescence. *Mar. Poll. Bull.*, **36**:429-436.

Ralph, P.J., R. Gademan, and W.C. Dennison, 1998. In situ seagrass photosynthesis measured using a submersible pulse-amplitude modulated fluorometer. *Mar. Biol.*, **132**:367-373.

Ralph, P.J., S.M. Polk, K.A. Moore, R.J. Orth, and W.O. Smith Jr., 2002. Operation of the xanthophyll cycle in the seagrass *Zostera marina* in response to variable irradiance, *J. Exp. Mar. Biol. Ecol.*, **271**:189– 207.

Raven, J.A., 1995. Photosynthesis in aquatic plants, pp. 299-318, *In*, Schulze, E.-D., and M.M. Caldwell (eds), Ecophysiology of photosynthesis, Springer-Verlag, Berlin.

Ressom, H., S. Fyfe, P. Natarajan, and S. Sriranganam, 2003. Monitoring seagrass health using neural networks, *Proceedings of the IEEE International Joint Conference on Neural Networks*, Portland, USA, July 20-24, 2003, Vol 2, pp. 1019-1024.

Richardson, A.J., and C.L. Wiegand, 1977. Distinguishing vegetation from substrate background information, *Photogramm. Eng. Remote Sens.*, **43**:1541-1552.

Richardson, L.L. 1996. Remote sensing of algal bloom dynamics. *BioScience*, **46**(7):492-501.

Richardson, L.L., D. Buisson, C-J. Liu, and V. Ambrosia, 1994. The detection of algal photosynthetic accessory pigments using airborne visible-infrared imaging spectrometer (AVIRIS) spectral data. *Mar. Tech. Soc. J.*, **28**(3):10-21.

Robbins, B.D., 1997. Quantifying temporal change in seagrass areal coverage: the use of GIS and low resolution aerial photography, *Aquat. Bot.*, **58**(3-4):259-267.

Robblee, M.B., T.R. Barber, P.R. Carlson, M.J. Durako, J.W. Fourqurean, L.K. Muehlstein, D. Porter, L.A. Yarbro, R.T. Zieman, and J.C. Zieman, 1991. Mass mortality of the tropical seagrass *Thalassia testudinum* in Florida Bay (USA), *Mar. Ecol. Progr. Ser.*, **71**(3):297-299.

Roberts, D.G., and D.J.W. Moriarty, 1987. Lacunal gas discharge as a measure of productivity in the seagrasses *Zostera capricorni*, *Cymodocea serrulata* and *Syringodium isoetifolium*, *Aquat. Bot.*, **28**(2):143-160.

Robertson, E.L., 1984. Seagrasses, pp. 57-122, *In*, Womersley, H.B.S. (ed.), The marine benthic flora of southern Australia, D.J. Woolman Govt Printer, South Australia.

Robinson, S.A., 1999. Sunlight: An all pervasive source of energy, pp. 380-397, *In*, Atwell, B.J., P.E. Kriedemann, and C.E. Turnbull (eds), Plants in action: Adaptation in nature, performance in cultivation, Macmillan Education Australia, Melbourne.

Robinson, S.A., and C.B.O. Osmond, 1994. Internal gradients of chlorophyll and carotenoid pigments in relation to photoprotection in thick leaves of plants with Crassulacean acid metabolism, *Aust. J. Plant Physiol.*, **21**:497-506.

Rock, B.N., J.E. Vogelmann, D.L. Williams, A.F. Vogelmann, and T. Hoshizaki, 1986. Remote detection of forest damage, *Bioscience*, **36**:439-445.

Rock, B.N., T. Hoshizaki, and J.R. Miller, 1988. Comparison of *in situ* and airborne spectral measurements of the blue shift associated with forest decline, *Remote Sens. Environ.*, **24**:109-127.

Roelfsema, C.M., S.R. Phinn, and W.C. Dennison, 2002. Spatial distribution of benthic microalgae on coral reefs determined by remote sensing, *Coral Reefs*, **21**:264-274.

Rollings, N., N. Doblin, and B. Light. 1998. CASI outperforms aerial photography, Daedalus 1268 and SPOT for mapping benthic assemblages in Port Phillip Bay, Victoria, *Proc. 9th Australasian Remote Sens. Photogramm. Conf.*, Sydney, Australia, 20-24 July 1998, CD-Rom.

Rollings, N., B. Light, N. Doblin, and T. Chiffings, 1993. An evaluation of remote sensing and associated field techniques for mapping the distribution of benthic habitats in Port Phillip Bay. Final Report, Port Phillip Bay Environmental Study, Task G2.1, November 1993.

Rouse, J.W., R.H. Haas, J.A. Schell, and D.W. Deering, 1973. Monitoring vegetation systems in the Great Plains with ERTS, *Proc. 3rd ERTS Symp.*, Vol. 1, pp. 48-62.

Rowan, K.S., 1989. Photosynthetic pigments of algae, Cambridge University Press.

Sabol, B. E. McCarthy, and K. Rocha, 1997. Hydroacoustic basis for detection and characterization of eelgrass (*Zostera marina*), *Proc. 4th International Conference on Remote Sensing for Marine and Coastal Environments*, Orlando, Florida, 17-19 March 1997, Vol I, pp. 679-693.

Sampson, P.H, P.J. Zarco-Tejada, G.H. Mohammed, J.R. Miller, and T.L. Noland, 2003. Hyperspectral remote sensing of forest condition: Estimating chlorophyll content in tolerant hardwoods, *Forest Science*, **49**(3):381-391.

Sand-Jensen, K., 1977. Effects of epiphytes on eelgrass photosynthesis. *Aquat. Bot.*, **3**:55-63.

Sathyendranath, S., and T. Platt, 1989. Computation of aquatic primary production: extended formalism to include effect of angular and spectral distribution of light, *Limnol. Oceanogr.*, **34**:188-198.

Saxena, A.K., V.K. Tewari, H.B. Triparthi, and J.S. Singh, 1985. Spectro-reflectance characteristics of certain plants of the Kumaun Himalaya and relationship of pigment concentration with leaf reflectance, *Proc. Indian Nat. Sci. Acad.*, **B51** (2):223-234.

Schiefthaler, U., A.W. Russell, H.R. Bolhàr-Nordenkampf, and C. Critchley, 1999. Photoregulation and photodamage in *Schefflera arboricola* leaves adapted to different light environments, *Aust. J. Plant Physiol.*, **26**:485-494.

Schimel, D.S., 1993. New technologies for physiological ecology, pp. 359-365, *In*, Ehleringer, J.R., and C.B. Field (eds), Scaling physiological processes: Leaf to globe, Academic Press, Sydney.

Schrieber, U., W. Bilger, and C. Neubauer, 1995. Chlorophyll fluorescence as a non-intrusive indicator for rapid assessment of in vivo photosynthesis, pp. 49-70, *In*, Schulze, E.-D., and M.M. Caldwell (eds), Ecophysiology of photosynthesis, Ecological studies 100, Springer-Verlag, Berlin.

Schrieber, U., and W. Bilger, 1987. Rapid assessment of stress effects on plant leaves by chlorophyll fluorescence measurements, pp. 27-76, *In*, Tenhunen, J.D., F.M, Catarino, O.L. Lange, and W.C. Oechel (eds), Plant response to stress: a functional analysis in Mediterranean ecosystems, NATO ASI series, vol. G15, Springer-Verlag, Berlin.

Seddon, S., R.M. Connolly, and K.S. Edyvane, 2000. Large-scale seagrass dieback in northern Spencer Gulf, South Australia, *Aquat. Bot.*, **66**:297-310.

Shephard, S.A., A.J. McComb, D.A. Bulthuis, V. Neverauskas, D.A. Steffensen, and R. West, 1989. Decline in seagrasses, pp. 346-393, *In,* A.W.D. Larkum, A.J. McComb, and S.A. Shepherd (eds), Biology of seagrasses: a treatise on the biology of seagrasses with special reference to the Australian region, Elsevier Science Publishers, Amsterdam.

Short, F.T., 1987. Effects of sediment nutrients on seagrasses: literature review and mesocosm experiment, *Aquat. Bot.*, **27**:41-57.

Short, F.T., and S. Wyllie-Echeverria, 1996. Natural and human-induced disturbance of seagrasses, *Environ. Conserv.*, **23**:17-27.

Short, F.T., D.M. Burdick, and J.E. Kaldy III, 1995. Mesocosm experiments quantify the effects of eutrophication on eelgrass, *Zostera marina*, *Limnol. Oceanogr.*, **40**:740-749.

Short, F.T., D.M. Burdick, S. Granger, and S.W. Nixon, 1996. Long term decline in eelgrass, *Zostera marina*, linked to increased housing development, pp. 291-298, *In*, Kuo, J., R.C. Phillips, D.I. Walker, and H. Kirkman (eds), Seagrass Biology: Proceedings of an International Workshop, Rottnest Island, Western Australia, 25-29 January 1996.

Shul'gin, I.A., V.S. Khazanov, and A.F. Kleshnin, 1960. On the reflection of light as related to leaf structure, *Academiya Nauk SSSR, Bot. Sci.*, **134**:197-199.

Sims, D.A., and J.A. Gamon, 2002. Relationships between leaf pigment content and spectral reflectance across a wide range of species, leaf structures and developmental stages, *Remote Sens. Environ.*, **81**(2-3):337-354.

Sinclair, T.R, M.M. Scriber, and R.M. Hoffer, 1973. Diffuse reflectance hypothesis for the pathway of solar radiation through leaves, *Agron. J.*, **65**:268-283.

Skoog, D.A., D.M. West, and F.J. Holler, 1992. Fundamentals of Analytical Chemistry, 6th edn, Saunders College Publishing, Fort Worth, USA.

Smith, G.M., T. Spencer, A.L. Murray and J.R. French, 1998. Assessing seasonal vegetation change in coastal wetlands with airborne remote sensing: an outline methodology. *Mangroves and Saltmarshes*, **2**:15-28.

Smith, R.D., W.C. Dennison, and R.S. Alberte, 1984. Role of seagrass photosynthesis in root aerobic processes, *Plant Physiol.*, **74**:1055-1058.

Smith, R.D., A.M. Pregnall, and R.S. Alberte, 1988. Effects of anaerobiosis on root metabolism of *Zostera marina* (eelgrass): implications for survival in reducing sediments, *Mar. Biol.*, **98**:131-141.

Spanner, M.A., L.L.Pierce, D.L. Peterson, and S.W. Running, 1990. Remote sensing of temperate coniferous forest LAI: the influence of canopy closure, understorey vegetation and background reflectance, *Int. J. Remote Sens.*, **11**(1):95-111.

S.P.C.C., 1981. The ecology of fish in Botany Bay, State Pollution Control Commission Report, BBS 23, 77 pp.

Sriranganam, S., H. Ressom, M.T. Musavi, S. K. Fyfe, and P. Natarajan, 2003. Neural Network based methods for estimation of photosynthetic efficiency for monitoring seagrass health, 9 pp.

Stone, C., L. Chisholm, and N. Coops, 2001. Spectral reflectance characteristics of eucalypt foliage damaged by insects, *Aust. J. Bot.*, **49**:687-698.

Stone, C., L.A. Chisholm, and S. McDonald, 2003. Spectral reflectance characteristics of *Pinus radiata* needles affected by dothistroma needle blight, *Canadian J. Botany*, **81**(6):560-569.

Stumpf, R.P., and M.L. Frayer, 1997. Temporal and spatial change in coastal ecosystems using remote sensing: example with Florida Bay, USA, emphasizing AVHRR imagery, *Proc. 4th International Conference on Remote Sensing for Marine and Coastal Environments*, Orlando, Florida, 17-19 March 1997, Vol I, pp. 65-73.

Stylinski, C.D., J.A. Gamon, and W.C. Oechel, 2002. Seasonal patterns of reflectance indices, carotenoid pigments and photosynthesis of evergreen chaparral species, *Oecologia*, **131**(3):366-374.

Supanawid, C. 1996. Recovery of the seagrass *Halophila ovalis* after grazing by dugong, pp. 315-318, *In*, Kuo, J., R.C. Phillips, D.I. Walker, and H. Kirkman (eds), Seagrass Biology: Proceedings of an International Workshop, 25-29 January, Rottnest Island, Western Australia.

Tageeva, S.V., A.B. Brandt, and V.G. Derevyanko, 1961. Changes in optical properties of leaves in the course of the growing season, *Akademiya nauk SSSR, Bot. Sciences*, **135**:266-288.

Tassan, S., 1996. Modified Lyzenga's method for macroalgae detection in water with non-uniform composition, *Int. J. Remote Sens.*, **17**(8):1601-1607.

Terrados, J., and C.M. Duarte, 2000. Experimental evidence of reduced particle resuspension within a seagrass (*Posidonia oceanica* L.) meadow, *J. Exp. Mar. Biol. Ecol.*, **243**:45-63.

Thayer, S.S., and O. Björkman, 1990. Leaf xanthophyll content and composition in sun and shade determined by HPLC, *Photyosynthesis Res.*, **23**:331-343.

Thenkabail, P.S., R.B. Smith, and E. De Pauw, 2000. Hyperspectral vegetation indices and their relationships with agricultural crop characteristics, *Remote Sens. Environ.*, **71**(2):158-182.

Thomas, J.R., and H.W. Gausman, 1977. Leaf reflectance vs. leaf chlorophyll and carotenoid concentrations for eight crops, *Agron. J.*,**69**:799-802.

Thomas, J.R., and G.F. Oerther, 1972. Estimating nitrogen content of sweet pepper leaves by reflectance measurements, *Agron. J.*, **64**:11-13.

Thomas, M., P.Lavery, and R. Coles, 1999. Monitoring and assessment of seagrass, pp. 116-139, *In*, Butler, A., and P. Jernakoff (eds), FRDC Project 98/223: Seagrass in Australia: strategic review and development of an R & D plan, CSIRO Publishing, Victoria.

Thomson, A.G., R.M. Fuller, T.H. Sparks, M.G. Yates, and J.A. Eastwood, 1998. Ground and airborne radiometry over intertidal surfaces: waveband selection for cover classification. *Int. J. Remote Sens.*, **19**(6):1189-1205.

Todd, J.S., R.C. Zimmerman, P. Crews, and R.S Alberte, 1993. The antifouling activity of natural and synthetic phenolic acid sulphate esters, *Phytochem.*, **34**(2):401-404.

Tomlinson, P.B., 1974. Vegetative morphology and meristem dependence: foundation of productivity in seagrasses, *Aquaculture*, **4**(2):107-130.

Tomlinson, P.B., 1980. Leaf morphology and anatomy in seagrasses, p. 7-28. *In*, Phillips, R.C., and C.P.McRoy (eds.), Handbook of seagrass biology: an ecosystem perspective. Garland Press, New York.

Touchette, B.W., and J.M. Burkholder, 2000a. Review of nitrogen and phosphorus metabolism in seagrasses, *J. Exp. Mar. Biol. Ecol.*, **250**:133-167.

Touchette, B.W., and J.M. Burkholder, 2000b. Overview of the physiological ecology of carbon metabolism in seagrasses. *J. Exp. Mar. Biol. Ecol.*, **250**:169-205.

Trocine, R.P., J.D. Rice, and G.N Wells, 1982. Photosynthetic response of seagrasses to ultraviolet-A radiation and the influence of visible light intensity. *Plant Physiol.*, **69**:341-344.

Tucker C.J., 1979. Red and photographic infrared linear combinations for monitoring vegetation, *Remote Sens. Environ.*, **8**:127-150.

Turner, S.J., S.F. Thrush, M.R. Wilkinson, J.E. Hewitt, V.J. Cummings, A.-M. Schwarz, D.J. Morrisey, and I. Hawes, 1996. Patch dynamics of the seagrass *Zostera novazelandica* (?) at three sites in New Zealand, pp. 21-31, *In*, Kuo, J., R.C. Phillips, D.I. Walker, and H. Kirkman (eds), Seagrass Biology: Proceedings of an International Workshop, Rottnest Island, Western Australia, 25-29 January 1996.

Udy, J.W., and W.C. Dennison, 1996. Estimating nutrient availability in seagrass sediments, pp. 163-172, *In*, Kuo, J., R.C. Phillips, D.I. Walker, and H. Kirkman (eds), Seagrass Biology: Proceedings of an International Workshop, Rottnest Island, Western Australia, 25-29 January 1996.

Udy, J.W., and W.C. Dennison, 1997a. Physiological responses of seagrasses used to identify anthropogenic impacts, *Mar. Freshwater Res.*, **48**:605-614.

Udy, J.W., and W.C. Dennison, 1997b. Growth and physiological responses of three seagrass species to elevated sediment nutrients in Moreton Bay, Australia, *J. Exp. Mar. Biol. Ecol.*, **217**:253-277.

Ustin, S.L., and B. Curtiss, 1990. Spectral characteristics of ozone-treated conifers, *Environ. Exp. Bot.*, **30:**293-308.

Ustin, S.L., M.O. Smith, and J.B. Adams, 1993. Remote sensing of ecological processes: a strategy for developing and testing ecological models using spectral mixture analysis. *In* J. Ehrlinger, and C. Field (eds.), Scaling ecological processes between leaf and landscape, Academic Press,

van Lent, F., J.M. Verschuure, M.L.J. van Veghel, 1995. Comparative study on populations of *Zostera marina* L. (eelgrass): in situ nitrogen enrichment and light manipulation, *J. Exp. Mar. Biol. Ecol.*, **185**:55-76.

Verhoef, W., 1984. Light-scattering by leaf layers with application to canopy reflectance modeling: the SAIL model, *Remote Sens. Environ.*, **16**(2):125-141.

Vermaat, J.E., N.S.R. Agawin, M.D. Fortes, J.S. Uri, C.M. Duarte, N. Marba, S. Enriquez and W. van Vierssen, 1996. The capacity of seagrasses to survive increased turbidity and siltation: The significance of growth form and light use, *Ambio*, **25**(2):499-504.

Virnstein, R., M. Tepera, L. Beazley, T. Hume, T. Altman, and M. Finkbeiner, 1997. A comparison of digital multispectral imagery versus conventional photography for mapping seagrass in Indian River Lagoon, Florida, *Proc. 4th International Conference on Remote Sensing for Marine and Coastal Environments*, Orlando, Florida, 17-19 March, Vol. I, pp. 57-.

Vogelmann, T.C., 1989. Yearly Review: Penetration of light into plants, *Photochem. Photobiol.*, **50**(6):895-902.

Vogelmann, T.C., J.N. Nishio, and W. K. Smith, 1996. Leaves and light capture: light propagation and gradients of carbon fixation within leaves, *Trends in Plant Science: Reviews*, **1**(2):65-70.

Vogelmann, T.C., B.N. Rock, and D.M. Moss, 1993. Red edge spectral measurements from sugar maple leaves, *Int. J. Remote Sens.*, **14**(8):1563-1575.

Walker, D.I., 1988. Methods for monitoring seagrass habitat. VIMS Working Paper #18, Melbourne, Australia, 26 pp.

Walker, D.J., and A.J. McComb, 1992. Seagrass degradation in Australian coastal waters, *Mar. Poll. Bull.*, **25**:191-195.

Walker, D., W. Dennison, and G. Edgar, 1999. Status of Australian seagrass research and knowledge, pp. 1-24, *In*, Butler, A., and P. Jernakoff (eds), FRDC Project 98/223: Seagrass in Australia: strategic review and development of an R & D plan, CSIRO Publishing, Victoria.

Walker, D.I., K.A. Hillman, G.A. Kendrick, and P. Lavery, 2001. Ecological significance of seagrasses: Assessment for management of environmental impact in Western Australia, *Ecological Engineering*, **16**:323-330.

Wang, J., L. Zhang, and Q. Tong, 1998. The derivative spectral matching for wetland vegetation identification and classification by hyperspectral data, p. 280-288. *In*, R.O. Green, and Q. Tong (eds), *Proc. of SPIE, Vol. 3502, Hyperspectral Remote Sens. Application*, September 1998, Beijing, China.

Ward, D.H., C.J. Markon, and D.C. Douglas, 1997. Distribution and stability of eelgrass beds at Izembek Lagoon, Alaska, *Aquat. Bot.*, **58**:229-240.

Watford, F.A. and R.J. Williams, 1998. Inventory of estuarine vegetation in Botany Bay, with special reference to change in the distribution of seagrass. NSW Fisheries Final Report Series No. 11, 44 pp.

Waycott, M., and D.H. Les, 1996. An integrated approach to the evolutionary study of seagrasses, *In*, Kuo, J., R.C. Phillips, D.I. Walker, and H. Kirkman (eds), Seagrass Biology: proceedings of an international workshop, Rottnest Island, Western Australia, January 1996.

West, R.J., 1995. Rehabilitation of seagrass and mangrove sites – successes and failures in NSW, *Wetlands (Australia)*, **14**:13-19.

West, R.J., 1997. Estuaries and estuarine management in south-eastern Australia-monitoring seagrasses and seagrass fish communities, unpub. *Proc. Qindgdao Workshop*, University of Wollongong, Australia.

West, R.J., A.W.D Larkum, and R.J. King, 1989. Regional studies - seagrasses of south eastern Australia, pp. 230-260, *In,* A.W.D. Larkum, A.J. McComb, and S.A. Shepherd (eds), Biology of seagrasses: a treatise on the biology of seagrasses with special reference to the Australian region, Elsevier Science Publishers, Amsterdam.

West, R.J., C.A. Thorogood, T.R. Walford, and R.J. Williams, 1985. An estuarine inventory for New South Wales, Australia, *Fisheries Bulletin 2*, Department of Agriculture New South Wales, Australia.

Westman, W.E., and C.V. Price, 1988. Spectral changes in conifers subjected to air pollution and water stress: Experimental studies, *IEEE Trans. Geosc. Remote Sens.*, **26**(1):11-21.

Wiginton, J.R., and C. McMillan, 1979. Chlorophyll composition under controlled light conditions as related to the distribution of seagrasses in Texas and the U.S. Virgin Islands, *Aquat. Bot.*, **6**:171-184.

Williams, S.L., and W.C. Dennison, 1990. Light availability and diurnal growth of a green macroalga (*Caulerpa cupressoides*) and a seagrass (*Halophila decipiens*), *Mar. Biol.*, **106**:437-443.

Wilzbach, M.A., K.W. Cummins, L.M. Rojas, P.J. Rudershausen, and J. Locascio, 2000. Establishing baseline seagrass parameters in a small estuarine bay, pp. 125-136, *In*, S.A. Bortone (ed), Seagrasses: monitoring, ecology, physiology and management, CRC Press, Florida.

Winer, B.J., 1971. Statistical principles in experimental design, McGraw-Hill.

Woolley, J.T., 1971. Reflectance and transmittance of light by leaves, *Plant Physiol.*, **47**:656-662.

Wright, S.W., and J.D. Shearer, 1984. Rapid extraction and high-performance liquid chromatography of chlorophylls and carotenoids from marine phytoplankton, *J. Chromatography*, **294**:281-295.

Wyllie, A., D. Lord, P. Collins, G. Kendrick, and H. Kirkman, 1997. Mapping historical changes in seagrass beds bordering Owen Anchorage, Western Australia, *Proc. 4th International Conference on Remote Sensing for Marine and Coastal Environments*, Orlando, Florida, 17-19 March 1997, Vol. 1, pp. 541-551.

Yoder, B.J., and R.E. Pettigrew-Cosby, 1995. Predicting nitrogen and chlorophyll content and concentrations from reflectance spectra (400-2500 nm) at leaf and canopy scales, *Remote Sens. Environ.*, **53**:199-211.

Young, A.J., 1993a. Carotenoids in pigment-protein complexes, pp. 72-95, *In*, Young, A., and G. Britton (eds), Carotenoids in Photosynthesis, Chapman Hall, London.

Young, A.J., 1993b. Factors that affect the carotenoid composition of higher plants and algae, pp. 160-2055, *In*, Young, A., and G. Britton (eds), Carotenoids in Photosynthesis, Chapman Hall, London.

Young, A., and G. Britton, 1990. Carotenoids and stress, pp. 87-112, *In*, Alscher, R.G., and J.R. Cumming (eds), Stress responses in plants: Adaptation and acclimation mechanisms, Wiley-Liss Inc, New York.

Young, A.J., D. Phillip, and J. Savill, 1997. Carotenoids in higher plant photosynthesis, pp. 575-595, *In*, Pessarakli, M. (ed.), Handbook of photosynthesis, Marcel Dekker Inc., New York.

Yu, B., M. Ostland, P. Gong, and R. Pu, 1999. Penalized discriminant analysis of in situ hyperspectral data for conifer species recognition, *IEEE Trans. Geosci. Remote Sens.*, **37**(5):2569-2577.

Zacharias, M., O. Niemann, and G. Borstad. 1992. An assessment and classification of a multispectral bandset for the remote sensing of intertidal seaweeds, *Canadian J. Remote Sens.*, **18**(4):263-274.

Zainal, A.J.M., D.H. Dalby, and I.S. Robinson, 1993. Monitoring marine ecological changes on the east coast of Bahrain with Landsat TM, *Photogramm. Eng. Remote Sens.*, **59**:415-421.

Zann, L.P. (ed.), 1995. Our sea, our future: major findings of the 'State of the Marine Environment Report for Australia', Dept Environment, Sport and Territories, Canberra, 112 pp.

Zar, J.H., 1984. Biostatistical analysis, 2nd ed. Prentice-Hall, Sydney.

Zarco-Tejada, P.J., J.R. Miller, G.H. Mohammed, T.L. Noland, and P.H. Sampson, 1999. Canopy optical indices from infinite reflectance and canopy reflectance models for forest condition monitoring: Applied to hyperspectral CASI data, *Proceedings IEEE International Geoscience and Remote Sensing Symposium*, Hamburg, Germany, 28 June – 2 July, 1999.

Zarco-Tejada, P.J., J.R. Miller, G.H. Mohammed, T.L. Noland, and P.H. Sampson, 2002. Vegetation stress detection through chlorophyll *a+b* estimation and fluorescence effects on hyperspectral imagery, *J. Environ. Quality*, **31**(5):1433-1441.

Zibordi, G., F. Parmiggiani, and L. Alberotanza, 1990. Application of aircraft multispectral scanner data to algae mapping over the Venice Lagoon, *Remote Sens. Environ.*, **34**(1):49-54.

Zieman, J.C., and R.G. Wetzel, 1980. Productivity in seagrasses: Methods and rates, pp. 87-116, *In*, Phillips, R.C., and C.P. McRoy (eds), Handbook of seagrass biology: An ecosystem perspective, Garland Press, New York.

Zimmerman, R.C., J.L. Reguzzoni, S. Wyllie-Echeverria, M. Josselyn, and R.S. Alberte, 1991. Assessment of environmental suitability for growth of *Zostera marina* L. (eelgrass) in San Francisco Bay, *Aquat. Bot.*, **39**:353-366.

Zimmerman, R.C., J.L. Reguzzoni, and R.S. Alberte, 1995. Eelgrass (*Zostera marina* L) transplants in San Francisco Bay: role of light availability on metabolism, growth and survival, *Aquat. Bot.*, **51**(1-2):67-86.

Appendices

Appendix 1.1. The short-term influence of low light stress on the photosynthetic and photoprotective pigment concentrations observed in *Zostera capricorni* leaves (n = 5 except for anthocyanins n= 4).

Variable	treatment	**initial day**	**day 7**	**day 13**	**day 86**
Tchls	control	1327.509 ± 200.377	1314.132 ± 418.966	1373.113 ± 610.203	1202.850 ± 312.763
	low light	1441.194 ± 390.113	1414.107 ± 494.904	1634.157 ± 358.487	1912.810 ± 446.585
chl *a*:*b*	control	3.137 + 0.068	3.289 + 0.079	3.536 + 0.129	2.855 + 0.447
	low light	2.986 + 0.135	3.182 + 0.164	3.230 + 0.095	2.732 + 0.308
lutein	control	117.658 ± 10.808	116.145 ± 36.374	123.675 ± 54.861	122.227 ± 25.930
	low light	133.889 ± 21.271	127.175 ± 40.153	141.467 ± 32.383	189.100 ± 40.676
L:Tcars	control	0.342 ± 0.020	0.366 ± 0.012	0.352 ± 0.015	0.355 ± 0.049
	low light	0.352 ± 0.025	0.392 ± 0.063	0.371 ± 0.023	0.363 ± 0.014
neoxanthin	control	58.830 + 13.869	56.787 + 22.315	60.675 + 26.195	65.401 + 14.917
	low light	64.688 + 12.160	60.092 + 26.000	66.064 + 16.058	94.643 + 22.171
N:Tcars	control	0.169 ± 0.023	0.176 ± 0.015	0.174 ± 0.008	0.188 ± 0.014
	low light	0.170 ± 0.020	0.173 ± 0.026	0.173 ± 0.012	0.182 ± 0.008
Tcars	control	345.152 ± 44.669	316.981 ± 97.148	350.228 ± 151.894	352.241 ± 100.802
	low light	382.058 ± 60.828	336.503 ± 125.849	381.036 ± 81.560	520.984 ± 114.508
Tcars:Tchls	control	0.263 + 0.038	0.242 + 0.023	0.258 + 0.023	0.293 + 0.040
	low light	0.272 + 0.038	0.235 + 0.020	0.234 + 0.017	0.273 + 0.015
T β-cars	control	80.608 ± 11.735	74.214 ± 22.169	82.922 ± 33.995	68.570 ± 36.126
	low light	85.638 ± 17.403	78.446 ± 31.809	86.347 ± 20.665	116.609 ± 27.830
T β-cars:Tcars	control	0.234 ± 0.017	0.235 ± 0.009	0.239 ± 0.009	0.185 ± 0.070
	low light	0.223 ± 0.014	0.230 ± 0.014	0.226 ± 0.021	0.223 ± 0.011
α-carotene	control	0.959 + 1.194	0.294 + 0.034	0.317 + 0.124	0.266 + 0.018
	low light	2.935 + 2.614	0.286 + 0.021	0.817 + 0.913	0.266 + 0.020
α-car:β-cars	control	0.013 ± 0.017	0.004 ± 0.001	0.004 ± 0.002	0.006 ± 0.006
	low light	0.038 ± 0.034	0.004 ± 0.003	0.009 ± 0.010	0.002 ± 0.001
VAZ pool	control	87.097 ± 15.760	69.540 ± 17.586	82.640 ± 38.321	95.776 ± 28.503
	low light	94.910 ± 17.729	70.505 ± 29.971	86.341 ± 15.882	120.365 ± 27.716

Appendix 1.1. (cont.)

Variable	treatment	**initial day**	**day 7**	**day 13**	**day 86**
VAZ:Tchls	control	0.067 + 0.014	0.054 + 0.009	0.060 + 0.009	0.080 + 0.010
	low light	0.067 + 0.008	0.048 + 0.010	0.053 + 0.005	0.063 + 0.008
VAZ:Tcars	control	0.252 ± 0.029	0.222 ± 0.019	0.234 ± 0.023	0.272 ± 0.016
	low light	0.248 ± 0.014	0.204 ± 0.029	0.228 ± 0.019	0.231 ± 0.017
V:Tcars	control	0.207 ± 0.048	0.203 ± 0.013	0.187 ± 0.045	0.233 ± 0.022
	low light	0.210 ± 0.024	0.186 ± 0.029	0.216 ± 0.019	0.201 ± 0.005
A+Z:VAZ	control	0.189 + 0.108	0.084 + 0.033	0.205 + 0.148	0.143 + 0.040
	low light	0.155 + 0.072	0.091 + 0.044	0.053 + 0.007	0.130 + 0.053
Z:VAZ	control	0.068 ± 0.044	0.022 ± 0.013	0.058 ± 0.063	0.023 ± 0.002
	low light	0.046 ± 0.018	0.022 ± 0.015	0.007 ± 0.004	0.024 ± 0.011
EPS	control	0.872 ± 0.073	0.947 ± 0.018	0.868 ± 0.105	0.917 ± 0.021
	low light	0.900 ± 0.044	0.943 ± 0.025	0.970 ± 0.005	0.923 ± 0.032
violaxanthin	control	71.276 + 18.689	63.743 + 16.627	69.559 + 38.717	82.689 + 27.602
	low light	80.125 + 16.826	64.593 + 28.508	81.843 + 15.395	104.692 + 24.608
antheraxanthin	control	10.343 ± 6.133	4.221 ± 2.556	9.677 ± 2.723	10.920 ± 2.645
	low light	10.472 ± 5.707	4.471 ± 3.195	3.932 ± 0.851	12.837 ± 6.579
zeaxanthin	control	5.479 ± 2.843	1.576 ± 1.219	3.404 ± 1.935	2.167 ± 0.587
	low light	4.313 + 1.692	1.441 + 1.118	0.566 + 0.272	2.836 + 1.562
anthocyanins	control	22.889 + 4.870	17.414 + 2.340	17.622 + 4.309	
	low light	22.889 + 4.870	19.276 + 5.220	17.606 + 4.236	

T = total, chl = chlorophyll, car = carotenoid, V = violaxanthin, A = antheraxanthin, Z = zeaxanthin, N = neoxanthin, L = lutein, EPS = (V + 0.5A)/VAZ.

Appendix 1.2. The short-term influence of high light stress on the photosynthetic and photoprotective pigment concentrations observed in *Zostera capricorni* leaves (n = 8 for days 1 and 17, n = 7 for days 7 and 13 except for anthocyanins n =8).

Variable	treatment	**initial day**	**day 7**	**day 13**	**day 17**
Tchls	control	1181.563 + 197.529	1145.203 + 208.108	1906.167 + 728.366	1848.805 + 444.357
	high light	1246.384 ± 477.773	1008.279 ± 351.346	1703.257 ± 373.223	1626.771 ± 541.095
chl *a:b*	control	3.880 ± 0.199	3.707 ± 0.680	4.151 ± 0.122	3.927 ± 0.164
	high light	3.854 ± 0.401	3.706 ± 0.837	4.174 ± 0.230	4.032 ± 0.291
lutein	control	120.470 + 20.473	99.973 + 18.511	142.498 + 55.147	145.952 + 31.308
	high light	194.021 + 222.791	88.431 + 23.358	134.321 + 29.987	133.110 + 48.613
L:Tcars	control	0.333 + 0.026	0.326 + 0.027	0.317 + 0.010	0.313 + 0.015
	high light	0.364 ± 0.116	0.310 ± 0.039	0.308 ± 0.020	0.304 ± 0.015
neoxanthin	control	52.199 ± 9.033	44.659 ± 8.686	80.521 ± 33.913	88.727 ± 17.978
	high light	55.133 ± 23.249	38.482 ± 11.671	77.108 ± 19.959	82.454 ± 30.415
N:Tcars	control	0.157 + 0.008	0.159 + 0.018	0.188 + 0.013	0.203 + 0.013
	high light	0.145 + 0.031	0.148 + 0.017	0.187 + 0.010	0.200 + 0.009
Tcars	control	359.955 + 43.229	307.400 + 55.687	447.930 + 163.470	466.439 + 96.843
	high light	455.773 ± 302.103	282.659 ± 61.295	437.870 ± 101.426	436.579 ± 153.972
Tcars:Tchls	control	0.308 ± 0.029	0.260 ± 0.037	0.236 ± 0.014	0.254 ± 0.012
	high light	0.353 ± 0.140	0.302 ± 0.089	0.259 ± 0.036	0.269 ± 0.017
T β-cars	control	78.497 + 12.865	69.275 + 15.458	112.487 + 32.043	116.127 + 24.376
	high light	81.653 + 28.469	68.812 + 12.954	105.494 + 20.483	102.140 + 32.807
T β-cars:Tcars	control	0.218 + 0.017	0.233 + 0.011	0.257 + 0.027	0.250 + 0.029
	high light	0.206 ± 0.053	0.248 ± 0.045	0.243 ± 0.014	0.236 ± 0.010
α-carotene	control	2.559 ± 1.659	0.592 ± 0.449	3.494 ± 2.375	1.575 ± 1.905
	high light	2.732 ± 1.509	2.176 ± 4.506	4.103 ± 1.237	2.878 ± 1.661
α-car:β-cars	control	0.031 + 0.017	0.008 + 0.004	0.033 + 0.023	0.013 + 0.015
	high light	0.036 + 0.026	0.032 + 0.068	0.042 + 0.015	0.031 + 0.018
VAZ pool	control	101.636 ± 10.220	86.300 ± 17.596	104.142 ± 43.093	108.298 ± 26.671
	high light	117.373 ± 54.031	80.972 ± 21.671	111.358 ± 32.383	110.601 ± 41.771

Appendix 1.2 (cont.)

Variable	treatment	**initial day**	**day 7**	**day 13**	**day 17**
VAZ:Tchls	control	0.089 + 0.019	0.073 + 0.014	0.054 + 0.004	0.059 + 0.007
	high light	0.094 ± 0.029	0.086 ± 0.026	0.066 ± 0.014	0.069 ± 0.009
VAZ:Tcars	control	0.285 ± 0.040	0.281 ± 0.022	0.229 ± 0.016	0.231 ± 0.023
	high light	0.276 + 0.056	0.284 + 0.029	0.252 + 0.024	0.254 + 0.021
V:Tcars	control	0.225 + 0.016	0.233 + 0.025	0.208 + 0.018	0.198 + 0.015
	high light	0.219 ± 0.052	0.219 ± 0.037	0.202 ± 0.034	0.198 ± 0.021
A+Z:VAZ	control	0.205 ± 0.068	0.165 ± 0.094	0.095 ± 0.032	0.139 ± 0.063
	high light	0.212 ± 0.066	0.230 ± 0.105	0.195 ± 0.135	0.217 ± 0.067
Z:VAZ	control	0.061 + 0.025	0.028 + 0.019	0.021 + 0.018	0.033 + 0.020
	high light	0.071 + 0.043	0.066 + 0.055	0.076 + 0.072	0.065 + 0.028
EPS	control	0.864 ± 0.046	0.903 ± 0.056	0.942 ± 0.022	0.914 ± 0.040
	high light	0.849 ± 0.049	0.852 ± 0.078	0.865 ± 0.102	0.845 ± 0.111
violaxanthin	control	80.476 + 7.503	71.602 + 14.132	94.992 + 41.789	92.921 + 23.265
	high light	93.138 + 42.516	63.343 + 21.411	89.804 + 26.570	87.889 + 39.500
antheraxanthin	control	14.964 ± 6.310	12.189 ± 7.073	7.154 ± 2.702	11.812 ± 6.157
	high light	16.462 ± 9.257	12.852 ± 5.452	13.665 ± 10.079	16.179 ± 5.068
zeaxanthin	control	6.196 ± 2.690	2.509 ± 1.694	1.996 ± 1.438	3.565 ± 1.919
	high light	7.772 + 6.648	4.776 + 3.325	7.889 + 6.293	6.533 + 2.028
anthocyanins	control	6.616 ± 1.464		9.013 ± 1.610	
	high light	6.782 ± 1.385		8.786 ± 2.868	

T = total, chl = chlorophyll, car = carotenoid, V = violaxanthin, A = antheraxanthin, Z = zeaxanthin, N = neoxanthin, L = lutein, EPS = (V + 0.5A)/VAZ.

Appendix 2.1. Correlation between the total chlorophyll content of *Zostera capricorni* samples grown in laboratory light stress experiments and spectral reflectance at individual visible wavelengths between 430-750 nm.

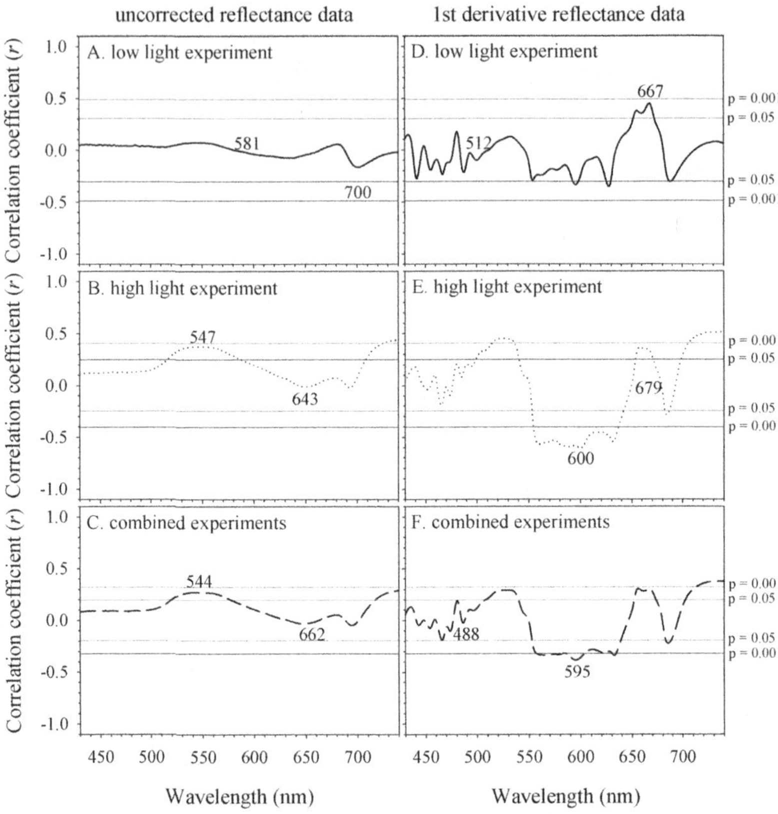

Appendix 2.2. Correlation between the chlorophyll *a*:*b* content of *Zostera capricorni* samples grown in laboratory light stress experiments and spectral reflectance at individual visible wavelengths between 430-750 nm.

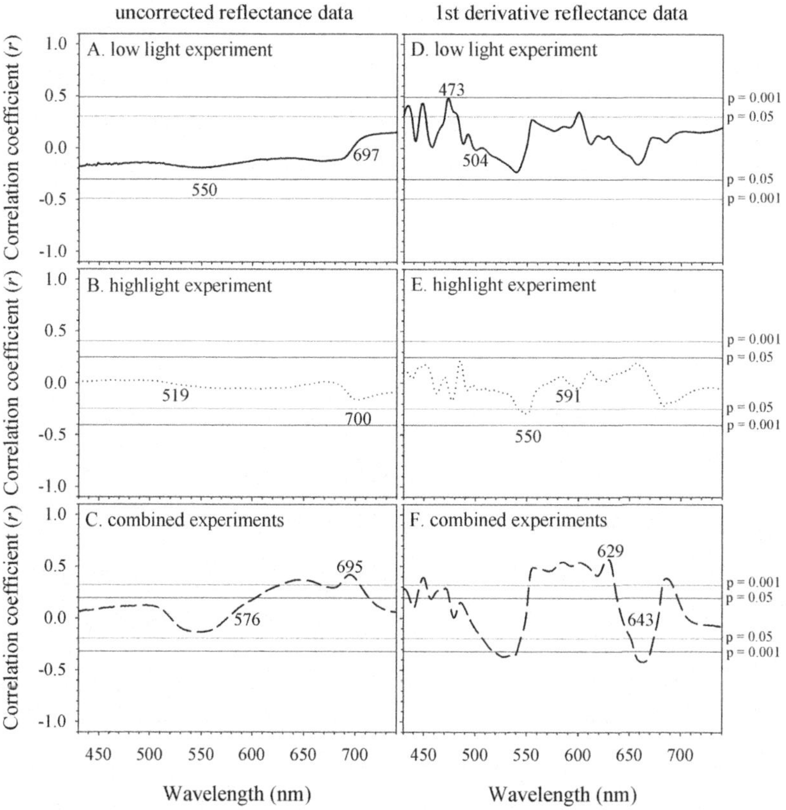

Appendix 2.3. Correlation between the VAZ:total chlorophyll content of *Zostera capricorni* samples grown in laboratory light stress experiments and spectral reflectance at individual visible wavelengths between 430-750 nm.

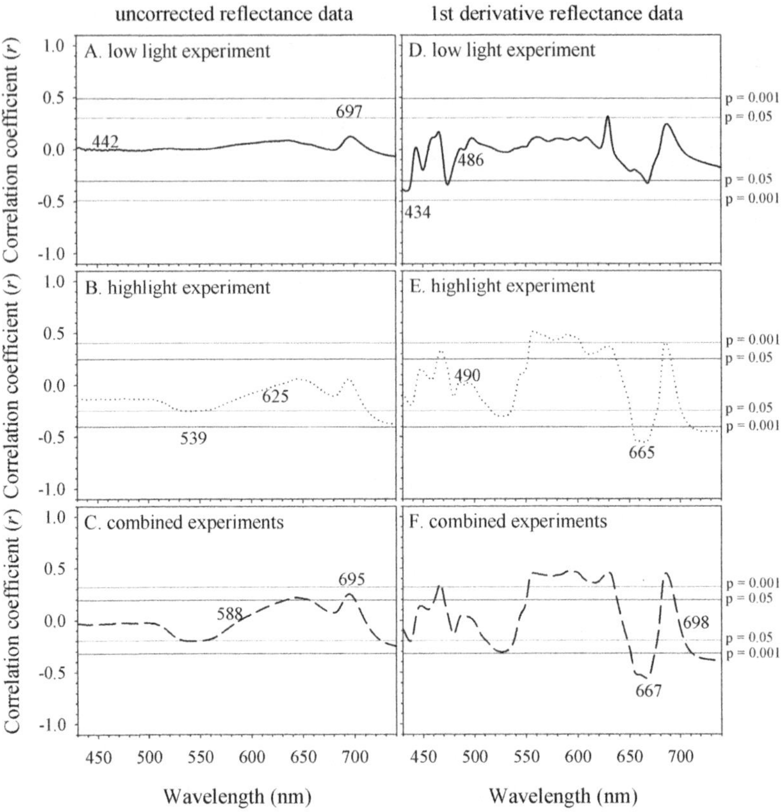

Appendix 2.4. Correlation between the VAZ:total carotenoid content of *Zostera capricorni* samples grown in laboratory light stress experiments and spectral reflectance at individual visible wavelengths between 430-750 nm.

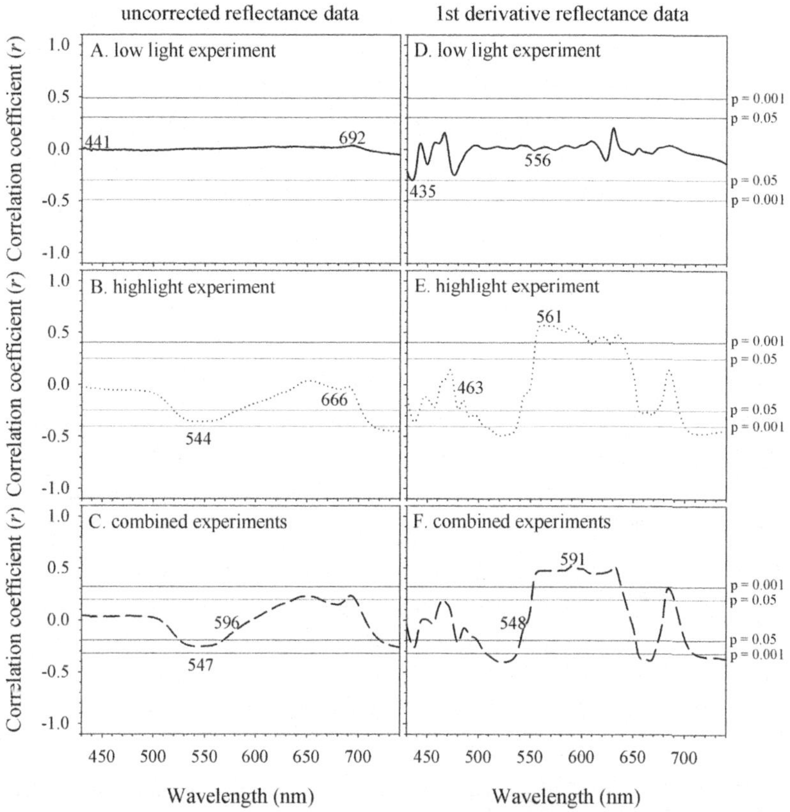

Appendix 2.5. Correlation between the Z:VAZ content of *Zostera capricorni* samples grown in laboratory light stress experiments and spectral reflectance at individual visible wavelengths between 430-750 nm.

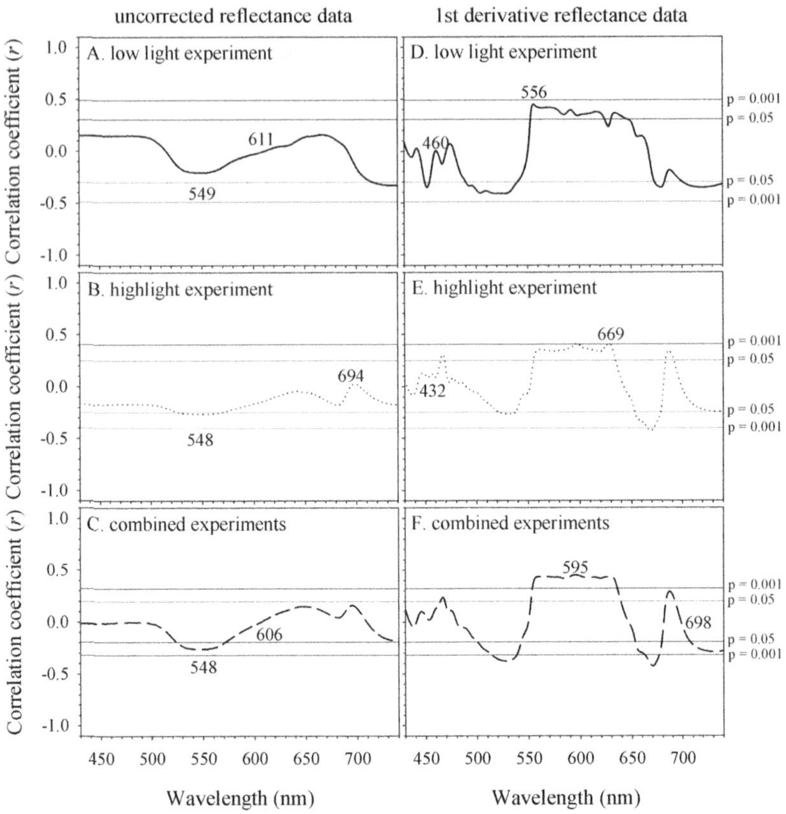

Appendix 2.6. Correlation between the photosynthetic efficiency (F_v:F_m) of *Zostera capricorni* samples grown in laboratory light stress experiments and spectral reflectance at individual visible wavelengths between 430-750 nm.

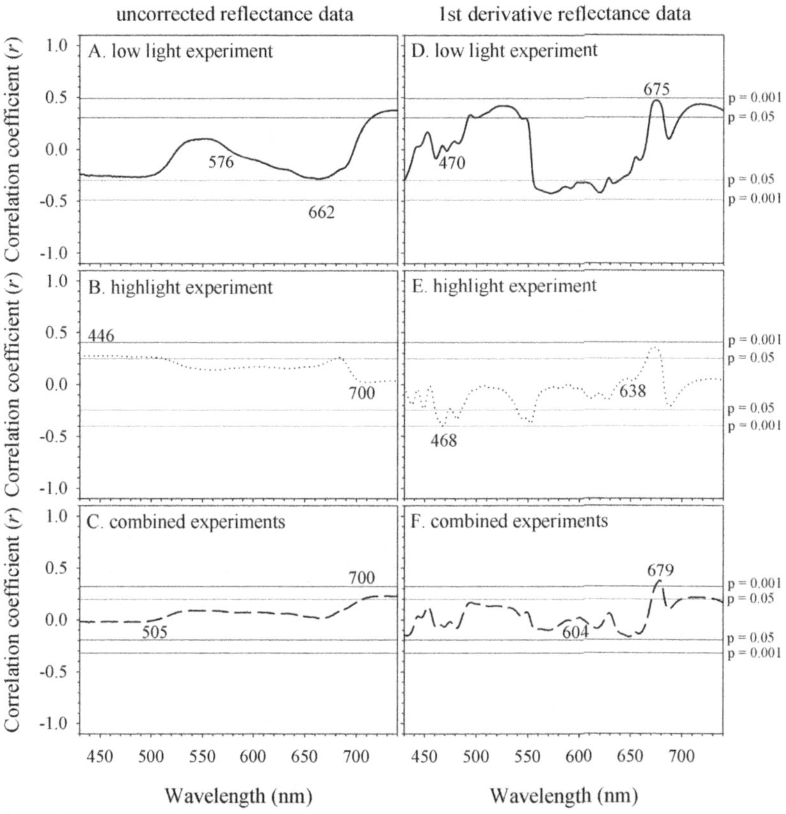

Made in the USA
Coppell, TX
11 July 2024

34534876R00226